DIVIDED EUROPE

Contents

Notes on contributors

Ray Hudson is Professor of Geography and Director of the Centre for European Studies of Territorial Development in the University of Durham. He is also a director of NOMIS (the National On-line Manpower Information System) and RCADE (the ESRC Resource Centre for Access to Data on Europe). His research interests are in geographies of production, work and employment and territorial development strategies in Europe. His recent publications include, *A Place Called Teesside: A Locality in a Global Economy* (with H. Beynon and D. Salder, 1994), *Towards a New Map of Automobile Manufacturing in Europe?* (with E. Schamp, 1995), and *Successful European Regions: Northern Ireland Learning from Others* (with M. Dunford, 1996). He is an editor of *European Urban and Regional Studies*.

Alan Williams is Professor of Human Geography and European Studies at the University of Exeter. His research interests are centred on uneven development in Europe, and he is currently undertaking projects on privatisation in Central Europe and tourism and migration in Southern Europe. He is author of *The European Community* (1994) and co-editor of *European Tourism: Regions, Spaces and Restructuring* (1995) and of *Turkey and Europe* (1993). He is an editor of *European Urban and Regional Studies*.

Adam Tickell is Professor of Geography at the University of Southampton. He has published widely in economic geography on the restructuring of finance, regulatory theory, local governance and regional development, and is the author of *Rogues, Regulation and the Culture of Finance*, to be published shortly. He has held appointments at Leeds and Manchester Universities and held fellowships from the Economic and Social Research Council and the Canadian High Commission. He has also acted as adviser to a number of policy making bodies including government departments, the European Parliament and local authorities in Britain. His current research is on the geography and culture of the financial industry.

Colin H. Williams is Research Professor in Sociolinguistics in the Department of Welsh, The University of Wales, Cardiff. He previously taught in Universities in Canada, England and the USA. His main scholarly interests are sociolinguistics and language policy in multi-cultural societies, ethnic and minority relations and political geography. Included in his publications are *Called Unto Liberty: On Language and Nationalism* (1994) and editor of *The Political Geography of the New World Order* (1993). He is currently engaged on a Welsh Language Board sponsored project on Community Language Planning and Policy, whose first major report was published as Williams and Evas, *Y Cynllun Ymchwil Cymunedol (The Community Research Project)* (1997).

Joe Painter is Lecturer in Geography at the University of Durham. His research interests lie mainly in political geography, and include the local state, local governance and regulation; the cultural aspects of political processes; the geographies of citizenship and democracy; and the relationships between geography and political theory. He is the author of numerous articles and chapters and of *Politics, Geography and 'Political Geography'* (1995).

Russell King is Professor of Geography and Dean of the School of European Studies at the University of Sussex. His research interests are in migration studies and the Mediterranean. Among his recent publications are two books on these themes: *The Mediterranean: Environment and Society* (1997), which he co-edited with Lindsey Proudfoot and Bernard Smith; and *Southern Europe and the New Immigrations* (1997), co-edited with Richard Black.

Marco Donati is Research and Information Officer at Associazione per la Pace, Rome. A political scientist with a special interest in security issues in the Mediterranean, he has done postgraduate work at the University of Sussex and has authored a number of articles in Italian on geopolitics and peace studies.

Vladimir Balaz is Principal Researcher at the Institute for Forecasting, Slovak Academy of Science, Bratislava. His research interests are in tourism and financial services. He has written several books on world capital markets in Slovakia and a book *Tourism During Transition* with Allan Williams (forthcoming, 1999), and a number of articles on tourism, regional development and emerging capital markets.

Diane Perrons is Lecturer in Geography at the London School of Economics and Political Science. Her research interests are in economic and social cohesion in Europe focusing on regional and gender dimensions of inequality. She has recently co-ordinated a cross national qualitative research project on flexible working and the reconciliation of work and family life for the European Union. Her recent publications include *Making Gender Work: Managing Equal Opportunities* (1995, edited with Jenny Shaw) and papers in Feminist Economics, *European Urban and Regional Studies*.

Paul White is Professor of Geography at the University of Sheffield. His research interests are in the field of comparative studies of European cities, particularly relating to social exclusion, urban population change, ethnicity, and neighbourhood change processes. He is also interested in international migration and migrant settlement issues. His recent publications include *Paris* (with Daniel Noin, 1997), *Writing Across Worlds: Literature and Migration* (edited with Russell King and John Connell, 1995) and *Europe's Population: Towards the Next Century* (edited with Ray Hall, 1995).

Anthony M. Warnes is Professor of Social Gerontology at the University of Sheffield, Research Director for the School of Health and Related Research in the Faculty of Medicine, and Chair of the British Society of Gerontology. He was formerly Professor of Human Geography at King's College London, where he was closely involved in the establishment of the first multi-disciplinary Institute of Gerontology in this country. His long-term research interests are in the demography of ageing societies, and migration and morbidity in later life. Major recent projects have been an ESRC-supported study of the retirement of British citizens to the Mediterranean (with Russell King and Allan Williams); a study of resource allocation for elderly programme services among Community Health Care Trusts in the Eastern Health and Social Service Board area of Northern Ireland, and a review of health and social care services for older people in London for the King's Fund Health Policy Institute. His books include *The Demography of Ageing in the United Kingdom* (1993) and *Homeless Truths: Challenging the Myths About Older Homeless People* (with Maureen Crance, 1997).

Lila Leontidou is Professor of Human Geography and European Studies and Head of the Department of Geography, University of the Aegean. She has held posts at King's College London, and at the Universities of Athens and Thessaloniki. She is currently co-ordinating a TSER programme of DG XII on 'Border cities and towns: cases of social exclusion in peripheral Europe'. She is author of *The Mediterranean City in Transition* (1990) and *Cities of Silence* (in Greek); she has also published articles on urban restructuring, modernism, local development and geographical epistemology, and book chapters on Mediterranean development.

Alex Afouxenidis is Adjunct Senior Lecturer at the Department of Geography, University of the Aegean. He is currently the co-ordinator of the Greek case study of a TSER programme of DG XII on 'Border cities and towns: causes of social exclusion in peripheral Europe'. He teaches sociology and political theory, European integration, qualitative methods of social research and medical geography. He is the author of articles and book chapters in Greek on labour relations, unemployment and the welfare state.

David Sadler is Reader in Geography at the University of Durham. His research interests are centred on questions of uneven regional development in Europe. He is managing editor of *European Urban and Regional Studies*, co-author of *A Place Called Teesside: A Locality in a Global Economy* (1994) and *Approaching Human Geography: An Introduction to Contemporary Theoretical Debates* (1991), author of *The Global Region: Production, State Policies and Uneven Development* (1992) and co-editor of *Europe at the Margins: New Mosaics of Inequality* (1995).

Preface

The 1990s have been a turbulent decade in Europe with the political changes in the East following the collapse of state socialism in 1989, the continuing processes of widening and deepening of the European Union, and the subsequent interaction between these two processes. What will the socio-spatial anatomy of Europe look like in the new millennium? This is a question of considerable intellectual interest and practical importance. There have been vigorous debates within the social sciences on how best to comprehend these changes within Europe and the broader changes in the character of contemporary capitalism, of which they form one part, and to work through their implications for the future of Europe.

Reflecting these concerns, the chapters in this book originated in a special session organized at the Annual Conference of the Royal Geographical Society with the Institute of British Geographers at Exeter in January 1997, focused on the theme of 'Europe in Crisis'. This sought to address the issues of Europe's changing place in a globalizing economy, the extent and character of changes within Europe, and the implications of these for re-shaping social and spatial divisions within Europe, and public policy responses to them; there was also a focus on the ways in which human geographers and other social scientists were seeking to understand these changes and, in some instances, influence policies towards them. In addition to revised versions of the papers given in Exeter, other chapters have been specially commissioned for this volume to broaden the coverage of issues.

In preparing the book for publication we have been greatly helped by the editorial staff at Sage, as well as our publisher's sponsorship of the special session at the Exeter conference. As always, the secretarial and technical staff at Durham and Exeter, particularly Arthur Corner and Joan Dresser at Durham and Jan Thatcher and Terry Bacon at Exeter, provided invaluable support and assistance for our efforts. We would also like to acknowledge the assistance of Mike Blakemore and Mike

Cuthbertson of *r.cade*, the ESRC Resource Centre for Access to Data on Europe, at Durham. Ray Hudson would also like to acknowledge the receipt of a Sir Derman Christopherson Foundation Fellowship from the University of Durham for 1997–8, which allowed time to work on the book.

Ray Hudson, *Durham*
Allan M. Williams, *Exeter*

INTRODUCTION

1

Re-shaping Europe: the Challenge of New Divisions within a Homogenized Political–Economic Space

Ray Hudson and Allan M. Williams

There is an apparent paradox in that, while Europe seems to be becoming an increasingly homogenized politico-economic space, it is also characterized by enduring and, in some instances, deepening social cleavages. After the paralysis of decision-making in the 1970s and 1980s, there has been a deepening of economic, and to a lesser extent political, integration symbolized by the single European market programme and the Single European Act, the three pillars of Maastricht, the Amsterdam Treaty and European monetary union (EMU). This has been paralleled by the *de facto* extension of European Union (EU) influence, and often decision-making, into central and eastern Europe. At the same time, however, there has been reproduction and recasting of the social and territorial divisions of Europe.

There is no shortage of media images for the new faces of poverty, deprivation and inequality in Europe: displaced families in the former Yugoslavia, the homeless on the streets of London, over four million unemployed in Germany, a spasm of racist attacks across Europe, the continued absence of women from the seats of power, and the failure of European welfare systems to keep apace with the increasing demands made upon them, particularly by those displaced in the labour market. These images, and others, are indicative of profound changes re-shaping the political economy of Europe, the place of Europe in a globalizing economy, the map of territorial differentiation within Europe, and the patterns of inequality along dimensions such as class, ethnicity and gender. There have been significant changes in the structure and performance of urban and regional economies, in living conditions and lifestyles between and within cities and regions, and in individual life chances and employment opportunities. There were some 52 million people living in poverty in the 12 European Union countries in 1988. And in 1993 there were over 19 million unemployed in the EU, over half of whom had been unemployed for more than a year, and an estimated three million homeless people (Rutherford, 1997). Extending the scope of enquiry to central and eastern Europe would greatly increase the extent of poverty and social division and as the EU extends progressively eastwards these problems will be internalized within it and make new demands for the Europeanization of welfare policies. There is also the challenge of creating and sustaining EMU, in the face of pessimistic scenarios of social tensions and civil unrest in response to increased unemployment in parts of the EU.

These divisions have to be seen against the longer-term evolution of the political economy of Europe since 1945, coupled with a genuine desire to ensure peace in western Europe and resist the advances of communism beyond the imaginary Iron Curtain. After the Cold War partition of Europe, a political consensus was constructed over much of western Europe around a Keynesian commitment to full employment and the role of the welfare state. There were more than two decades of sustained economic growth based on Fordism, and gradual internationalization of both capital and labour (see Chapter 2). This provided the pre-conditions for the establishment and subsequent enlargement of the EU, creating an increasingly homogenized economic space (A.M. Williams, 1994). The consensus as to the redistributive and regulatory role of the state allowed for the expansion of national systems of welfare provision. Although there were some major differences among the competing models of the welfare state (Esping-Andersen, 1990), relatively low levels of unemployment and high rates of GDP growth allowed expansion of welfare provision in all parts of north-western Europe (although there was significant divergence from it over much of Mediterranean Europe). This is not to claim that there were no social inequalities or that poverty had been eliminated. Rather, there was a long phase of convergence in terms of incomes (Atkinson, 1995) and regional GDP per capita levels (Dunford, 1994) until the late 1960s and early 1970s, even if other social cleavages remained little altered in this period. Thereafter, the triad of growth, welfarism and social convergence began to unravel.

Some of these processes of change from the 1970s, leading to more sharply etched contours of socio-spatial differentiation within Europe, relate to hotly contested claims about an epochal shift in the character of contemporary capitalism from one dominant growth model to another (see Chapter 2) and related claims about the rise of a global economy (see Chapter 3). The perceived enhanced salience of globalization is partly linked to recognized limits to national state regulation, associated with the crisis of a Fordist regime of accumulation in its various western European variants. The existence of different models of welfare capitalism meant that there were variations in the form and timing of crises, and the responses to them, but they could not be avoided (Esping-Andersen, 1994; Rhodes, 1996).

Recognition of the limits to state capacities and the pressures for a 'lean welfare' state (Drèze and Malinvaud, 1994) stimulated a search for new neo-liberal macro-scale regulatory models that accepted the limited powers of national states to counter global market forces. This 'subversive liberalism' (Rhodes, 1995) led to a re-definition of the boundaries between private and public sectors, and a cutting back of the welfare state and the scope of welfare provision. These changes had a direct impact upon patterns of service provision, employment opportunities and individual life chances, with differential effects on different cities, regions and social groups. To some extent, it has led to convergence between the contested social models of welfare capitalism (Rhodes, 1996) but national states within Europe remain as key sites of regulation, governance and identity.

This presaged the export of new neo-liberal regulatory models from North America and western Europe to the countries of eastern Europe. Social and territorial differences have also become more clearly demarcated in central and eastern Europe. From the late 1960s, the highly centralized state planning system began to unravel, with Hungary and Poland, in particular, moving towards more decentralized economic decision-making, putative private property rights and increasing trade with and investment flows from the West (Landesmann and Székely, 1995). The process of transition post-1989 was subject to external pressure to adopt a neo-liberal 'shock therapy' model of reform (Gowan, 1995; see also Gowan, 1996; Lloyd, 1996), the key elements of which were stabilization, market liberalization and the development of market-supporting institutions, privatization, currency convertibility and trade liberalization. However, in practice the process of transition was path-dependent–path-creating (Nielsen et al., 1995) and was conditional upon national economic and political structures. As a result, the transition has been marked by differences in the timing and form of economic reforms, culminating with the July 1997 recommendation of the European Commission to prioritize the future accession of some applicants: Poland, Hungary, the Czech Republic, Slovenia and Estonia (and Cyprus). There has also been the creation of a new rich class and the polarization of incomes, and above all property ownership, within the countries of central and eastern Europe (Duke and Grime, 1997; see Chapter 7).

The perceived loss of national economic sovereignty has also stimulated a debate on whether the EU, or some other supra-national body, should and could replace the regulatory role of the national state, that is 'Europeanize' regulation. To some extent this has occurred, but integration has been largely negative, involving the removal of barriers to trade, capital and labour mobility. In contrast, the role of the EU in social welfare has been severely constrained. Although the Commission has proposed social action programmes in the 1990s and has had an effect on some areas of workers' rights, such as limiting working hours, its role has been strictly confined. There are manifold barriers to the development of any such wider role due to the EU's lack of means to pursue an employment strategy, intergovernmentalism in its decision-making and budgetary limits, as well as the difficulties of harmonizing deeply culturally embedded welfare institutions.

The combined effects of these changes in and to the political economy of Europe in the 1980s and 1990s has been to deepen and modify the map of socio-spatial differences. The pattern of social exclusion has been modified by the differential engagement of social classes, men and women, migrants and ethnic groups, and those at different stages of the life course, in the redistribution of income, wealth and power which has accompanied changes in European capitalism. The geographies of social exclusion have become increasingly complex and range from inner cities to Europe's border regions (see Chapter 11). Prevailing notions of citizenship have been re-defined in the face of changing social expectations, the emergence of a proto-European citizenship and the challenges posed by intra- and extra-European refugee movements, and changes in governance (Chapter 5). The remainder of this chapter explores the seeming paradox that the persistence and deepening of many axes of social inequality have coincided with a period of increasing homogenization of the political–economic space of Europe. But first we consider the debate about globalization.

The character of contemporary capitalism: Europe in a global economy, globalization and state power

There is widespread agreement that the character of contemporary capitalism has shifted in significant ways but much less agreement as to the theoretical and political significance of these changes. This is registered in competing claims, such as those claiming a transition to late capitalism (Mandel, 1975), late modernity (Giddens, 1990), reflexive modernity (Beck, 1992) or post-modernity (Harvey, 1989a). A common thread running through these competing accounts, however, is a recognition of the significance of the dramatic reduction in the tyranny of distance, a world in which time–space has shrunk dramatically (albeit very unevenly) as a result of technological innovations in transport and communications, especially information technologies. This in turn has been associated with claims about the transition to a global political economy and processes of globalization and these implicitly or explicitly relate to debates about changes in the regulatory capacities of national states and the 'hollowing out' of national states, upwards, downwards and sideways (Jessop, 1994). The collapse of the Bretton Woods system of fixed exchange rates between national currencies in the early 1970s is seen as a particularly critical moment in the transition towards a global economy since this had profound implications for the national regulation of national economies. One of the particularly important aspects of the move 'upwards' of the regulatory powers of national states has been the creation of macro-regions such as NAFTA and the EU as a result of agreements between national states. These have been critical formative moments in the emerging macro-geography of globalization and the EU has emerged as one of the major regions of the triad in this global economy (Ohmae, 1990, 1995).

There are difficult policy and political choices within Europe as to whether to seek to resist or to encourage globalization processes: the dilemmas certainly are not new, but they have become more difficult to deal with precisely because there is no theoretical consensus on the meaning and impact of globalization. Globalization is a hotly contested concept, not least in terms of perspectives on the relationship between the local, the national and the global and the degree to which national states are seen as having been 'hollowed out' (and by implication the extent to which embryonic supra-national states such as the EU could and should extend their reach to fill the resultant policy vacuum). Beyond that, however, there is debate on the extent to which globalization is seen to encompass more than just a shifting geography of regulation as the local and the global inter-penetrate and influence one another in new and novel ways.

At the risk of some over-simplification, four broad analytical positions can be identified, although those who subscribe to them often draw very different political implications from them (for a fuller review, see Amin, 1997a; for a slightly different perspective, see Weiss, 1997). First, there are those who claim that little in fact has changed at all, other than that there has been expanded reproduction of the capitalist economy as an international one, with national states continuing to behave, as before, as regulators of national economies. Hirst and Thompson (1996a: 47–8; see also 1996b) assert that the notion of globalization is 'just plainly wrong' and that the idea of a new, highly internationalized, virtually uncontrollable economy based on world market forces is wide of the mark.

A second position is diametrically opposed to such a view, and its proponents (such as Robinson, 1996: 13–14) see the global economy as one in which transnational capital scours the globe in search of profit, with virtually no constraint on its activities. National states are fatally weakened and deprived of regulatory capacity as one facet of a radical process of 'hollowing out' of the national state which shifts power decisively to 'footloose' transnationals, thus facilitating the emergence of a 'borderless' world. The implication is that there is a pressing need for the EU to develop rapidly into a heavy-duty political actor with the weight and influence to deal with hyper-mobile transnational capital.

A third view (for example, Boyer and Drache, 1996) accepts that there have been fundamental shifts in the organization of the global political economy, and that globalization does indeed register a qualitative change in the character of the political economy of contemporary capitalism. This does not, however, entail the emasculation of the national state; announcements of the death of the national state are premature (Anderson, 1995). The implication of this view is that the regulation and governance of economy and society will reflect the distribution of political powers, competencies and resources between the national and supra-national levels.

Finally, a fourth view accepts that there are qualitative processes of change, which can be caught in the concept of globalization, but argues that these permeate the crevices of everyday life and individual experience of contemporary capitalism. It thus emphasizes that globalization not only leads to a re-definition of relations between a global economy and national states but between a global economy and national and local civil societies, and by implication a transnational European civil society. Held (1995: 20), drawing on Giddens' (1996) concept of time–space distanciation, suggests that globalization can be taken to denote the stretching and deepening of social relations and institutions across space and time. As a result, on the one hand, day-to-day activities are increasingly influenced by events happening on the other side of the globe and, on the other hand, the practices and decisions of local groups can have significant global reverberations. Individuals will simultaneously identify with the local, national and European scales as new forms of cultural hybridity help shape identities. European citizens will bear multiple identities (not all of which, of course, will be territorially defined) and structures of governance within Europe will need to acknowledge and reflect this.

The perspective adopted here in seeking to understand the relationships between Europe, the EU and processes of globalization is one that combines elements of the third and fourth of the above positions. Globalization is conceptualized as a process of linkage and interdependence between territories and of 'in here–out there' connectivity (Amin 1997b). There have certainly been processes of 'hollowing out' of the national state upwards (notably to the EU, which has become a significant site of political authority and power), downwards (to the regional and local levels) and outwards (into a variety of organizations within civil society at sub-national, national and supra-national scales), and significant related changes in the mode of regulatory activities of national states. Equally, the national state form retains a continuing salience, co-existing alongside the EU, with complex reciprocal links between the two entities, rather than it being replaced by the EU. National states are active subjects in shaping processes of globalization rather than being simply passive objects and victims of them. So too is the EU. Consequently, globalization is not conceptualized as a process somehow external to national economy, society and state for national states have played a key role in bringing about regulatory and other changes that have facilitated the emergence of processes and agencies of globalization (for example, see Cerny, 1990). Consequently, both the EU and the national states of Europe are key sites of power in mediating the relations between processes of globalization and territorial patterns of socio-economic change within national spaces, although decentralization of state power to local and regional levels has altered the forms of this mediation. While there are reciprocal relationships between changes at local, regional and global scales, the impact of globalization has been deeply uneven, both between and within national territories and within the wider space of the EU and Europe.

Creating a new unified and homogenized political–economic space: political and policy changes in Europe

The changing political–economic map

The political–economic map of Europe which has evolved in the context of globalization in the post-war period has been highly contested. Different states have pursued particular versions of capitalism (Lash and Urry, 1987; Esping-Andersen, 1990) and contested the supra-national institutions which have been imposed on the European space. The EU was discursively constructed as a model of liberal capitalism, and 'its originators drew on a prevailing ideology which was economic and political . . . In essence, this was the ideology of liberal capitalism, or the assumption that the self-interest of enterprise could be harnessed in the public interest through a liberalisation of trade, capital and labour markets' (Holland, 1980: 4). Its main triumphs in the 1960s were the creation of a customs union and a common external tariff. Its major collective policy, the common agricultural policy (CAP), not only dominated its expenditure (precluding a Europeanization of welfare policies) but also contributed to the widening of income differences within and between rural areas. It also sat uneasily with claims about the superiority of market resource allocation.

Over the next three decades there was an ongoing re-definition of the boundaries between the two parts of western Europe, which at one time had been crystallized into the competing economic spaces of the EEC and EFTA; each of the enlargements of the EU in the 1970s, 1980s and 1990s effectively drew the two sets of member states together, driven on by the logic of international competition and trade, mass production and market access. There was also a deepening as well as widening which was evident in the increasing role of the EU in regulation, the development of new areas of (weak) welfare policies, and the upward drift of power from the member states. The progressive expansion of the EU, widening the scope for market forces to shape territorial development trajectories, has increased the scale of spatial inequalities within the EU. Further expansion eastwards as part of the process of seeking to establish capitalism and liberal democracy more firmly in eastern Europe – heralded initially by the free-trade Europe Agreements and latterly by the Commission's 1997 opinions on applications for membership – will undoubtedly change the character and exacerbate the depth of inequalities.

While significant national and regional variations therefore remain in the forms in which these economic and political relations are constituted within Europe, the project to widen the European Union represents an attempt to reduce this variability. At the same time, and in many ways in conflict with this process of widening, there have been powerful pressures further to deepen and transform the character of the Union itself. The single European market programme, European monetary union and the prospect of a single 'euro' currency are indicative of the drive towards the creation of a unified economic space in which market forces will have much greater scope to influence the socio-spatial distributions of economic activities, resources and income. Thus these changes in the character of the EU can also be regarded as bringing about a homogenization of its space, seeking to establish the hegemony of capitalist social relations over its entirety. At the same time, giving wider and freer play to market forces has led to increasing territorial differentiation within the EU. Seemingly paradoxically, these processes of homogenization are enhancing the significance of differences between places in influencing the locations of economic activities and the quality of people's lives within Europe. A corollary of this is that, *ceteris paribus*, territorial inequalities will widen further, undermining cohesion within the Union.

The most far-reaching political–economic change within Europe has undoubtedly been the re-definition of the relation between East and West into one between central and eastern and western Europe. This transition to capitalism in central and eastern Europe was intended to help insert it into at least the margins of the wider global economy. Within central and eastern Europe the possibilities for the longer-term success of the establishment of new regulatory regimes and successful capitalist development seem greatest in the Czech Republic, Hungary and Poland (Commission of the European Communities, 1996b: 26) but equally there is evidence that the processes of transition are having strongly territorially differentiated effects within these countries (see Chapter 7). The process of reform and transition is thus creating new inequalities within the East, and re-defining the map of uneven development within Europe. Opening up significant swathes of territory largely denied for decades to capital has had major implications for geographies of uneven development within this enlarged European space. It has created new opportunities for some companies, people and places; this is perhaps most evident in the various forms of direct foreign investment via new greenfield factories, joint ventures and the acquisition of former state enterprises. Conversely, it has generated a serious threat to others.

While much of the attention of the EU since 1989 has been focused on its eastern neighbours, there has also been a long process of re-definition of relationships with the Mediterranean region (see Chapter 6). Some but not all parts of the northern Mediterranean have, of course, been formally integrated into the EU, while others are enmeshed with it through trade, capital and labour flows. The Mediterranean region, as a whole, has long provided a pool of reserve labour for the EU, and one which it drew on in the 1960s in particular as a critical ingredient in the long-sustained post-war boom of northern Europe. Subsequently, uneven development within the region, as well as the enlargement of the EU, have fragmented the region as a source of migration, with the member states of the EU – Italy, Spain, Greece and Portugal – becoming destination countries for immigrants from the other side of the Mediterranean. But the divisions within the Mediterranean region are not all north–south, as is evident in the political and economic disintegration of Albania and, especially, the former Yugoslavia. The collapse of the old regimes in central and eastern Europe, ushering in the potential to create new forms of democracy and governance (see Chapter 5) has also unleashed the forces of nationalism and racism. This was most starkly evident in the former Yugoslavia, where the political crisis in the state(s), exacerbated in part by the palpable failure of the proto-EU common foreign and security policy (Carter et al., 1996), led to violent conflict. This had massive humanitarian and economic costs, measured in terms of the victims of 'ethnic cleansing', the massive outflow of refugees, homelessness, and the virtual collapse of production and international trade in several of the new republics, thereby adding a new and particularly tragic twist to the deepening of social and territorial cleavages in what was also becoming an increasingly complex area located beyond the homogeneous European economic space.

Public policy responses in Europe

Until the 1970s western European welfare states were able to maintain their traditional goals of redistribution, welfarism and full employment, even in the face of an intensification of international competition. Four distinctive strands of welfare capitalism developed in western Europe in the post-war period and, following Esping-Andersen (1990) and Rhodes (1996), these are summarized here (see also Chapter 10):

- The Scandinavian model characterized by a high degree of universality, corporatism and consensus between capital, unions and the state on the need for an active employment policy and a strategy for rationalization and technical change.

There has been some narrowing of per caput income inequalities between member states, largely as a result of faster income growth in Greece, Ireland, Portugal and Spain (Commission of the European Communities, 1996b). The southern European enlargements of the EU in the 1980s successively increased the magnitude of regional inequalities in per caput GDP within it (Hudson and Lewis, 1985) and, while the impact of the Scandinavian enlargement of the 1990s was much more muted, the accession of Finland did bring further significant regional problems within the EU (Commission of the European Communities, 1995a: 15). The enlargement that had the most profound impact upon and led to the most marked re-definition of the intra-EU map of regional differentiation was the re-unification of Germany as the new Eastern *Länder* became some of the most problematic regions of the Union (Commission of the European Communities, 1995a: 151–6). Moreover, within countries (with the exception of the Netherlands) regional income differences have widened. In addition, the incidence of unemployment across the Union as a whole has become much more uneven, spatially and socially (Figures 1.1 and 1.2). As a result, 'even though a process of convergence between the Member States is apparent, economic and social cohesion *within* most Member States seems to have experienced a setback during the 1990s in the form of widening disparities in income and unemployment' (Commission of the European Communities, 1996a: 49), though such a view may understate the extent to which cohesion *within the EU* has also experienced a setback. This suggests that the various policy interventions to narrow territorial inequalities and promote cohesion have not been able to prevent a widening of regional and social disparities internally within the member states (Commission of the European Communities, 1996a: 57) and have had, at best, limited efficacy. It remains an open question whether such disparities would have been even wider without these policies. On the other hand, it is clear, as we have argued above, that the effects of other policies have been to undermine the attainment of reduced regional differentials and this conflict between policy objectives is certainly one reason why inequalities have grown within member states.

Conclusion: implications for the future

There is no doubt that the patterns of social and spatial inequality within Europe are being re-shaped by a range of political and economic processes, some global in their scope, others more specifically European. Not only are these social divisions reflected in gendered, racialized, life-course and territorial differences in access to power and wealth, but these dimensions are fundamental to the way in which an increasingly homogenized but divided Europe is being constructed. The ending of the Cold War has not, of course, eliminated East–West differences: not

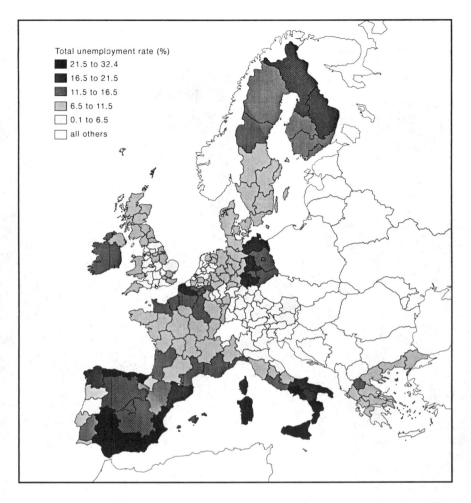

FIGURE 1.1 *Percentage unemployment in the EU15, 1996 (ESRC Resource Centre for Access to Data on Europe)*

only has it added a new layer of social and territorial inequalities to the European map, but it has affected and will affect the way in which social and territorial divisions are constituted elsewhere in the macro-region.

What will be the future outlines of a divided Europe? There are those who argue for a future map of Europe in which market forces and national states remain the dominant cartographers and in which sharpened inequalities are regarded as necessary and desirable. Others argue for a more powerful role for the European Union and for a more social conception of the market, with a much greater emphasis on the goal of enhanced cohesion and with socio-spatial inequalities being held within narrower bounds. Yet, recognizing both the need for cohesion policies and their limited effectiveness to date, the Commission of the European

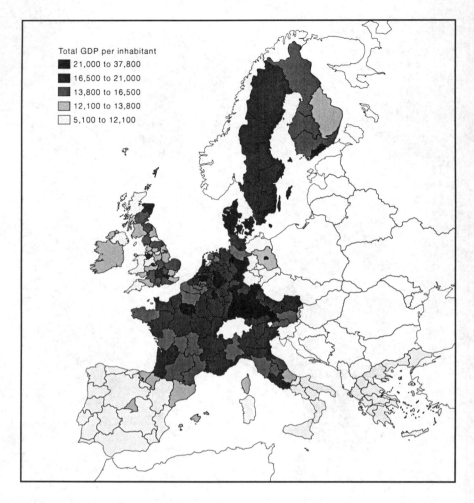

Total GDP per inhabitant
- 21,000 to 37,800
- 16,500 to 21,000
- 13,800 to 16,500
- 12,100 to 13,800
- 5,100 to 12,100

FIGURE 1.2   *GDP per capita in the EU 15, 1996 (ESRC Resource Centre for Access to Data on Europe)*

Communities (1996a) poses questions as to how their effectiveness could be enhanced. It is suggested that concentration is the key principle underlying the effectiveness of cohesion policies (Commission of the European Communities, 1996a: 116) and that there needs to be a greater degree of selectivity in future. More generally, there is a suggestion that with just over 50 per cent of the EU population eligible for the four regional objectives of the structural funds, 'there may be a case for a more determined application of the principle of concentration' (Commission of the European Communities, 1996a: 117). This tends to suggest a need to trim the extent of the structural funds for budgetary reasons rather than a prioritization of the goals of cohesion, however. There is recognition that the general climate of financial rigour in Member States has implications for the Union's policies. A major theme will be how to combine, *in a balanced way*, fiscal discipline with solidarity both within the poorest Member States and regions and with the most disadvantaged regions and people in the more prosperous Member States. (Commission of the European Communities, 1996a: 11, emphasis added)

The key issue here will surely be the way in which 'in a balanced way' is translated into practice: the extent to which competitiveness and cohesion can be made compatible goals in practice, and which will take priority if they cannot be made compatible. The way in which this issue will be addressed undoubtedly holds the key to the sort of Europe that will emerge in the new millennium.

At one level, the debate on the determinants of the future map of Europe and the future anatomy of European society is cast in terms of a struggle between the two broadly opposing perspectives of national states/market forces versus stronger EU role/more social construction of markets. In both cases, however, there is an acceptance that the dominant social relations of the economy will remain capitalist. As a result, it may be that deeper structural forces will both constrain the form and extent of state involvement and redraw the map of Europe according to the latest cartographic version of the principles of combined and uneven development. There are those who argue for a 'third way', with a greater emphasis on devolution and decentralization of political power to the institutions of civil society at local and regional levels (Cooke, 1995, for example). Some see this as a form of associationalism linked to a thin and procedural state (Hirst, 1994), others regarding a strong and enabling state as a critical pre-condition for such decentralized associationalist approaches to be compatible with notions of social cohesion and socio-spatial justice (Amin, 1996). Even so, tackling social exclusion via such a third way still has to confront the inequalities endemic in capitalism. This suggests that the issue may well be the extent to which various forms of exclusion can be tackled and what forms of socio-spatial inequality can be legitimated rather than abolishing inequality in its various forms.

It may also be the case, however, that incompatible policy objectives cannot be reconciled and that hard political choices will have to be made as to which policy goals are to be prioritized. There is certainly a theoretical literature that suggests that the contradictions of a capitalist economy cannot be abolished by state intervention, that as a result such contradictions are internalized within the operations of the state itself and in due course emerge as crises in the state's mode of crisis management (Habermas, 1975; Offe, 1975). In these circumstances, priority tends to be given to those policy objectives most closely linked to the core relationships of capitalism, and the reinvigoration of economic growth and the accumulation process are given priority over the pursuit of goals of social cohesion, equity and territorially balanced development. This suggests that the future for Europe will be one of continuing divisions and that the future map of Europe will continue to be marked by deep qualitative and quantitative socio-spatial inequalities at a variety of scales.

2

The New Economy of the New Europe: Eradicating Divisions or Creating New Forms of Uneven Development?

Ray Hudson

The past two decades within Europe have seen major changes in European economies, organizationally, sectorally and spatially. These restructuring processes have taken place in the context of profound changes in the geo-political map of Europe and in the character of contemporary capitalism. Perhaps the most significant of the latter changes relates to hotly contested claims as to an epochal shift from one dominant growth model to another and related claims about the rise of a global economy (see Amin and Thrift, 1994; Amin, 1997a; also Chapter 1). The growing significance of processes of globalization reflects, *inter alia*, decisions by national states to change the international regulatory framework by participating in institutions such as the IMF, the World Bank, GATT, and the Multi-Fibre Agreement. Despite such growing tendencies, to date there are very few firms and sectors that are organized on a truly global basis, even in the financial sector, often seen as the prime exemplar of globalization (Weiss, 1997; see Chapter 3). A corollary of this is that commentators such as Ohmae (1990, 1995) see the contemporary economy as organized around three global regions, of which Europe is one. The growing significance of processes of globalization is also partly linked to perceived limits to national state regulation, associated with the demise of a Fordist regime of accumulation. The quantitative and qualitative expansion of state involvement in the economy and society of much of western Europe in the post-war period transformed and internalized economic crisis tendencies within the state. In due course, they erupted as crises of the state itself and its mode of crisis management. Paralleling this crisis of state management and regulation of the economy in the capitalist West was the more profound crisis and collapse of state socialist economies in the East.

Recognition of the limits to state capacities stimulated a search in western Europe (and the USA) for new neo-liberal macro-scale regulatory models that both accepted and further eroded national states' limited powers to counter global market forces. These in turn were exported to the countries of eastern Europe as a form of 'shock therapy' which conditioned the ways in which they were inserted into the wider global economy (Gowan, 1995; see Chapter 7) and re-defined the map of locational possibilities for production within Europe. It is, however, important not to over-state the decline in the significance of the national state. The processes of 'hollowing out' (Jessop, 1994) or 'unbundling' of national state competencies (Ruggie, 1993) may have gone further in the countries of the EU than elsewhere, but claims of the 'death of the nation-state' are premature (Anderson, 1995), although there have been important changes in its mode of operation and its relationship to economies and civil societies within Europe. These various neo-liberal and 'post-Fordist' regulatory experiments led to the advanced capitalist countries of the West becoming characterized by a greater degree of socio-spatial inequality, while 'shock therapy' created new inequalities in the East. These changing intra-national and global conditions in turn served to define the parameters within which corporate and industrial restructuring took place within Europe.

The most far-reaching geo-political change within Europe has been the re-definition of the relation between East and West – a deep political–economic and ideological divide – into one between eastern and western Europe, a difference between parts of a common capitalist space in Europe. Opening up significant swathes of territory largely denied to capital for decades has major implications for geographies of economies within this enlarged European space. Although the full repercussions have yet to be revealed, it has created new opportunities for some companies and places, while generating a serious threat to others. As well as these changes, however, others were underway in western Europe, originating in the 1950s with the creation of the EEC and taken a stage further in 1960 with the establishment of EFTA. The next two decades witnessed a continuing re-definition of the boundaries between the two parts of western Europe and the deepening as well as widening of the EC as it became the EU. One consequence of these changes was that the influence of the EU was enhanced, both as a macro-region in the globalizing economy and, linked to a degree of 'hollowing out' of national states, as a new site and level of regulation. Thus the EU simultaneously encourages globalization and is a site of resistance to it, and this is reflected in tensions between its industrial policies which seek to promote globally competitive companies (via support for R&D, a permissive attitude to intra-EU acquisitions and mergers, especially in key high-tech sectors), its competition policies which seek to create a unified single market, and those policies which seek to encourage social and spatial cohesion and equity within the EU. This tension has had an important influence on corporate restructuring and patterns of territorially uneven development (Ramsay, 1990). In addition, with the prospect of the single European market, inward investment from North America and South-east Asia to secure market access within the EU became an increasingly significant influence on the organizational and territorial pattern of production.

In summary, these macro-scale changes – globally, within western Europe and between East and West – set the context in which economic restructuring has occurred within Europe. The structure of the rest of the chapter is as follows. First, some aggregate indicators of economic change within Europe are briefly examined. Then sectoral changes and their labour marker implications are analysed. This is followed by consideration of corporate strategies of Europeanization and globalization. In the next section, different models of production organization are summarized. Following this, tendencies to spatial concentration and decentralization within the European economy are discussed. Then some of the implications of these various changes for trade unions are considered. Finally, some conclusions are drawn about the implications for future geographies of economies and restructuring tendencies within Europe.

Indicators of political–economic change: jobless growth, job shedding growth and rising unemployment in Europe

There are three basic points that are of particular relevance here. First, during much of the 1950s and 1960s the economies of western Europe, especially those of the six founding members of the EEC, which grew consistently at unprecedented rates of around 5 per cent per annum, expanded rapidly. So too did the economies of eastern European countries, although it is difficult to draw precise comparisons with those of the West as the goals of economic policy and the statistical data available as a result are not comparable. From the second half of the 1960s, however, it became increasingly clear that the European variants of the post-war 'Fordist' growth model (for example, see Lash and Urry, 1987) were becoming crisis prone. The conjunctural effects of the 1973–4 oil crisis intensified the underlying structural contradictions (at micro- and macro-scales) of this growth model and produced a deep recession, leading to sharp absolute falls in industrial production and output (for example, see Mandel, 1978). Following the second round of oil price rises in 1979–80, a recovery in national economic growth rates did occur in western Europe from the early 1980s, in part linked to a growing adoption of neo-liberal national economic policies. Nevertheless, average

TABLE 2.3 *Sectoral distribution (%) of gross value added in the EU15 (current market prices and exchange rates) 1985–94*

	Manufacturing		Market services		Non-market Services		other	
	1985	1994	1985	1994	1985	1994	1985	1994
Austria	30.4[3]	26.1	41.6[3]	44.0	16.1[3]	15.8	11.9[3]	14.1
Belgium	22.8	19.6	49.3	55.1	15.2	13.9	12.7	11.4
Denmark	19.5	19.1	44.6	46.9	22.1	22.4	13.8	11.6
Finland	25.5	25.6	37.3	42.2	18.6	21.7	18.6	10.5
France	21.6	18.8	46.1	52.0	17.8	17.9	14.5	11.3
Germany	29.8	24.6	44.1	51.4	14.0	13.4	12.1	10.6
Greece	18.5	15.3[1]	37.3	44.8[1]	16.2	15.7[1]	28.0	23.2
Ireland	28.5	30.6[2]	36.6	41.9[1]	18.1	16.3[1]	16.8	11.2
Italy	24.6	20.9	47.0	51.6	12.9	13.5	15.5	14.0
Luxembourg	26.3	22.1[2]	53.3	52.7[2]	11.4	14.8[2]	9.0	10.4
Netherlands	17.4	16.8	49.5	57.2	12.2	10.8	20.9	15.2
Portugal	27.8	24.4[2]	43.4	45.5[2]	12.1	15.9[2]	16.7	14.2
Spain	25.2	18.8	43.2	48.4	12.7	14.7	18.9	18.1
Sweden	n.a.	n.a.	n.a.	n.a.	n.a.	n.a.	n.a.	n.a.
UK	23.8	20.2	42.9	54.2	15.4	13.2	17.8	12.4

[1] 1993; [2] 1992; [3] 1989.

Source: Eurostat (1996)

unemployed were in the 15–24 age group (according to the 1996 Labour Force Survey: Commission of the European Communities, 1996a). The distribution of unemployment also continues to be related to gender (see Chapter 8) and ethnic (see Chapter 9) differences. Labour productivity consistently grew more rapidly than output so that much of western Europe experienced 'jobless' or even 'job shedding' growth as companies strove to respond to global competitive challenges and depressed national economies in Europe. In many areas unemployment rose to very high levels as a direct consequence of major industrial closures or job shedding in an attempt to restore competitiveness (see Commission of the European Communities, 1997o; and Chapter 8). Such processes were given an additional twist as national states strove to meet the EMU convergence criteria in anticipation of the launching of the single European currency (probably from 1999).

The transformation in eastern Europe was even more dramatic. The rapid growth in unemployment there after 1989 was unprecedented as the state socialist economies had been organized in such a way as to abolish unemployment (albeit at the cost of considerable under-employment). National unemployment rates in excess of 10 per cent became common (though not universal), with rates in particular locali-ties far in excess of this (for example, in those affected by the collapse of armaments production: see A. Smith, 1996a, b). These dramatic increases in unemployment led to severe competition between places – at

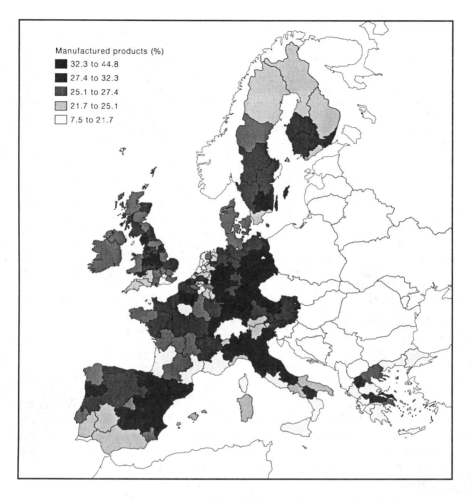

FIGURE 2.1 *Percentage employed in manufacturing products in the EU15,*
1996 (ESRC Resource Centre for Access to Data on Europe)

national, regional and local levels – for fresh inward investment to help
provide alternative sources of employment, generally with little success.

Sectoral change: the growth of the service sector, feminization and flexibilization of the labour market?

Without any doubt, the most significant aggregate structural change in
the European economy has been the growth of the service sector. As
well as the overall expansion of the service sector, however, there have
been important changes within it. Initially, service sector expansion was
closely linked with the growing role of the state in many north and

TABLE 2.4   *Unemployment in the EU15, 1985–95*

	Number of long-term (12+ months) unemployed			Numbers (1000) unemployed			Rate (%)		
	1985	1990	1995	1985	1990	1995	1985	1990	1995
Austria	n.a.	n.a.	n.a.	121	115	159[3]	3.6	3.2	4.3[3]
Belgium	304	186	244	449	283	391	11.3	7.2	9.3
Denmark	67	69	55	215	242	196	7.8	8.3	7.0
Finland	n.a.	n.a.	n.a.	130	88	429	5.1	3.4	17.2
France	1,041	873	1,189	2,436	2,259	2,977	10.3	9.4	11.9
Germany	890	673	1,505[1]	1,932	1,491	3,179	6.9	4.9	8.2
Greece	131	138	194	304	282	381	7.8	7.0	9.1
Ireland	145	118	129[2]	234	186	206[1]	18.0	14.1	14.6[1]
Italy	1,360	1,559	1,676	2,154	2,314	2,663	9.5	9.8	11.8
Luxembourg	2	1	1	5	3	5	3.0	1.6	2.9
Netherlands	380	233	238	601	526	552	10.5	7.7	7.5
Portugal	216[2]	900	164	408	229	336	8.8[2]	4.7	7.1
Spain	1,698	1,241	1,928	2,925	2,439	3,533	21.3[2]	16.3	22.7
Sweden	n.a.	n.a.	n.a.	124	76	333	2.8	1.7	7.6
UK	1,500	667	1,073	3,151	2,008	2,468	11.5	7.0	8.7
EU15	n.a.	n.a.	n.a.	n.a.	n.a.	17,805	n.a.	n.a.	10.7

[1] 1994; [2] 1996; [3] 1993.

Source: Eurostat (1996)

western European states (for example, see Esping-Andersen, 1990). Subsequently, there was some expansion in southern European states following their transitions from dictatorship to democracy. More recently, the dominant pattern has been of a shrinkage of the public sector and the expansion of private sector services, with the mass privatization programmes of central and eastern Europe providing a significant boost to the privatizing tendency (see Chapter 7). Within these broad categories, however, there is considerable heterogeneity, encompassing local consumer services, educational, health and social services, service provision for international tourists, and major international financial and business services, with varying geographies of spatial dispersal and concentration (Figure 2.2). In terms of its relationship to broader processes of economic change, the most significant feature has been the growth of private sector business and financial services, the latter often linked to financial product innovations and the deregulation of financial markets (see Chapter 3). In part, the growth of the service sector reflects increased out-sourcing, the introduction of new forms of inter-firm relationships and a re-definition of the social division of labour.

The increasing importance of the service sector has also been associated with important changes in European labour markets. Several features of significance can be identified. The first of these is the further

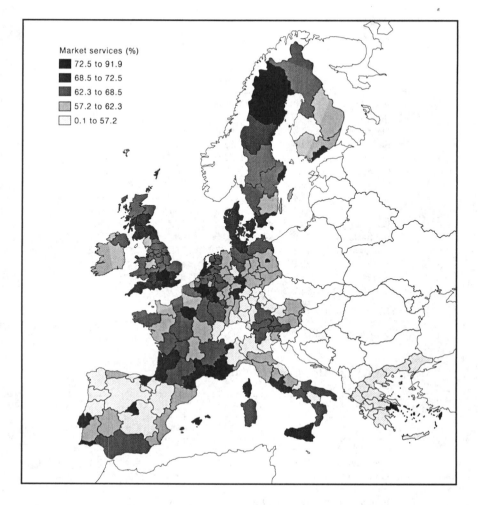

FIGURE 2.2 *Percentage employed in market services in the EU15, 1996*
(ESRC Resource Resource Centre for Access to Data on Europe)

'feminization' of the labour market. The employment rate for women has increased markedly, and more rapidly than that for men in all countries and the great majority of regions of the EU (Table 2.5). As a result of this growth, on average in the EU 70 per cent of women's jobs are in the service sector, while half of the service jobs carried out in the EU are held by women (though with significant national variations around these averages: see Chapter 8). There has been some growth of well-paid, full-time secure professional employment for women but, more generally, growth has been in less desirable jobs, in terms of pay and conditions of employment. Paradoxically, the female unemployment rate has simultaneously risen as the growth in demand for female labour has not kept up with that in supply (Table 2.6).

TABLE 2.5 *Employment rates (%) for men and women in the EU15, 1985–95*

	Female			Male		
	1985	1990	1995	1985	1990	1995
Austria	49.0	53.2	55.8	77.7	77.5	76.6
Belgium	37.0	40.8	45.4	68.7	68.1	66.9
Denmark	67.4	70.7	67.0	80.3	80.1	80.7
Finland	69.8	70.5	58.3	74.6	75.6	62.2
France	49.5	50.9	52.0	72.1	71.0	67.3
Germany	47.3	54.0	55.3	76.3	78.7	73.9
Greece	36.1	37.5	38.0	75.9	73.4	72.2
Ireland	31.4	35.5	39.8	67.7	67.8	65.3
Italy	33.6	36.4	35.6	73.6	72.0	65.7
Luxembourg	39.7	41.4	42.2	77.4	76.5	74.3
Netherlands	36.0	46.7	53.5	68.2	75.2	74.4
Portugal	47.3	53.3	54.3	76.9	71.6	71.2
Spain	25.1	30.7	31.2	63.3	68.2	60.8
Sweden	76.9	81.0	70.8	83.5	85.1	73.5
UK	54.2	61.7	61.4	75.9	80.5	74.8
EU15	n.a.	n.a.	49.5	n.a.	n.a.	70.0

Source: Eurostat (1996)

TABLE 2.6 *Unemployment rates (%) for men and women in the EU15, 1985–95*

	Female			Male		
	1985	1990	1995	1985	1990	1995
Austria	3.6	3.6	4.5[3]	3.7	3.0	4.1[3]
Belgium	17.9	11.5	12.3	7.4	4.6	7.4
Denmark	9.6	9.0	8.6	6.5	8.0	5.7
Finland	4.6	2.8	16.8	5.6	4.1	17.7
France	12.7	12.2	14.2	8.6	7.2	10.1
Germany	8.6	6.2	9.7	5.9	4.1	7.1
Greece	12.1	12.0	14.1	5.8	4.4	6.4
Ireland	19.5	15.2	14.8[2]	17.8	14.0	4.8[2]
Italy	15.5	15.8	16.3	6.4	6.5	9.3
Luxembourg	4.4	2.4	4.4	2.3	1.2	2.1
Netherlands	12.4	10.9	9.2	9.5	5.7	6.4
Portugal	11.9[1]	6.7	8.1	7.0[1]	3.4	6.8
Spain	25.4[1]	24.4	30.5	19.8[1]	12.1	18.2
Sweden	2.9	1.6	6.9	2.8	1.7	8.4
UK	11.1	6.7	7.0	11.9	7.3	10.2
EU15	n.a.	n.a.	12.5	n.a.	n.a.	9.6

[1] 1986; [2] 1994; [3] 1993.

Source: Eurostat (1996)

TABLE 2.7 *Women employed part time as a percentage of all employed women in the EU, 1985–95*

	1985	1990	1995
Austria	14.8	18.9	20.3[3]
Belgium	21.1	25.8	29.8
Denmark	43.9	38.4	35.5
Finland	12.1	10.2	11.1
France	21.8	23.6	28.9
Germany	29.6	33.8	33.8
Greece	10.0	7.6	8.4
Ireland	15.5	17.7	21.7[2]
Italy	10.1	9.6	12.7
Luxembourg	16.3	16.5	20.3
Netherlands	51.6	59.6	67.2
Portugal	10.0[4]	9.4	11.6
Spain	13.9[1]	12.1	16.6
Sweden	43.7	40.0	39.8
UK	44.8	43.2	44.3
EU15	n.a.	n.a.	31.0

[1] 1987; [2] 1994; [3] 1993; [4] 1986.

Source: Eurostat (1996)

TABLE 2.8 *Women employed part time as a percentage of all part-time employed people in the EU, 1985–95*

	1985	1990	1995
Austria	87.6	87.8	88.3[3]
Belgium	86.1	88.6	87.5
Denmark	81.0	75.7	73.3
Finland	71.0	67.8	65.1
France	83.0	84.0	82.0
Germany	90.5	89.7	87.4
Greece	64.7	64.9	62.7
Ireland	74.3	72.1	71.5[2]
Italy	61.6	67.2	70.6
Luxembourg	76.7	82.2	91.1
Netherlands	77.5	70.6	73.6
Portugal	65.8[4]	66.2	69.1
Spain	71.5[1]	78.0	76.3
Sweden	86.7	83.4	80.1
UK	88.0	86.2	82.3
EU15	n.a.	n.a.	81.1

[1] 1987; [2] 1994; [3] 1993; [4] 1986.

Source: Eurostat (1996)

The second feature to note relates to new forms of employment contract, and the growth of part-time and casual work. The expansion of female employment in services has also been closely related to the growth of new forms of employment contract and work organization so that women form a disproportionately large part of the so-called 'flexible workforce' (Tables 2.7 and 2.8). Employers may vary numbers of staff by the hour, day or week or use temporary contracts over longer time periods; increasingly there is a substitution of part-time for full-time work in many segments of the service sector (Townsend, 1997: 21). By 1992 over 20 per cent of employment in the EU involved such 'flexible' forms of work (Table 2.9; see also Commission of the European Communities, 1994b). Employers may combine strategies of 'numerical flexibility' with those of 'functional flexibility' (Atkinson, 1985), varying both the tasks that workers perform as well as the number of people performing them. However, 'flexibilization' for employers is often experienced as risk and uncertainty by employees, most of whom are women. Indeed, Beck (1992) interprets the growth of 'flexible employment' as a new kind of Taylorism of time and employment contracts, a point explored more fully below.

Much of the expansion of female employment has therefore been on a part-time, temporary or casualized basis, with irregular hours or intermittent contracts, in poorly paid service sector jobs. One-third of all women employed in the service sector work on a part-time basis.

European Communities, 1996b: 38). Their success in winning inward investment further reinforces their attraction as a destination for private sector investment and widens territorial inequalities within central and eastern Europe.

While acquisition and merger have led to companies becoming bigger within Europe, there has also been an increasing prevalence of strategic alliances among these bigger companies in search of economies of scale and scope as part of strategies to remain competitive. These have been particularly prominent in manufacturing industries characterized by high entry costs, the continuing importance of economies of scale, rapidly changing technologies and substantial operating risks (such as IT, biotechnologies, automobiles and new materials industries: Dicken and Oberg, 1996). In recent years, a parallel set of processes of acquisition, merger and strategic alliances has increasingly swept through parts of the service sector as diverse as banking, financial services and retailing in response to new competitive pressures and possibilities created by technological changes and deregulation of national markets (for example, see Townsend, 1997).

In the past decade or so, transnationals based in North America and South-east Asia have been increasingly actively involved in creating strategic alliances with European companies. There has also been a surge in acquisition activity by non-EU transnationals in Europe, along with increased direct inward investment into Europe. Securing access to the EU market, one of the three key macro-regions of the global economy, is a key element in corporate strategies of 'glocalization'. As with intra-EU flows, there was a surge in such inward investment in anticipation of the completion of the single market, with the gross inflow between 1986 and 1991 estimated at 120 billion ECU (Commission of the European Communities, 1994b: 12). Much of the inward FDI to the EU has been in banks and financial services, very heavily concentrated in existing major metropolitan centres, such as the City of London (Hudson and Williams, 1995). At the same time, there has been a tendency to locate some new types of service activity, serving international markets, in more peripheral locations. A combination of technological advances in computing and telecommunications, deregulation of many previously nationally protected markets in services and resultant pressures on companies to reduce costs to remain competitive has led to the location of international call centres and customer service centres in a diverse range of activities (such as airline billing and ticketing operations, direct banking and other financial services) in Ireland and the UK, often on the fringes of major urban areas such as Dublin (Coe, 1997) and metropolitan areas in the United Kingdom outside of the south-east region (Richardson and Marshall, 1996). The combination of relatively low labour costs (Table 2.10), ample available labour and English language skills has been critical in shaping the geography of these latter activities.

TABLE 2.10 *Average hourly labour costs in industry in the EU15, 1988–94 (ECU)*

	1988	Rank 1988	1994	Rank 1994	Change in rank (1988–94)
Austria	14.75	7	22.77	3	+4
Belgium	16.97	3	24.26	2	+1
Denmark	15.51	5	21.07	5	–
East Germany	n.a.	n.a.	16.44	n.a.	n.a.
Finland	17.56[2]	2	17.82	9	–7
France	15.27	6	19.12[1]	7	–1
Greece	5.35	14	7.64	14	–
Ireland	10.62	12	13.22	12	–
Italy	14.24	9	18.09[1]	8	+1
Luxembourg	13.61	10	19.29	6	+4
Netherlands	16.37	4	21.33	4	–
Portugal	2.98	15	5.53	15	–
Spain	9.13	13	14.13	11	+2
Sweden	14.43	8	16.05	10	–2
UK	10.97	11	12.65	13	–2
West Germany	18.27	1	26.14	1	–

[1] 1992; [2] 1988–94.

Source: Eurostat (1996)

There has also been further inward investment in manufacturing, but with a rather different geography. There has been some investment in activities such as European HQ offices and R&D in core regions, but it has been generally very limited in extent, as such activities typically remain deeply embedded in home countries (Dicken, 1994). Most inward investment has been in routine production activities. Acquisition of existing companies and factories and inward investment in new greenfield manufacturing plants have often been a way of guaranteeing access to EU markets in key consumer goods sectors such as automobiles, electronics and 'white goods'. New greenfield factories have typically been in peripheral locations in both eastern and western Europe, characterized by high unemployment (permitting great selectivity in recruitment strategies), generous state financial subsidies to cut the costs of capital investment and sometimes the absence of pollution regulations or laxity in their implementation. This in many ways reproduces the familiar pattern of branch plant investment in peripheral locations in Europe (which is discussed more fully below; see also Turok, 1993). There are, however, claims that these new factories (especially those of Japanese and South-east Asian origin) may represent a higher quality of investment and be more 'embedded' in the surrounding regional economies, with greater regional developmental potential than the earlier generation of branch plants (Cooke, 1995), although others contest this (for a fuller discussion of these issues, see Hudson, 1995b).

Heterogeneous production: the simultaneous persistence of different forms of production in Europe

Rather than a linear sequence of different forms of production organization sequentially succeeding one another, as 'better' ways of producing are discovered, the European economy is characterized by the
simultaneous persistence of a range of forms of production organization.
These include continuous flow production of basic chemicals and
metals, mass production of consumer goods (with automobile production the prime exemplar), new forms of high-volume production and
small-batch production of diverse commodities ranging from oil rigs to
haute couture fashion items in both large and small firms and factories.
This persistence of different forms of production reflects the fact that
there are social and material limits to the ways in which different
products can be viably produced. Differing product market conditions,
for example, can have a decisive influence on choice of production
technology. While consumer goods such as automobiles and refrig-
erators can be profitably mass produced, or produced in high volume in
other ways, products such as *haute couture* female clothing can only be
produced in, at most, small batches, while major items of fixed capital
equipment such as power stations or oil rigs are only produced to order,
usually one at a time. These points are explored in a preliminary way in
this section and the discussion is further extended in the following
section which considers the co-existence of different geographies of
production in Europe.

Mass production in Europe was and is characterized by the pursuit of
economies of time and scale in large factories (for example, see Best,
1990) and a high degree of vertical integration within the firm, *a fortiori*
in the former *combinats* of eastern Europe. It is marked by a deep
technical division of labour within companies and their, typically large,
production units. There is a sharp distinction between occupations
requiring mental and manual labour, informed by Taylorist views of
scientific management, with a strong vertical hierarchy of control;
individual workers are restricted to single, specialized, often deskilled,
tasks (Braverman, 1974). This can lead to problems – for employers and
managers – of lack of workers' motivation and innovation for it fails to
capture the knowledge that workers develop through doing their jobs.
One consequence of alienation is a tendency for the production line to
go down for long periods, with a direct impact on productivity (for
example, see Beynon, 1984). Another consequence is that there are often
problems with product quality, usually concealed on the line and then
dealt with at the end via fault rectification sections. Capitalist companies
typically produce the same components or products in different plants
to minimize the risks of labour unrest disrupting production.

Mass production is also characterized by a particular social division of
labour, regulated politically in state socialist economies and primarily

via prices and market relations between customers and suppliers in capitalist ones. Typically, capitalist companies pursue dual (or multiple) sourcing strategies to minimize the risks of problems with their suppliers. Even so, the risks of regulating relations between companies in this way generates requirements for large and expensive buffer stocks, giving rise to the characterization of production as 'just in case'. The uncertainties of the eastern 'command economies' generated similar contingencies, leading to the hoarding of parts and materials. Despite extensive stocks, problems of maintaining an adequate balance in component flows along the line, which interrupt its smooth flow, remain.

Mass production became increasingly crisis prone in western Europe from the late 1960s because of contradictions internal to this model of production organization, changes in the international division of labour and deep systemic crises in the political economy of capitalism. Within the factories, continuous attempts to speed up the line and intensify workload were increasingly resisted by workers, with an explosion of often unofficial strikes and other forms of industrial action – notably in the 'hot autumn' of 1968 – as the mass collective worker spontaneously organized to challenge managerial imperatives (Hudson, 1997b). Companies responded to this by seeking a series of 'spatial fixes', switching production first to parts of the European periphery and increasingly to areas of the Third World as a constituent moment in the emergence of a new international division of labour. Within western Europe, national states were finding it increasingly difficult to manage the economy in a way that preserved the class compromises of the welfare state, the balance between the growth of labour productivity and mass production and that of living standards and mass consumption. In the east, this form of production collapsed spectacularly and precipitately with the political changes of the post-1989 era, although there had been some reforms towards a market economy from the 1960s in Hungary and Poland. Such forms of production nevertheless remain within Europe. Although they are certainly much less prevalent than two decades ago, they are far from absent.

In the past two decades, in response to the perceived shortcomings of Taylorist mass production (and of successive spatial fixes to contain them, which are discussed more fully below), there have been a variety of experiments with new ways of producing in large quantities within Europe (Hudson, 1994a, 1997a). Perhaps their key common characteristic is a concern to combine the benefits of economies of scope, small batch craft production and a greater flexibility in responding to consumer demand with those of economies of scale. This is most sharply exemplified by mass customization with its goal of batch sizes of one: that is, uniquely customized commodities. Experimenting with new forms of high-volume production (HVP) can be seen as corporate attempts to embrace Schumpeterian concepts of competition via linked product and process innovations. But while HVP methods are in many respects more

flexible than 'just-in-case' production, their specific requirements introduce their own rigidities into the organization of production within and
between firms (Sayer and Walker, 1992). As forms of *high-volume*
production, they constitute variations around the basic mass production
theme. There are strict limits, deriving from the material and social
requirements of commodity production, as to the range of industries
and products in which these new HVP approaches *could* be applied (see
Hudson, 1997a).

While these new HVP approaches have common characteristics, there
are also differences between them. The choice of a particular HVP
approach is linked to labour and product market conditions. HVP production strategies can be connected with concepts of segmenting markets
by territory, culture and taste and then 'regionalizing' or 'glocalizing'
production in response to this (see, for example, van Tulder and Ruigrok,
1993). Companies have moved away from very high levels of automation
to more labour-intensive high-volume production technologies within
Europe in the 1990s (see, for example, Conti and Enrietti, 1995, for a
discussion of Fiat's strategy in the 1990s compared with the 1980s).
Lower fixed capital intensity offers greater scope for combining flexibility with profitability as aggregate levels of demand have declined
and/or as labour market conditions have changed (see Hudson and
Schamp, 1995).

The introduction of HVP has required some re-definition of the social
relations of production, between companies and between employers and
employees. HVP incorporates producing 'just in time' and close relations between customers and suppliers; for example, between manufacturers and their suppliers or retailers and their suppliers (Crewe and
Davenport, 1992). Relations between companies are often said to be
based on trust and linked with single supplier deals. An important
consequence of producing 'just in time' is that there are minimal buffer
stocks. Quality has to be 'built in' from the outset. Furthermore, problems of balancing flows of parts on the line are reduced since components are only ordered and delivered in response to orders from
customers (either other companies or final consumers) or from workers
further up the line. If, however, such imbalances do arise, production
quickly grinds to a halt.

New forms of HVP necessarily require workers to perform a wider
range of tasks than on the Fordist assembly line. There is considerable
debate on the implications of this change for those carrying out the tasks
of production. Some see workers as multi-skilled, empowered and more
creatively involved in the production process and place great stress on
teamwork. These new forms of work are seen by some to provide better
jobs, recombining the mental and the manual which Taylorism had torn
asunder (for instance, see Wickens, 1986; Cooke, 1995). Others dispute
this. For them, jobs in these factories are no better than those on the old
mass production lines of earlier years, involving not multi-skilling but

multi-tasking, a search for new ways of intensifying the labour process (Garrahan and Stewart, 1992; Beynon, 1995). From the point of view of labour, the new HVP approaches may involve greater intensification of work and stress than before. These alternative interpretations stress not the qualitative differences from the old Taylorist model but rather the continuities between the old and the new.

In order to introduce new ways of working, there have been corresponding changes involving greater selectivity in recruitment, putting more emphasis on 'appropriate' attitudes, making sure companies hire the 'right' people on the labour market, workers who will accept and adapt to new ways of working (for example, see Aviss, 1995). For those already employed in existing mass-production plants being reorganized around new HVP concepts, the fear of redundancy and unemployment is a powerful persuasive force in securing changes in established ways of working. The broader macro-economic and geographical conditions in Europe, especially generalized and persistent high unemployment, spatially concentrated, provide the context in which the texts about employment practices can be re-written within Europe.

In assessing the extent to which there have been significant changes in the dominant forms of organization of production in an (allegedly) post-Fordist European economy, it is important to recognize the variety of forms and relations of production that were to be found within the (allegedly) monolithic forms of mass-production organization in East and West. Put simply, not all production was, or could be, organized as mass production. In its western capitalist variant, there was a proliferation of small and medium-sized enterprises (SMEs), and the evolution of these in some circumstances into industrial districts is discussed below. There was also 'one-off' and small batch production of the means of production by big firms, continuous flow production of chemicals and metals, and widespread evidence of out-sourcing and sub-contracting alongside a high degree of vertical integration in mass-production sectors. In the eastern state socialist variant, there were various forms of 'informal' production between and around the formal state firms, including 'nomenklatura privatizations' in Hungary and Poland, and vertically integrated combinats, while mass production existed alongside other forms of small-batch production within the state sector.

Considerable emphasis has been placed upon the transformation of the role of former mass-production manufacturers and 'first-tier' component suppliers to a new role as 'systems integrators', co-ordinating the activities of a range of suppliers to which they have sub-contracted various production activities. Capital goods producers have operated in this way for decades (Vaughan, 1996). This particular organizational model is not, therefore, new. Often each new order or contract involved the creation of a new web of suppliers and sub-contractors for the precise purpose of fulfilling *that* order and attempts to introduce mass-production principles (for example into shipbuilding) met with at best very partial success. Moreover, such companies have 'traditionally' produced to order, 'just in time', and continue to do so, precisely because of the characteristics of the commodity and product markets. This system of production developed temporally alongside that of the mass production of consumer goods. Much of the allegedly radical change in forms of HVP was therefore prefigured in the era of mass production.

Finally, it is important to emphasize the extent to which Taylorist forms of work organization are becoming prevalent in many parts of the service sector, one response to pressures of cost competition (often intensified by market deregulation) and enabled by a combination of technological change and labour market conditions. There has been an expansion of 'traditional' Taylorization of work in many office and service occupations and the creation of 'new' forms of Taylorism associated with corporate strategies of flexibility (Beynon, 1995; Allen and Henry, 1997). While there are widespread proclamations heralding the end of Taylorism in manufacturing (though these are contested and by no means universally accepted, as we have shown), it is worth noting that commentators such as Beck (1992) have seen the creation of new forms of work in the service sector as heralding the ascendance of a new kind of Taylorism of time and employment contracts. As he puts it, referring to the individualization of employment relations and the destandardization of labour (1992: 147):

> In this sense, one could say that Taylor's 'philosophy of dismemberment' is transferred . . . from the substantive aspects of labour to the temporal and contractual relations of employment. The starting points for this new 'Taylorism of employment relations' are no longer situated in the combination of labour and machine but in the temporal limitation, legal (non)-protection and contractual pluralisation of the employment of labour.

Such processes are most marked – and not necessarily new – in much of the service sector; they have, however, been further intensified in occupations as diverse as serving 'fast food', cleaning offices and providing temporary secretarial help. While Taylorization may be shrinking in manufacturing, it is undeniably expanding rapidly over much of the service sector in Europe.

New geographies of European economies? Simultaneous tendencies towards spatial concentration and spatial dispersal

In contemporary Europe, companies are seeking new routes to competitiveness in an increasingly stretched-out and 'mobile' economy, with accelerating flows of capital and commodities as time–space compression reaches new levels. Capital is becoming even more (hyper)mobile but, seemingly paradoxically, place has become more significant to it as globalization has proceeded apace. Exploitation of place-specific concentrations of people in search of work, often reproduced in social and cultural settings that shift the costs of reproducing labour power on to local societies and states (via the payment of social benefits and meeting the costs of training as well as through unpaid domestic labour), has assumed an enhanced significance. This is reflected in Europe in simultaneous tendencies of agglomeration and deglomeration, dispersal and concentration of economic activities, with varying implications for urban and regional development strategies.

Spatial concentration and the (re)regionalization of the European economy

For many, the crisis of Fordism in Europe opened the door to a re-regionalization of the economy and new developmental possibilities for peripheral regions as a result. In part, this renewed focus on territorial agglomeration reflected a belief that the sorts of regional re-concentration of production associated with 'just-in-time' methods of production in Japan and their transplantation into parts of North America would also be transposed into Europe as a necessary corollary of the introduction of new HVP methods (Mair et al., 1988). Inward investment from Japan into Europe quickly revealed, however, that there is no precise correlation between 'just in time' and 'in one place'. For *some* components in the car industry there *is* undoubtedly evidence of 'in-one-place' 'synchronous production'. These are typically high-bulk, low-value components, with frequent deliveries and changes of specification, often located on purpose-built industrial parks adjacent to assembly plants (for example, see Ferrão and Vale, 1995). Furthermore, in *some* locations, especially where the transport infrastructure is very poor, as in the former German Democratic Republic, there is of necessity a greater degree of spatial clustering. Even so, this is often a 'pseudo just in time', with buffer stocks held in warehouses near to assembly plants. The absence of a firm correlation between 'just in time' and 'in one place' in what is typically seen as the prime exemplar industry (Hudson and Schamp, 1995) suggests that it is reasonable to conclude that this will also be the case in other sectors in which 'just-in-time' approaches can be implemented in Europe. Indeed, more generally the key issue is predictability and regularity of deliveries. The crucial variable is 'temporal certainty' rather than spatial proximity *per se*. In such cases, the vehicles used in transport function as mobile warehouses in which inventories are held.

There is no 'obvious' geography to other new HVP methods in the way that there initially seemed to be with 'just in time'. As a concern with economies of scale remains central in many industries, the introduction of lean production almost certainly means fewer factories.

Nevertheless, *where* these factories might be sited remains an open question, subject to locations satisfying labour market and other production requirements. There is, however, one way of interpreting the processes at work as leading to spatial agglomeration of production in existing core regions, amplifying rather than narrowing regional uneven development as a result. From this point of view, heavily automated assembly plants, requiring very high levels of fixed capital investment, will be drawn to existing 'core' regions, near to the main markets of the EU, further reinforcing the concentration of economic activities there. The more that such factories produce according to the principles of mass customization, the greater will be the propensity to locate in core regions as the increased inputs of mental labour into production, the greater weight attached to the role of R&D, and the much larger emphasis placed upon post-sales services as part of the product could help pull such production to them.

The strongest claims for the re-emergence of regional economies are not, however, associated with large firms and factories but with the (re)discovery of industrial districts based around networks of small firms in a variety of European locations. Over the past couple of decades, heavily influenced by Bagnasco's (1977) seminal study revealing the existence of a 'Third Italy', several empirical studies have been carried out that are claimed to lend credence to the notion that industrial districts are (re)emerging in parts of Europe. In some cases, high technology and certain sorts of micro-electronics production are seen as pivotal (Isaksen, 1994). Commonly cited examples include the M4 corridor, the M25 ring and the area around Cambridge in the UK and Grenoble in France. Baden-Wurtemburg in Germany represents a different case, in which the emphasis is more on the applications of new production technologies within existing networks of inter-firm relations, in which large firms are typically prominent. A third example is constituted by various sorts of areas specializing in niche market consumer goods, such as designer clothing and craft products, such as central and north-east Italy and parts of Jutland in western Denmark (see Dunford and Hudson, 1996). This variation suggests that a variety of processes may underlie such territorial clustering of production, and indeed Garofoli (1986) identifies significant variation within the Third Italy in this respect, suggesting a need for caution in categorizing regions with very different social relations of production into a uniform category of 'new industrial space' (as Scott, 1988, does).

There has certainly been a growth in the number of SMEs in Europe (not least because a switch to small firms and endogenous growth has characterized national and regional development strategies). This tendency has been present in eastern as well as western Europe. Prior to 1989, the emergence of SMEs was linked to '*nomenklatura* privatizations' in Poland and Hungary, and post-1989 there has been a rapid growth of new small firms over much of eastern Europe. Growing numbers of small firms and, more specifically, industrial districts and locally agglo-merated flexible production systems are seen to represent a new departure and a progressive response to the crisis of mass production (Storper and Scott, 1989). Industrial districts are characterized as relatively self-contained, product-specialized regional economies of linked small firms, which form internationally competitive nodes in a global economy. They evolve in circumstances in which there has allegedly been a significant and irreversible growth in consumer sovereignty, which requires that production be flexibly organized to meet volatile consumer demands for better quality and more differ-entiated goods, with increasingly reduced life-cycles. Consequently, satisfaction of market demands necessitates decentralized co-ordination and control, a horizontal division of labour between independent but inter-linked producers, numerical and task flexibility among workers who at the same time are required to display greater ingenuity and innovatory 'on-the-job' capacities, and elimination of time and wastage in delivery and supply. Such forms of production organization are claimed to be particularly evident in industries in which volatility and product innovation in niche markets is most pronounced. The new market conditions allegedly require a radical transformation of the production system towards flexible intra-firm and inter-firm arrangements which simultaneously combine economies of agglomeration, scope and versatility.

Moreover, it is argued that such a transformation requires a spatial concentration of the different agents involved in a production *filière*. Such agglomeration offers a series of Marshallian benefits, upon which a system of vertically disintegrated and 'knowledge'-based production can draw. This includes reductions in transaction and transport costs; the production of a local pool of expertise and know-how; a culture of labour flexibility and co-operation, grounded in associative social relations of trust and dense localized social interaction within civil society (see Cooke, 1995; Storper, 1995); and the growth of a local infra-structure of specialized services, distribution networks and supply structures. The resultant geography of production becomes organized around a set of internally cohesive regional economies which each grow in a virtuous circle of self-reproducing and self-regulating territorially based social relationships.

Small is therefore no longer necessarily tainted with connotations of technologically backward or archaic forms of production organization. Small firms may deploy quite technically sophisticated production methods, have sophisticated design capabilities and possess consider-able autonomy in marketing strategies. On the other hand, many SMEs in western Europe do lack such design and marketing capabilities and as a result produce 'rigidly' rather than 'flexibly', locked into either sub-contracting arrangements whereby they produce to the qualities and quantities specified by their customers at the controlling retailing end of the production *filière* or, alternatively, into the routine production of standardized components for those manufacturing companies further up the *filière* which form their markets. Much of the growth of both new small firms and new service sector activities, especially those denoted as producer services, in western Europe has been precisely linked to the new sub-contracting and out-sourcing strategies of major companies. Moreover, many small firms in eastern Europe, which emerged in the interstices of the former state socialist economies and continue to exist there, are characterized by qualities of obsolescence and archaism (see Smith, 1996b).

Most small manufacturing firms in Europe are *not* therefore embedded into the social and technical production structures of industrial districts. In sharp contrast, most small firms exist in the niches that are unattractive to bigger companies, are often subservient to them and established as a direct consequence of their strategies of concentrating on core competencies, and are engaged in fierce price competition with one another. In brief, their market position is based on profoundly asymmetrical power relations between companies in the production *filière*, characterized by fierce competition between SMEs for the markets which bigger companies provide. Even in the exemplar industrial districts of central and north-east Italy, many small firms are locked into dependent relationships with larger producers. The emphasis upon egalitarian networked relationships in industrial districts underplays the extent to which SMEs have been a persistent organizational form rather than a sudden arrival on the European corporate scene and overplays the extent to which SMEs are linked into the form of industrial districts with their connotations of socially progressive and territorially embedded production structures. Such forms of regional industrial structure are rare and will remain so. Moreover, they by no means have a guarantee of continuing economic success (for example, see Sunley, 1992). They may well be susceptible to the same sort of 'institutional lock-in' that has led to a switch from economic success to decline in other European regions (see Hudson, 1994b) as they fail to generate the sorts of institutional innovation that will allow them to become continuous 'learning regions' (see Hudson, 1996).

While recognizing that regional concentration of economic activities has perhaps increased recently, it is important to acknowledge three other points. First, industrial districts have a long history in Europe, pre-dating Fordism. Secondly, in the Fordist era, the early stages of mass production of consumer goods such as automobiles were characterized by spatial concentration of production. Production was concentrated in those major industrial conurbations in which mass production had sprung up in the decades between the 1920s and 1960s. Thirdly, there were forms of production characterized by territorial concentration in other industries. In shipbuilding, lead companies acted as 'system integrators' in a way that presaged certain notions of 'post-Fordist' production and supplier companies tended to be clustered around the assembly yards (Hudson, 1997a). There was also a powerful tendency towards spatial agglomeration in highly automated continuous process industries such as bulk chemicals and metals. This form of territorial clustering was and is based on input–output linkages (often between separately owned companies) because of a continuous interchange of products as one plant's or company's output becomes another plant or company's input. This territorial clustering was often associated with a move to new locations dotted around the coasts of Europe as these maritime production complexes increasingly relied upon imported raw materials and exported products. Often they were located in peripheral problem regions in which generous state financial aid was available to offset the massive fixed capital investment costs involved in establishing such complexes on greenfield sites (Hudson, 1983; Hudson and Sadler, 1989). They were not necessarily permanent features of the economic landscape of such regions, however. The extent to which investment costs were subsidized through regional and other state policies significantly reduced the turnover time of fixed capital from the perspective of private sector investors.

Spatial decentralization and the continuing spatial fragmentation of production processes

The paradigmatic form of spatial organization of production in the era of mass production became – and remains – the siting of different stages of the production process in different locations. In the state socialist period in eastern Europe, geographies of mass production were determined politically, both intra-nationally and internationally within the Comecon bloc, with particular areas assigned roles in spatial divisions of labour (for example, see Smith, 1996b; Swain, 1996). In western Europe, such geographies were determined by an interplay of corporate decisions and state policies. Initially, production was concentrated in those major industrial conurbations in which mass production of consumer goods had become established in the decades leading up to the 1960s. As this approach to production reached its limits in the 1950s and 1960s, companies sought new 'spatial fixes' to preserve the viability of the old production model. Thus new spatial divisions of labour emerged, first intra-nationally, then internationally (Hudson, 1988). While key strategic decision-making and R&D remained in the old core areas, various peripheral locations (previously unindustrialized agricultural regions and deindustrialized former workshops of the world) became the location of unskilled and/or deskilled work in component production and/or routine assembly. Companies were persuaded to locate such functions there by the attractions of abundant available labour, along with often generous financial incentives in the form of grants and loans. Labour-intensive component production, organized on classic Taylorist mass-production principles, continues to be located in the cheap labour and high-subsidy peripheries of Europe. The opening up of eastern Europe to capital investment from the West has again re-defined the ways in which Europe's peripheral regions are being incorporated into the mass-production chain. It has created fresh opportunities for corporate use of peripheral spaces to preserve Taylorism and intense competition between peripheral locations in Europe, both eastern and western, for such investment. Some companies therefore continue to find that Taylorist mass production remains viable within Europe via exploitation of favourable localized labour market conditions. It is clearly premature to announce the death of Taylorism and mass production in Europe. Moreover, in developments that raise further questions about the reproduction of industrial districts, firms within industrial districts are increasingly internationalizing their production strategies via sub-contracting labour-intensive operations and/or direct investment to exploit the spaces of eastern Europe recently re-opened to capital and in which Taylorist production is being established (Dunford and Hudson, 1996). The opening up of new economic spaces in central and eastern Europe is also having other implications and impacts, particularly in conditioning capital–labour relations in other parts of western Europe, especially in adjacent border areas in countries such as Austria (see Tödtling, 1997).

New forms of HVP now co-exist alongside more traditional forms of small firm and mass production within Europe, however. While some interpretations of the immanent geographies of HVP emphasize a tendency to spatial concentration and a re-regionalization of production, others point out the extent to which these new forms of HVP involve the use of spatial differentiation within Europe that bears a marked resemblance to the geographies of mass production. Key decision-making functions and R&D tend to remain located in core regions as existing companies introduce HVP methods, underpinned by both national government and EU policies intended to enhance the international competitiveness of European companies (for example, see Commission of the European Communities, 1996a). When inward-investing companies from South-east Asia establish decision-making and R&D activities in Europe, they too tend to follow this pattern; Nissan, for example, has its European HQ in Amsterdam and its R&D facilities at Cranfield, in the south-east of England. In contrast, both routine production and experimental plants are located in peripheral locations in western Europe with little or no history of automobile production, the latter so that companies can experiment with new ways of producing without endangering production in their main core plants (Hudson, 1994a, 1997a,b). New Fiat plants in the Mezzogiorno in the 1990s support such an interpretation, for example (see Conti and Enrietti, 1995). The locational strategies integral to the penetration of capital into parts of eastern Europe illustrate, *a fortiori*, a variant of such an approach; for instance, General Motors' new plant at Eisenach and Volkswagen's at Mosel (see Schamp, 1995). The availability of substantial regional policy grants in these parts of the European periphery, plus the ready availability of skilled engineering labour which is none the less amenable to the introduction of new working practices, make *these* locations attractive for the introduction of new HVP methods (for example, see Swain, 1996). Indeed, a large proportion of inward investment into the automobile sector in eastern Europe has been in the form of acquisitions of existing plants (for example, Volkswagen's investment in Skoda), in areas with experienced workers willing to accept new ways of working for wages that are a fraction of those paid in comparable western European plants.

There is also evidence of similar geographies of production being created in the service sector, linked in considerable measure to the Taylorization of work. This is reflected in the ways in which geographies of service provision are changing within Europe, for a range of both producer and consumer services (for example, see Townsend, 1997). On the one hand, key decision-making functions and R&D remain located in existing core regions (see Chapter 3). New service activities, especially business and financial services, have also been concentrated in major metropolitan areas, particularly those aspiring to and having attained 'world city' status (see Hamnett, 1994), although there is evidence that capital cities are generally attracting less of these activities than other major cities of western Europe (Moulaert and Tödtling, 1995). Nevertheless, this reinforces the gap between major metropolitan centres and other European locations. Similar tendencies are observable in eastern Europe, with FDI in banking and business services heavily concentrated in capital cities (see above). Their growth has been linked to increasing spatial inequality both between these areas and other cities and regions but also within the metropolitan areas themselves.

While the highest-level decision-making functions are increasingly concentrated in the central metropolitan areas, there is some evidence of the decentralization of 'back office' functions, which routinely process large volumes of paper and/or electronic transactions within the metropolitan areas and their immediate hinterlands, to peripheral locations on the fringes of cities (Illeris, 1989) and on occasion to more distant locations in peripheral regions (for example, see Beynon et al., 1994). There are also important changes taking place re-shaping the geographies of the service sector at an international scale within Europe and beyond. Changing forms of organization of the labour process within parts of the service sector, coupled with technological changes in electronic and remote communications which have radically reduced the tyranny of distance, and regulatory changes that have opened previously protected national markets to the forces of global time–space compression, have led to new geographies of service production. These echo the geographies of decentralized branch plants of Taylorized mass production in manufacturing as part of what was once hailed as a new international division of labour (Fröbel et al., 1980). Driven by a competitive necessity to cut production costs by reducing labour costs, companies in sectors such as financial services have re-located 'back office' operations that deal with routine processing of data on a mass basis to peripheral locations, both intra-nationally and internationally, in which cheap, typically female, labour is available. More recently, the English-speaking parts of the EU have become the location of a range of international call centres and service centres dealing with a range of operations from tele-marketing to sophisticated computer support (as noted earlier). Paralleling earlier developments in manufacturing, such service centre companies are now establishing sophisticated international divisions of labour within the EU, within Europe more widely (for example, Russia is a major centre for sub-contracting computer programming) and beyond (with India competing with Russia in this market, for example).

New threats to and opportunities for organized labour and trade unions

Historically, trade unions developed from a plant or local basis to national organizations; sometimes these took the form of industry-specific unions, sometimes more 'general' unions representing workers in a range of industries. The national increasingly became the territorial basis on which trade unions as institutions of organized labour were built, especially in the Fordist era of national regulation within western Europe (although there were many trade unions in which important sub-national divisions remained). As well as trade unions being involved in agreeing national corporatist bargains, workers developed other sources of power and influence within workplaces. Above all, the mass collective worker, in the big Taylorized factories in urban labour markets, was able to wield considerable bargaining power over wages, terms, conditions and 'grievances' procedures, though not over the control of the labour process. As companies sought to respond to competitive pressures by increasing line speeds and labour productivity, there was growing shop-floor unrest and a sharp increase in 'unofficial' industrial action, especially marked in some sectors of manufacturing (notably automobile production) and some countries, such as France , Italy and the UK in the late 1960s. Many other sectors remained much less heavily unionized, with trade unions relatively under-represented in much of the private service sector economy and in small firms, though with exceptions to this general trend (for example, small firms in parts of the Third Italy such as Emilia Romagna were heavily unionized).

The growing power of the mass collective worker during the 1960s in the First World led to companies re-assessing their production strategies and ushered in a series of spatial fixes, initially intra-nationally (for example, Fiat within Italy; see Amin, 1985), but increasingly internationally (Dicken, 1994) to try to preserve the competitive viability of mass production. This divided workers within companies and sectors. Attempts to restore unity via 'combines' and in other ways met with little success in the face of companies' ability to divide and rule via relocation and capital flight. The organizational and institutional bases of trade union organization were further challenged by the growing tendency to globalization, further calling into question the salience of 'the national'.

The subsequent increasing adoption of new production concepts – new forms of high-volume production, flexibly specialized small firms – further undermined traditional trade union methods of organization. An increasing reliance upon out-sourcing and sub-contracting further divided workers from one another. Along with sectoral change in the economy – above all the expansion of the traditionally little unionized service sector – this posed growing challenges to established forms of trade union organization and representation of workers' interests. Moreover, the growing neo-liberal turn in state policies in parts of western Europe sought to re-define and restrict the scope of trade union activities, in turn facilitating such changes.

The changing structure and increasing multiple segmentation of the labour market posed further problems for trade union organization. HVP approaches both require and permit the shattering of old forms of trade unionism and the institutions of labour and their recasting in new moulds. The new forms of work organization and union responses to them can further serve to divide workers from one another. While many workers were worse off as a result of sectoral and organizational changes, left in marginal, precarious jobs as a result of such changes, other workers seemed to be more secure and materially better off. New labour recruitment, retention and work organization and remuneration practices, such as performance-related pay and individually tailored contracts for key 'core workers', further served to create divisions within workforces. Trade unions often responded to the threat of non-union workplaces (for example, with works councils rather than trade unions) by agreeing to single-union deals, a return to 'new' forms of company unionism by trade unions desperate for a chance to recruit new members to compensate in part for losses as a result of decline in other sectors and companies. They were often only too eager to trade off sole bargaining rights for various 'sweetheart' deals to combat their own falling membership.

On the other hand, if many of the changes challenged established trade union modes of operation, they also – so some argue – opened up new possibilities for influence and action. New flexibly networked 'just-in-time' production models are very susceptible to any stoppage in production. Companies are acutely vulnerable to interruptions to production in approaches built around vertical disintegration, 'just-in-time' principles and minimal stock levels. This was sharply emphasized by the strike in October 1991 at Renault's engine and gearbox plant at Cleon, which soon ground production to a halt in many of Renault's other plants for which it was the sole JIT supplier. Companies therefore have to strive to ensure that individual workers and organized labour remain co-operative and compliant, to re-form the institutions of organized labour and re-define the culture of capitalism in Europe in a way that is consistent with their interests. They do so on a terrain which is favourable to this project. For, in contrast to the era of 'full employment' in the 1960s, the 1990s are marked by high unemployment in many industrial areas. This is the venomous sting in the tail from the point of view of labour. For the new models of work organization are undeniably predicated on their being *no* return to 'full employment'. Such labour market conditions would destroy companies' capacity to be so selective in deciding which individuals to employ.

It is also suggested that there are new opportunities for transnational trade union organizations and co-operation as the scale of regulation shifts upwards; for example, to the EU and initiatives such as the social chapter of the Maastricht Treaty (see Wills, 1997). On the other hand, there is a fierce competition in the place market for investment and jobs, and trade unions have often become involved in territorially based cross-class alliances in defence of 'their place' (for example, within the steel industry; see Hudson and Sadler, 1983). Others suggest that there is scope for trade unions to take on a broader role than one simply confined to employment issues and plant and industrial politics. Unions can, for example, become involved in providing a wider range of services to their members, re-define their role in terms of greater community involvement in issues such as training and employment provision (as, for example, ISTC (Iron and Steel Trades Confederation) is seeking to do in the UK). It is also important to remember that in some places trade unions have long had such an embedded role in local society and community – for example, in Emilia Romagna in the Third Italy – but that this is based on long-established tradition. It may not, therefore, be easy to transfer the lessons of this experience to other places.

Conclusion: the future of the European economy and its geographies

The recent restructuring of the European economy – sectorally, spatially and organizationally – has been closely linked to broader processes of globalization as well as processes of fundamental political–economic change within Europe. It is by no means obvious that Europe will continue to be successful in a global struggle with the other constituent parts of the triad, in particular a resurgent USA. The EU will surely continue to expand and evolve, but Europe's position as one of the three core regions of the global economy is by no means automatically assured. The drive towards economic and monetary union can be seen as central to a political project that seeks to underpin Europe's competitive position within the triad. Ensuring the future economic success of Europe, and more specifically of European-based companies in the global economy of the triad, requires the creation of a more deeply integrated European economic space (although successfully completing EMU looks to be far from unproblematic). It remains an equally open question whether – and if so how – creating a more deeply economically integrated European Union can be made compatible with further significant enlargement of the European Union. What is clear, however, is that further economic integration and the homogenization of the European economic space will both pose threats to and create opportunities for economic development within Europe. It will remain a critical market for major transnationals (even if Europe slips somewhat in the pecking order within the triad), and as a result they will seek to invest within Europe.

There is no doubt that the European economy has experienced fundamental changes over the past two decades; these include continuing deindustrialization, tertiarization, growing female employment, experiments with a variety of new forms of production of goods and services, and a growth in flexible forms of working. It is equally clear that further changes will follow. There are, however, limits to the extent to which the growing relative importance of the service sector and of the employment of women, often in poorly paid and 'flexible' jobs, can proceed. There are also grave doubts about the longer-term social sustainability of continuing high levels of unemployment as these pose serious challenges to the 'European model' of society and to sociospatial cohesion within Europe. Discovering a new macro-economic growth model that can address problems of environmental sustainability, and social exclusion, is a major policy and political challenge for Europe in the twenty-first century (see Hudson, 1995b; Hudson and Weaver, 1997).

Mass production in both eastern and western Europe is now less significant than it once was but it has by no means disappeared completely. Indeed, even in the era of mass production, there was a plethora of other approaches to the production of goods and services, both in large and small firms that co-existed alongside Taylorism in large workplaces. Mass production still co-exists alongside other organizational models, now including new forms of high-volume production based on large firms and more flexible forms of small firm production associated with industrial districts. Economic evolution is more complicated than a simple sequential change so that these differing forms of production organization co-exist in Europe as elements in a more complex mosaic of methods of production. The range of options open to private sector companies is certainly now wider than it has ever been. The possible combinations of choices of location, both inside and outside Europe, production technologies and products varies with forms, types and sizes of companies. Overall, however, the range of available choices is unprecedented. This poses immense problems for trade unions seeking to create effective labour organizations in the face of this complexity and fragmentation which effectively outflanks their traditional models of organization and action, especially in a context of generalized high unemployment. On the other hand, new constraints, such as those of ecology, may be becoming more important. How, then, will the economic geography of Europe evolve in the foreseeable future?

The continuities with the past will be at least as strong as the breaks with it. The construction of Europe as an increasingly homogenized economic space will enhance opportunities for companies both to create and exploit differences between places in the organization of production. The future will continue to be one of uneven development, in all probability more so than in the past. There is little in the past record of national and EU industrial, regional and cohesion polices to suggest otherwise, especially as the policy challenges of combined and uneven development within the EU will intensify with successive eastward enlargements. Those involving the southern European countries in the 1980s, with much less advanced economies (Hudson and Lewis, 1985), led to much greater heterogeneity within the EU in terms of national and regional economic structure and performance, forms of corporate organization and the balance of large and small firms. Enhanced socio-spatial inequality posed serious challenges to emerging cohesion policies within the EU. These will pale into insignificance in relation to the effects of enlargements involving the eastern European countries and the policy challenges that these will pose. The re-unification of Germany provides graphic evidence of the implications and impacts of expanding eastwards, the more so as the former GDR was the most economically advanced of the eastern European countries and the former FRG was able to channel vast resources into the new eastern *Länder* on a scale that will not be repeatable elsewhere.

Scenarios which suggest a future of generalized regional regeneration around flexibly specialized ensembles of firms linked into industrial districts are fundamentally misconceived. On the one hand, it is by no means obvious that such districts will be able to avoid the dangers of institutional lock-in and continue to evolve as economically dynamic and successful learning regions. On the other hand, there have always been sectors and areas within capitalism within which small batch production for niche markets has been present. But there are strong systemic pressures within capitalism towards high-volume production as a consequence of competitive pressures, both between capital and labour (to force down wage costs) and between companies (over the distribution of surplus value). The economic geography of Europe will continue to be dominated by major transnationals (many controlled from outside Europe) and by high-volume production and consumption. Moreover, there will be strong public policy pressures for this to be so, at national and EU levels, in order that European firms can successfully compete in key sectors of the global economy.

The future economic map of Europe will equally be one of growing concentration of corporate power, high-level business and banking services, R&D and advanced manufacturing producing the technically sophisticated new products in existing 'core' regions. There is some evidence of the pressures of time-compression in product and process development forcing firms in advanced or growth sectors to rely on spatial propinquity as a key element in their competitive strategies but in so far as these pressures have an impact, it is in further concentrating such activities in existing core regions. Furthermore, EU and national research, technology and development policies have 'an *in-built* tendency to favour the richer regions of the Community' (Commission of the European Communities, 1996a: 7, emphasis added), reinforcing the advantages of 'islands of innovation' and the qualitative gap in developmental potential between the regions of the EU. In recognition of this, there have been attempts to promote such activities in peripheral regions and countries, but the funds allocated to this have been limited and, in any case, there are powerful pressures which would constrain such activities to location in core regions (for example, in terms of access to key financial and political decision-makers or skilled workers). In so far as there is spatial deconcentration of such activities, it will generally be limited to locations in and around the fringes of the major metropolitan regions.

Consequently, the suggestion that Europe will experience a generalized reconstruction of regional economies around clusters of HVP industries producing 'just in time' and 'in one place' is no more tenable than that of a future of industrial districts of networked small firms. This will not be the future of the vast majority of peripheral regions. In strong contrast, their future will continue to be characterized by routine production in both manufacturing and services. For many in the margins of Europe (Hadjimichalis and Sadler, 1995) the future will offer even less: precarious work in parts of the service sector such as retailing and tourism or, for many, some combination of informal work, illegal activity and long-term unemployment. The European industrial economy will remain one that is spaced-out, and although the specifics of the spatial patterns will continue to alter, there will be strong threads of continuity with the spatial divisions of labour of mass production in both East and West. There will, of course, be favoured locations, in which well-paid, skilled jobs of high quality will be found. Even so, these will tend to be in deeply segmented metropolitan labour markets, with sharp socio-spatial divisions within them. The policy and political challenges of socio-spatial inequality will intensify rather than subside as Europe seeks a new, more socially inclusive and sustainable model of economic development.

3

European Financial Integration and Uneven Development

Adam Tickell

The 1990s has been a momentous decade for the European financial sector. At its start, Europe's financial system consisted of: (a) the separate national systems of the member states of the European Union (EU) which, although economically linked, were subject to very different political–economic circumstances, regulations and cultures; (b) the separate systems of the market economies on the periphery; and (c) the largely centrally controlled, state-owned systems of the former Warsaw Pact countries. By the end of the decade, the whole of Europe's financial system will be dominated by the single European currency, the euro; there will be a European central bank and financial regulations will be largely compatible. This is as true for countries not in the EU as for those that are. This chapter examines the development of an increasingly integrated European financial space during the 1990s, particularly during the '1992 process', as the European Commission strove to turn the rhetoric of economic community into economic reality with its attempts to force nation-states to harmonize regulations and create a single market, and in the development of European monetary union (EMU) after the Treaty of Maastricht in 1992. Although the Commission had neglected the financial sector for most of its existence, during the mid-1980s it came to believe that financial restructuring, particularly through the promotion of competition and regulatory harmonization, was more important than any other sector. The Commission believed that the European financial system was 'inefficient' on a global scale because the presence of a plethora of smaller institutions hampered its ability to compete with US and Japanese institutions and, more importantly, that inefficiencies impacted negatively on the economy more generally.[1] Indeed, somewhat hyperbolically, the Cecchini Report (Cecchini, 1988) claimed that harmonization of regulation in the financial sector could bring half the total efficiency savings of the whole of the

1992 process: 'The integration of financial markets across Community borders is uniquely important, however, in the sense that it will not only have important effects on the efficiency of the sector itself, but also on the efficiency and resource allocation of sectors using financial markets' (Commission of the European Communities, 1988). However, it is argued in this chapter that the harmonization of regulation in the financial sector has had a significantly smaller impact than the Commission anticipated and that much of the restructuring that took place in the EU during the early 1990s was a response to existing processes of globalization in financial markets, while eastern and central European countries were responding to the challenges of moving to a new economic system. The second section of the chapter goes on to consider the likely effects of the single currency on European finance. It is argued that this will re-shape its geography, both within the EU and elsewhere in the continent, far more fundamentally than regulatory harmonization has done, particularly by exacerbating the already heavy geographical concentration of activity.

Creating a single market?

The global financial industry went through a transformation during the 1970s and 1980s. The relatively stable financial order of the post-war period was undermined when President Nixon began to print dollars to pay for the escalating costs of the Vietnam War and as financial institutions increasingly took advantage of the 'euromarkets' where they could trade in currencies away from regulatory control (Strange, 1987; Thrift and Leyshon, 1988; Leyshon and Tickell, 1994). The global financial order gave way to a volatile and integrated international financial system. Nation-states found themselves increasingly unable to control the activities of the financial markets as banks internationalized and made use of their relative mobility to circumvent regulations, effectively forcing governments to 'deregulate' (Cerny, 1993). National regulators increasingly found that the informal and clubbish regulatory styles appropriate for nationally based financial systems had less and less resonance and influence (Moran, 1991). Yet, with the notable exception of London and, possibly, Frankfurt, the European financial system remained relatively untouched by such pressures. While international finance was being reformed and increasingly concentrated in global centres, the overwhelming majority of EC member states retained significant control over their national financial systems through regulatory particularism, state ownership of financial institutions, exchange controls and legally imposed barriers to entry by new institutions.[2]

As Europe became increasingly integrated during the 1980s, the Commission became concerned that Europe's financial system was not only one of the areas where progress towards economic integration was slowest, but that it was incapable of meeting the needs of European commerce and industry (Vives, 1991; Llewellyn, 1992). Although the problem was a multi-faceted one, the Commission believed that the principal problem was that the plethora of national regulations, the majority of which tended to restrict intra-European trade, inhibited the development of competition within the sector. Furthermore, the breakdown of the post-war economic settlement during the 1970s (Tickell and Peck, 1992) undermined the European financial arena even further. As Leyshon and Thrift (1997) have pointed out, the European regulatory framework was ineffective in an era of globalizing credit provision for two reasons. First, fragmentation allowed for pan-European speculation by financial institutions wishing to capitalize on different regulations and different prices and, secondly, European states became subordinate to Japan and the USA in credit provision, allowing Japanese and US companies to dominate world banking.

The publication of the Cockfield Report in 1985 laid the foundations for the development of a harmonized financial system, arguing that member states needed to liberalize capital movements entirely, to standardize most regulations and to introduce a single banking 'licence'. This meant that if an institution was authorized to carry out business by one member state, it would be able to carry out business in all member states as long as the activity was legal. The Commission acted swiftly after this and between 1985 and 1990 it issued 21 directives affecting banking, capital markets and securities, including the Directives on Capital Movements (1988), the Second Banking Directive (1989) and the Investment Services Directive (1993). The common aim of these initiatives was to create a single European market in financial services by:

- abolishing exchange controls;
- developing a single securities market;
- creating the freedom for any European institution to locate anywhere else in the Community;
- establishing the right for consumers to buy financial services from anywhere in the Community;
- establishing the freedom for companies to supply financial services without specific authorization from each member state (see Vives, 1991; Llewellyn, 1992; Warner, 1992).

Although seemingly technical and arcane, the potential impact of the changes being wrought by the Commission was immense. As Tsoukalis (1997: 98) has argued, 'Although some parts of the territory were still left uncharted, the significance of the [Second Banking Directive] cannot easily be overestimated. With one stroke, the EC countries went further than a federal country such as the United States in creating the conditions for an integrated banking sector.' By the time the Second

Banking Directive was in force on 1 January 1993, for example, an institution that was authorized by its home supervisor effectively had a licence to operate in all member states, as long as it complied with local regulations (*The Banker*, 1990).[3] In the run up to 1992, there was largely a consensus that the changes in regulation would transform European finance (although see Llewellyn, 1992). The Commission believed that increased competition and efficiency would remove internal cross-subsidies, lead to reduced prices, increase the number of mergers in the industry and increase diversification. Liberal commentators such as Xavier Vives (1991: 10) argued that '*the main effect of integration will be to change the focus from collusion and regulatory capture to competition . . . The weight of the history of the industry, with its restrictive practices linked to a heavily regulated environment is being erased by the "1992" idea*' (original emphasis).

In the event, however, the introduction of a set of rules which *formally* create a single market have been insufficient to produce a truly competitive market and the 1992 process has had a relatively minor impact on European finance. The European financial system continues to remain dominated by a set of national systems with variable approaches to the financial sector. During the 1990s, for example, significant state aid to troubled banks was made in France (when FFr 5bn, together with a FFr 18.4bn guarantee for real estate, were loaned to the state-owned Credit Lyonnais in 1994 to stave off the threat of bankruptcy),[4] while in Sweden in 1992–3 the government provided 90.5bn kronor to help banks overcome bad debts. However, the convergence of policy around Anglo-American norms has contributed to the further concentration of wholesale financial services in London and, to a much less developed extent, Frankfurt. For example, London's share of branches of non-European banks in the EU has continued to rise during the 1990s from 43 per cent in 1991 to 47 per cent in 1996 (calculated from *The Banker*, November and December 1991, 1996a) and continental European banks increasingly transact their wholesale activities there.

Retail banking was, perhaps, least touched by the 1992 process. If, for example, the number of bank branches per head of population is used as a proxy for the efficiency of the retail banking sector,[5] Table 3.1 shows that for Europe as a whole there was no appreciable change in the number of branches during the 1990s, while only the UK and the Scandinavian countries saw significant falls in branch numbers.

Implicit in the Commission's strategy was a desire to see the number of financial institutions in Europe fall in order to help to create globally competitive financial institutions, principally through mergers and acquisitions. However, again, the numbers of mergers and acquisitions has remained quite small and, in every country in the Union, has *fallen* since 1991–2. Part of the explanation for this is that the rhetoric of 1992 stimulated banks to think about their competitive position in Europe *before* the single market was completed (see, for example, Thomson and

TABLE 3.1 *Number of bank branches (000s) in the EU10, 1980–95*

	1980	1990	1995	Change 1980–95 (%)	Branches per 1,000 people (1995)
Belgium	7.8	8.3	7.8	0.0	0.77
Finland	3.4	3.3	2.1	−38.2	0.42
France	24.3	25.7	25.5	+4.9	0.44
Germany	39.3	39.8	37.9	−3.6	0.46
Italy	12.2	17.7	23.9	+95.9	0.42
Netherlands	5.5	8.0	7.3	+32.7	0.47
Norway	1.9	1.8	1.4	−26.3	0.36
Spain	25.8	35.2	36.0	+39.5	0.92
Sweden	3.7	3.3	2.7	−27.0	0.31
UK	20.4	19.0	16.6	−18.6	0.28
EU 10	144.3	162.1	161.2	+11.7	0.43

Source: McCauley and White (1997: table 16)

Taylor, 1994) and research carried out by the Bank of England in 1993 identified 247 cross-border mergers which were stimulated by the fear of it (Bank of England, 1993). However, as Altunbas et al. (1997: 324) have recently demonstrated, a more convincing explanation for the relatively limited number of mergers and acquisitions in financial services during the early 1990s was that 'only limited opportunities for cost savings from big-bank mergers [exist and also] that such mergers are more likely to result in an increase in total costs'. However, monetary union is likely to increase the pressures on institutions in countries which fully participate in the euro, in other members of the European Union and in other European states, which will increase the likelihood of mergers in the financial system (Warner, 1997).

Furthermore, in some countries in Europe the – geographically variable – institutional structure of ownership has acted as a barrier to change, particularly as far as state ownership is concerned. While state-owned banks control over 70 per cent of the banking market in Greece, and similarly high levels in other southern countries, privately owned groups are dominant in Britain, Germany and the Netherlands (where just five groups controlled over 80 per cent of the market for all financial services in 1991; Bakker, 1992). In Italy, for example, there are approaching 1,000 banks which are less profitable than banks elsewhere in Europe and have poor credit ratings. Mergers and take-overs have been slow to develop because the majority are owned either by the Italian state or by not-for-profit foundations and also because the Bank of Italy was, until 1995, publicly opposed to hostile take-overs (Lane, 1994; Moore, 1996). Yet it is worth remembering that, contrary to the liberal theories which see a large number of banks as an inevitable drag on efficiency, the localized and not-for-profit nature of banking systems, such as in Italy or in Jutland, have traditionally supported small and medium-sized enterprises and provided them with shock absorbers during cyclical down-turns (for example, Carnevali, 1996; more generally, see Dow, 1994; Martin, 1995; MacKay and Molyneux, 1996). One of the dangers of creating an 'efficient' and 'competitive' financial system is that smaller businesses will be less able to ride out the vagaries of the economic cycle.

The attempts during the early 1990s to create the single market in financial services foundered because they were carried out on the basis of regulation alone. Regulatory change is a necessary, but in itself insufficient, stimulus to restructuring because its impacts are tempered by pre-existing and continuing economic and cultural factors. Economically, and ironically, the existing over-capacity in some European financial markets has, in some cases, prevented restructuring of the industry from taking place for two reasons. First, while it may be beneficial for the banking sector as a whole to reduce its capacity, if any institution were to act unilaterally to do so their competitors would gain market share at their expense. Secondly, chronic over-capacity has acted as a barrier to more efficient foreign institutions entering markets in the way that the Commission originally envisaged (Llewellyn, 1992; Dermine, 1996; McCauley and White, 1997).

There also remain significant cultural differences within Europe which have affected the competitiveness of the markets, both in terms of the range of products used by consumers (D. Miles, 1994) and in terms of the methods used by companies to raise finance (historically ranging from the bank-based system in Germany to the equity-based system in the UK, see Pratt, 1998). Furthermore, the intangible nature of money means that consumers of financial services have been reluctant to move their custom to names that they do not trust. Finally, although the regulatory changes introduced significant harmonization, there also remain significant differences in legal and tax frameworks, and while financial regulations and supervisory cultures are converging, they remain different (see Gardener and Molyneux, 1990; Begg and Green, 1995).[6]

This, however, is not to say that there were no changes in the European financial system during the early part of the 1990s but that the 1992 process probably had less impact than the transformation of the *international* financial system (Leyshon and Thrift, 1997). In this, perhaps the central development was the international normalization of Anglo-American financial structures and mores. One of the tensions that emerged during the negotiations in the Commission was the extent to which the new rules reflect practices in different countries in Europe. For example, German banks have long been 'universal banks' which operate investment banking operations in-house, whereas in the UK the same function has, until recently, been performed by independent merchant banks and similar institutions. Regulations in each country to ensure the stability of the financial markets reflected these differences: whereas UK merchant banks were subject to relatively low capital ratios,

ratios were higher in Germany because equivalent work was carried out within banking groups (for a discussion of the general characteristics of competing European economic and welfare models, see Chapter 1). Accordingly, in the negotiations over the Investment Services Directive, Germany argued that low ratios would penalize their institutions, whereas British representatives argued the opposite (see, for example, Tsoukalis, 1997). The directive, which was due to become effective in 1996, is closest to Anglo-American practices and reflects a more general standardization of policy around liberal ideals.

Creating a single financial space?

Twelve years after it was established with the signing of the Treaty of Rome (and nearly two decades after the Treaty of Paris established the European Coal and Steel Community), the EEC first became committed to economic and monetary union at the Hague summit in 1969. The Werner group reported a year later with concrete proposals to achieve union in three stages during the 1970s (Federal Trust, 1995). These proposals were premissed on the maintenance of the international financial order of fixed exchange rates guaranteed under the Bretton Woods settlement that was crumbling and by 1973 it was clear that economic and monetary union was going to be a long-term aim rather than a short-term possibility. In March 1979 the launch of the European monetary system (EMS) attempted to establish a system of fixed, but adjustable, exchange rates between member states, and in the period up to 1992 exchange rates became increasingly stable as inflation rates converged. Between 1987 and 1992, for example, there was only one minor adjustment – of the margins of fluctuation for the Italian lira – and by April 1992 the only EU member state which had not joined the system was Greece. However, the effect of the development of rigidities that had grown up within the system, the asymmetrical relationship between the Deutschmark and other currencies, and speculative attacks on currencies that were thought to be weak led to crisis in September 1992, when sterling and the lira were forced out and the peseta and escudo were systematically devalued within it.

Despite the 1992 crisis in the EMS, the EU continued with plans for economic and monetary union. These had been given fresh impetus in 1989 when the Committee for the Study of Economic and Monetary Union reported (Delors, 1989), calling for irrevocably fixed exchange rates, the eventual introduction of a single currency for the Community and the establishment of the European Central Bank. At Maastricht in 1992 the member states of the EU agreed that the single currency would begin for countries which satisfied economic convergence criteria on 1 January 1999, that it would start later for others and that the UK and Denmark could opt not to join without jeopardizing their role in Europe.

The rest of this chapter considers the potential impact of EMU on the European financial sector and examines the effects that it is likely to have on the geography of European finance. It will be argued that EMU will lead to a significant restructuring of the industry, particularly in the bond markets and in banking, in both the medium and long term, with some withdrawal and concentration of capital. As Dermine (1996: 1) argues, 'A single European currency will change fundamentally and permanently the sources of competitive advantage of financial institutions.' Although the single currency is not designed principally to bring about this restructuring, the Commission has rehearsed the same arguments that it used about the completion of the single market, arguing that EMU will lead to a reduction in transaction costs, a reduction in risk, increased competition and the ability of Europe to compete head to head with the dollar (Emerson, 1990).

Although the onset of monetary union will reduce uncertainty and costs for companies and individuals, for the financial sector it is likely to be a brutal and costly exercise. The costs of the transition will disproportionately fall on the financial sector and the long-term impacts will also see a reduction in revenues, increased competition, falling margins and some increase in uncertainty. The period of transition from national currencies to the euro will give rise to increased costs for banks, as computer systems have to be changed, as staff need to be trained, as banks bear much of the costs of collecting old banknotes and replacing them with euro notes, and as banks provide their customers with support and advice. Estimates of the cost to banks of the transition range from 8 to 15 billion ECU, or added costs of 1.5 per cent of total revenues every year for four years (Lee, 1996; McCauley and White, 1997).

The impacts of EMU on the European financial sector will be felt most significantly in the longer term because they will directly lead to reduced income and increased competition, while the deflationary macro-economic effects are also likely to impact negatively on financial institutions.[7] First, financial institutions will see a fall of 1 per cent of all their revenues, as the need for foreign exchange transactions and travellers' cheques within Europe disappears (Dermine, 1996). Secondly, there will be a large fall in trading in foreign exchange derivatives as European companies no longer need to hedge against adverse movements in exchange rates, and if the euro develops an international role comparable with that of the dollar there will also be less need to hedge against non-European currencies. Thirdly, the deflationary effect of the single currency, and the effective Europeanization of the Bundesbank's obsession with low inflation (see Dyson et al., 1995), will mean that the need to hedge against interest rate changes will be reduced.

At the same time, the euro may lead to a significant increase in the competitive pressures in European financial markets (see, for example, Emerson, 1990; Bugie, 1996; Dermine, 1996; Lanno, 1996). In this reading, 'the existence of separate currencies must divide the market in money even more than those in other goods and services' (Johnson, 1996: 51). A single currency, therefore, will lead to much greater awareness of the differences in interest rates and other fees being offered by banks from different countries in the EU at the same time as the risk of adverse movements in the exchange rates disappears. Together with the single licence rules introduced during the 1992 process and the increased use of information and communication technologies in retail banking, this would mean that users of financial services will both borrow from, and deposit money with, banks from other countries and also exert pressure on domestic banks to reduce charges.

With the exception of bond trading in the City of London, Europe's government and corporate bond markets are largely nationally segmented and the onset of monetary union is likely to transform these markets. Bond underwriters, who buy, trade and market bond issues, draw their competitive advantage from relations with issuers (either governments or companies), from their general understanding of the local financial markets and from their relationships with end investors. Of these, the final two depend in Europe to a significant degree on national currencies. For example, Italian banks have an advantage in understanding the Italian financial markets and have a large pool of Italian investors willing to buy Italian bonds. This is true across Europe and is reflected in the fact that 86 per cent of French borrowers – and 77 per cent of all borrowers – use French companies for FFr bonds, while 84 per cent of guilder issues are managed by Dutch companies and even in the highly competitive and internationalized markets of the City of London, 44 per cent of sterling issues are managed by British firms (McCauley and White, 1997; see also Dermine 1996). As firms and governments begin to issue debt in the euro, specific knowledge of the currency and interest rate climate will disappear and the competitive advantage of national institutions will be eroded (Brown, 1996). One upshot of this is likely to be that, as competition intensifies, raising finance will become cheaper for both governments and companies yet the corollary of this is that financial institutions will see their profits fall.

The extent to which intra-European competition will really be stimulated by the introduction of the euro is, however, a matter of some debate (see Hoschka, 1993; Salomon Brothers, 1996). In retail banking, reputation and market profile is at least as important as price to the majority of consumers, and in the United Kingdom, for example, the introduction of lower rates of interest by telephone-based mortgage lenders has been slow to make an impact on the market share of established companies. Furthermore, new entrants to markets will not have the specialist knowledge of the local markets that existing institutions have. However, as McCauley and White (1997) argue, a single currency may change the behaviour of financial institutions and induce them to internationalize both corporate and retail activities as they identify

TABLE 3.2 *Employment in European banking (000s)*
in selected countries, 1980–94

	1980	1990	1994	Change 1980–1994 (%)
Belgium	68	79	76	+11.8
Finland	42	50	36	−14.3
France	399	399	382	−4.3
Germany	533	621	658	+23.5
Italy	277	324	332	+19.9
Netherlands	113	118	112	−0.9
Norway	24	31	23	−4.2
Spain	252	252	245	−2.8
Sweden	39	45	42	+7.7
UK	324	425	368	+13.6
EU 10	2011	2344	2274	+13.1

Source: McCauley and White (1997: table 17)

profitable niche areas. This would result in intensified competition in key markets and exert a downward pressure on prices and profitability.

The onset of economic and monetary union will, therefore, almost certainly lead to an increase in costs, a reduction of revenues and increased competitive pressures for European financial institutions. At a macro-level, there is a very real danger that competitive pressures may lead to an increase in the likelihood of a contagious bank failure, as institutions attempt to gain market share by taking on riskier customers, although this will be offset by the reduction in derivatives trading as the single currency provides a more stable macro-economic environment. For individual institutions, competitive pressures will, in the medium term, contribute to widespread, but geographically variable, restructuring in European financial services. At least one estimate has suggested that EMU will eventually lead to a 50 per cent reduction in employment in the major European banks (Economist Intelligence Unit, cited in *The Banker*, August 1996b: 28). The extent to which such apocalyptic forecasts will be realized will depend on the geographically variable extent to which jobs have already been cut (see Table 3.2). However, during this process it is clear that the skill composition of employment will change in two directions as retail financial service employment becomes polarized between routinized, clerical functions and more specialized sales functions.

The introduction of the single currency also raises a series of governance issues in Europe. EMU irrevocably fixes exchange rates between member states and interest rate policy is effectively transferred to an independent European Central Bank (ECB) modelled on the Bundesbank and charged with the maintenance of low inflation (but with no responsibility for unemployment or economic growth). This means that some of the most important tools in a government's armoury are being transferred away from nation-states to the supra-national level (Dyson et al., 1995). In some EU member states, at least, this has precipitated debates about the very nature of the European project. However, it is important to separate form from substance. The neo-liberal political–economic agenda pursued since the late 1970s, which stimulated the growth of global financial markets, undermined or abolished exchange controls and promoted a liberal world trade regime, simultaneously undermined the effectiveness of nation-states' ability to follow independent economic policies (Peck and Tickell, 1994; Tickell, 1998b; but see Hirst and Thompson, 1996b; and Chapter 1). In this context, the pooling of economic sovereignty implied by economic and monetary union may allow European nation-states to recapture some control over global financial markets.

Nevertheless, the levers of economic policy in Europe remain far from the people of Europe. The very independence of the ECB and the mechanisms of economic and monetary policy (which, through 'Ecofin', the Council of Finance Ministers, and the 'Euro Club', which includes the finance ministers of euro member countries, are only obliquely democratically accountable) mean that there is very real danger of a European democratic deficit (see Chapter 5). This is exacerbated by the influence that the EU has beyond its borders, as it prescribes a liberal economic medicine for the former Warsaw Pact countries.

EMU and uneven development

This chapter has argued that one outcome of EMU will be a widespread and fundamental restructuring of the financial services industry in Europe. Although nowhere will be untouched by these processes, the effects of monetary union will be geographically variable, both between and within the nation-states of the EU. In the short term, the liberalization of capital markets engendered during the 1992 process and in the run-up to monetary union has led to a flurry of activity on the financial markets of countries such as Greece and Ireland (see, for example, J.M. Brown, 1997; Hope, 1997). In the longer term, however, the aggregate impact of monetary union is likely to be a concentration of activity in core locations and a reduction of employment and branches in peripheral ones.

Monetary union will have its earliest, and arguably its greatest, impact on the corporate and wholesale financial markets. Financial institutions and large companies have better access to information than do individuals, while the costs of even a small differential in interest rates are significant on large amounts of money. Furthermore, as the introduction of the single currency will reduce the need for interest and exchange rate hedging, there will be a reduction in overall demand for derivative instruments by European companies. Institutionally, these developments are likely to result in both a concentration of activity in a smaller number of large banks and an increase in the extent to which companies and pension funds deal directly in the financial markets. Geographically, increased competition will strengthen the existing pressures to concentration of wholesale financial services in a small number of centres.

There is, however, some debate about where this concentration will take place. As one of the three pre-eminent global financial centres, alongside Tokyo and New York, London has long been the centre of the European financial services industry and, although there are few significant British-owned institutions left, London has increased its dominance over other centres and most large French and German banks have their main dealing desks in London rather than in Paris or Frankfurt. London's dominance is both a regulatory construction, as the British regulatory authorities have had a relaxed attitude to international finance, and the result of the development of the information networks and expertise increasingly necessary for the financial sector (see Amin and Thrift, 1992; Thrift, 1994). These networks will remain important after monetary union and, even if Britain is not a fully participating member, the Governor of the Bank of England has argued that London will remain dominant in the European financial system:

> The City of London thrives on broad, liquid markets *regardless* of currency, and provided we are prepared – as we will be from 1 January 1999 – to quote, trade and settle in euro, as we are now in dollars or yen or DM, then I am convinced that the broader markets in euro instruments will represent an opportunity rather than a threat to the City. The fact is that overseas-controlled financial institutions are continuing actively to build their presence in the City, notwithstanding expectation that the UK will not be among the initial euro-member countries.[8]

And yet, despite London's current strengths, the City's position is not set in stone. The ECB will be based in Frankfurt, taking at least some economic decision-making and discourse-making with it, while there have been conscious attempts to lure business elsewhere. In the derivatives markets, for example, although the London International Financial Futures Exchange (LIFFE) has a pre-eminent position within Europe, the main German, French and Swiss exchanges announced a link-up in 1997. This gave the three exchanges a joint list of products, single screen trading and a unified clearing and settlement scheme which would give the unified exchange a combined turnover which exceeded LIFFE and threaten London's dominance of the European derivatives markets at the same time as turnover will fall (*Financial Times*, 18 September 1997; although see LIFFE, 1997). The decision to delay British entry into EMU may damage the prospects of the City further, as institutions in continental centres are believed to have a competitive advantage in assessing the euro (see, for example, Bourcier, 1996; Johnson, 1996), although the extent to which this occurs will depend on whether the retention of sterling is a short-term expedient or a long-term political and economic strategy (see, for example, Munchau, 1997b).

Although there are questions about the extent to which monetary union will lead to a rapid transformation of retail financial services, price transparency will work in conjunction with the '1992 rules' and, in the medium term, contribute to a widespread restructuring. At the sub-national level, Europe will see an erosion of regional banking systems across much of continental Europe, as mergers between institutions undermine the local bases of banks, and geographical concentration in major cities will increase in much the same way as happened to the United Kingdom's regional banking system during the period between 1890 and 1914 (see Kindleberger, 1974). The more recent history of British retail banking is also instructive here. The intensification of competition in British financial services since the mid-1980s, which was stimulated by regulatory restructuring and the Lawson–Thatcher credit boom, has seen a progressive fall in both employment and branch numbers, while mutual companies have disappeared and institutions merged in order to build up major financial institutions. In this process, job losses and branch closures have been concentrated in inner cities and poorer rural areas (Pratt et al., 1996; Tickell, 1997). Such patterns have a brutal economic logic to them, in that banks are increasingly viewing each branch as a cost centre which has to meet given profit targets. Although the embedded nature of financial institutions in some parts of Europe will ameliorate the effects of this, monetary union will increase pressures towards the generalization of such processes across Europe.

If the broad outlines of the effects of integration and EMU on EU member states are clear, it would be a mistake to underestimate their impacts on countries on the European periphery. The creation of an increasingly coherent and powerful financial system in Europe, together with the relentless globalization and concentration of financial markets (see Tickell, 1998b), is exerting powerful pressures on countries as diverse as Switzerland and the former communist countries to the East. While Switzerland, for example, has a robust financial sector with world class institutions (Danthine, 1993; Swoboda, 1993), in order to maintain its presence as an international centre, regulators have had to emulate EU directives on investment funds. Furthermore, the euro is likely to have a more indirect impact on Switzerland as fund managers, who need to maintain a diverse portfolio of currencies, buy Swiss francs and, in the process, maintain an artificially high level for the currency.[9]

The problems of the former Warsaw Pact countries are likely to prove more intractable than for other countries within the euro's sphere of influence. It is only since the beginning of the 1990s that eastern countries have started to create financial systems which resemble the systems which have been evolving in the rest of Europe for at least two centuries. However, the desire to join the European Union has acted as a powerful incentive to attempts to emulate western European regulatory systems and institutional structures, as the Union has made membership conditional on having 'properly functioning market systems' with developed financial markets (Commission of the European Communities, 1997d). Since 1989, independent or quasi-independent central banks have been created in Poland, the Czech Republic, Slovakia, Hungary, Estonia, Romania, Latvia, Lithuania and Slovenia, while Bulgaria has a formally independent central bank which is effectively tightly controlled by the state (Commission of the European Communities, 1997f–n).

At the same time as creating independent central banks, there have been attempts to create financial sectors which both serve the needs of industry (by developing commercial banks which intermediate between lenders and borrowers and by developing stock markets in which to raise new equity finance) and help to provide a stable home for savings. In all countries, the prescription has been similar: initially developing state-owned commercial banks and stimulating capital markets through privatization of state-owned companies, and later privatizing the banks and welcoming foreign participation in financial markets. As in the European Union, a sub-text of many policies is to reduce the total number of small, co-operative banks and introduce a more commercial banking sector. Formally, the financial systems of eastern Europe have been transformed, but the effectiveness of the reforms has varied markedly, and bad debts, mismanagement and poor supervision precipitated banking crises in Latvia, Lithuania and Estonia, while the European Commission has rather dryly noted that in Bulgaria 'Imprudent lending, poor supervision, and an absence of "hard budget constraints" – the idea that debts have to be repaid – resulted in mounting losses' (Commission of the European Communities, 1997l: 23).

Even in those countries which have had most success at transforming their financial systems, the transition from a monolithic to a more complex, western-style financial sector has been slow and difficult. Banking systems remain dominated by state-owned institutions (in Poland, for example, approximately half of bank capital remains in government hands); while in every country capital markets remain weak, illiquid and fragmented. In recommending that five former Warsaw Pact countries be accepted for membership of the EU, the Commission argued that 'the Czech Republic, Estonia, Hungary, Poland and Slovenia . . . can be considered functioning market economies, *even if in all these cases some important features, such as capital markets, still need to mature and develop further'* (Commission of the European Communities, 1997d: 46; see also Kim, 1997). Accession to membership of the EU will be a double-edged sword for the financial institutions of these countries. While companies will get better access to the developed financial markets of other member states, the competitive pressures on banks will be intense and are likely to lead to substantial consolidation (C. Jones, 1997), exacerbating the tendencies towards uneven development in the wider European financial system and leading to a further concentration of activity in Europe's major centres.

Conclusion

The European financial sector is part way through its most fundamental restructuring since the Treaty of Rome. This chapter has argued that financial integration has been slow to develop and the stimulus of the single currency will be the main political catalyst for change in the sector. In the medium term, the widening and deepening of financial integration will result in intensification of competition in the sector, with reduced prices for users of financial services and a commensurate reduction of institutions and employment. And yet this will not be a painless project. The failures of financial institutions in East and Southeast Asia during the closing months of 1997 should serve to remind users and regulators of finance alike that 'restructuring' is often used as a euphemism for collapse, that there are very real losers from financial failures and that too rapid a transformation of the financial system unleashes forces which threaten the viability of the system itself. Furthermore, although still an incomplete project, it is increasingly clear that financial integration will strengthen the position of Europe's major financial centres. Whether, in the long run, London or Frankfurt or Paris emerges as the premier financial centre for the continent is, in some respects, less important than the fact that it will be one of these (or, most likely, all three). Smaller financial centres, such as Milan or Manchester, will see their influence wane in spite of the lobbying efforts of bodies such as the Association of European Regional Financial Centres. As this will take place at the same time as the enlargement of the Union, which will mean that many places lose their eligibility for regional aid, the impact on peripheral regions could be severe.

ACKNOWLEDGEMENT

Thanks are due to Allan Williams and Ray Hudson for incisive comments on an earlier draft. I would like to acknowledge ESRC's support for the research fellowship 'Regulating finance: the political geography of financial services' (award number H52427001394). The content remains, as ever, my responsibility.

NOTES

1 In fact, after the short period during the 1980s when the Commission was formulating policy, Japanese financial institutions were *less* competitive than

most European ones. During the mid and late 1980s, Japanese banks could borrow money more cheaply than European ones (and so could gain market share and higher profits), but as the depth of the crisis in the Japanese financial system has become clearer the situation has reversed and Japanese banks now have to pay a hefty premium on their borrowing (compare Zimmer and McCauley, 1991, with Tett, 1997).

2   Although neo-liberal commentators are wont to describe such regulatory fragmentation and protectionism as being entirely detrimental, it is important to remember that it emerged in an era of relative national autarchy and as part and parcel of the response to the financial crises of the 1920s and 1930s which had fuelled the growth of fascism.

3   Although during the run-up to the completion of the single market, rhetoric stressed that 'fortress Europe' was being created, there was no attempt in financial services to replace national barriers with a European barrier. This reflected the increasingly global nature of financial markets. Indeed, the Directive (1993) on Capital Adequacy merely put into European law 'voluntary' agreements signed by international bank regulators under the auspices of the Basle-based Bank for International Settlements.

4   In fact, by the end of 1997 it had become clear that the total bill was closer to FFr 150bn (Tucker and Jack, 1997: 2).

5   While, of course, it is a fair proxy for efficiency and competitiveness, it is important to remember that efficiency is itself used as an excuse for closures of branches in poorer neighbourhoods (Pratt et al., 1996).

6   This has been compounded by some resistance by member states to implementation of EU directives. The Investment Services and Capital Adequacy Directives, for example, were due to come into force in all member states by January 1996 but by the middle of 1997 Germany still had not complied with the rules and one German banker was reported to have said to foreign competitors that they should 'relocate here and enjoy the advantages while you can' (quoted in Shirreff, 1997: 1).

7   As McCauley and White (1997: 21) argue, 'Many of the sovereign participants in EMU should eventually gain the benefits of faster growth, lower inflation and lower market volatility that EMU will bring. However, for banks, this will be a mixed blessing. Banks will share in the benefits of faster growth, but lower inflation could well lead to a reduction in interest rate spreads [and] also (over the medium term) exacerbate problem loans associated with current low property prices, although (over the longer term) it might make a repetition of such loans less likely.'

8   Eddie George in a speech to the British Chamber of Commerce in Hong Kong, 23 September 1997 (see also Dermine, 1996).

9   This is likely to occur to other 'strong' currencies in Europe and elsewhere which are outside the euro zone.

POLITICAL DIMENSIONS

4

Nationalism and its Derivatives in post-1989 Europe

Colin H. Williams

Classical nationalism

The year 1989 represents a watershed in European history comparable to 1945, 1918, 1848 and 1789, marking a transition from one epoch to another. The demise of state communism, it is claimed, has heralded the re-emergence of pan-European features such as the market economy, a semblance of local democracy, the renewal of ethnic ties and the re-establishment of trans-frontier communication routes and commercial links and, most significantly, the revival of nationalism in central and eastern Europe. Such developments are profoundly ironic as commentators in western Europe and North America have been predicting the imminent disappearance of nationalism as a political force in an increasingly inter-dependent world economy (Latawski, 1995).

In this chapter I examine the role of nationalism and its derivatives in the perpetuation of a divided Europe. Far from being moribund or consigned to the pages of history, I argue that varieties of nationalism are sufficiently flexible to make common cause with other political and social movements and have been remarkably successful in realizing the goals of the project for national self-determination. However, despite the emergence of 'new' states in central and eastern Europe, and the establishment of national and regional assemblies in unitary states in western Europe, such as Spain and the United Kingdom, we should be wary not to attribute such sea-changes to nationalism as an autonomous force. Complex structural changes should not be ascribed to a single ideological force, although elements of contemporary nationalism have pervaded many other political movements, often for prosaic and instrumental reasons, always to gain influence and wield power.

In analysing *nationalism*, geographers have tended to focus on the role of territory, symbolic places, boundaries and frontiers, access to resources, population movement and strategic locations, rather than the ideological and processual aspects of nationalist mobilization and conflict resolution. Changes in the inter-relationship between globalization and localism, together with holistic and ecological ideas, and reformed conceptions of space occasioned by telematic revolutions, have altered our conception of the appropriate context for analysing political–geographical phenomena. There has also been a concern to differentiate between nationalism and other forms of political territorial assertion, such as *regionalization*, defined as the regional application of state policy, and *regionalism*, understood as the attempt to optimize the interests of a region's population through the manipulation of the political process. Commonly, the three forms merge into one collective movement dedicated to the advancement of a particular nation's cause and political well-being. Always hard to define, nationalism's enduring quality lies in its flexibility and ability to recast itself to suit the vagaries of each generation, as befits one of the three major political ideologies of the modern world.

In common with liberalism and socialism, classical nationalism has lost some of its allure. Since the events of 1989, it is argued that the academic debate is no longer one between nationalists and anti-nationalists about the virtues and vices of nationalism, but rather one conducted almost exclusively between different kinds of anti-nationalists. Knight (1997: 178) asserts that:

> conventional liberals promoting an ostensibly value-free but pointedly debunking social science of nationalism are challenged in one way, by neo-republican liberals advocating a constitutional patriotism that ignores cultural identity, and in another, by post-modernists opposing all traditional identities. The present absence of real debate poignantly illustrates the sterility of contemporary social science when faced with a meaningful and value-laden subject.

In this chapter I argue that Knight's assessment is correct in respect of classical or primary nationalism but very wide of the mark in relation to sub-state or secondary nationalism.

Nationalism and its derivatives

The power of nationalism and regionalism lies in their ability to mobilize people on the basis of their historical occupation of a cherished environment. Despite nationalism's potential for destruction, as witnessed in the struggle for Yugoslav succession, it can also provide a beguilingly complete socio-cultural framework for political and economic action. A.D. Smith (1993: 11) has phrased its allure thus:

> nations derive their profound hold over the feelings and imaginations of the people because they are historically embedded. They are rooted in older and more long-lasting ethnic ties, myths and sentiments from which these modern nations draw much of their emotional and cultural sustenance and much of what makes them distinctive, even unique. If nationalism is the normalisation of the unique, then we should not be baffled by its global power. It satisfies the dual craving to preserve what is felt to be a collective self and all its special cultural values, while inserting that self as a political community into the community of nations by endowing it with the standard attributes of the nation.

The defence of territory figures both as a context for socio-political processes and a repository for a threatened group identity. Hence the concern with an accurate definition of the nation and its territory so as to realize a new basis for political legitimacy, usually in the call for some form of autonomy. The most virulent form of political regionalism in the EU is ethnic separatism, as manifested among elements (usually a fraction of the elite) of the Basques, Bretons, Catalans, Corsicans, Flemish, Scots and Welsh. These enduring national separatist movements act as a counter to the general thrust of globalization and integration so redolent of the so-called 'post-modernist, New World Order' (Williams, 1993a). But they also signify an awareness that such political pressure is essential if historically subordinated nations are to benefit from the opportunities within the new European order. Notwithstanding this trend, I argue that the mobilization of regional interests at the European level threatens nationalism's pristine ideological and emotional appeal. European structural changes will force nationalist movements to adopt more accommodating, pluralistic policies and thus they may cease to be effective articulators of the 'desire of nations'.

Theories of nationalism describe the nationalist claim as a search for collective equality which involves grievances packaged under headings such as territorial defence, language recognition, economic development and social justice. National separatism stems from an acute concern over the erosion of a group's identity and resource base which can only be halted through territorial separation to form a sovereign state co-equal with all other states. Regionalists assert that one need not seek the break up of the state but insist that its internal arrangements should reflect its pluralist character. Regionalists and national separatists thus pose significant challenges to the territorially fixed nature of monopolistic sovereign space.

However, there is a sting in the tail. Nationalism has fascinated many because of its perverse and enigmatic character. Despite its apparent demise, it is still capable of stirring warring passions and fuelling inter-ethnic conflict in capitalist economies and former socialist republics. Increasingly, the two systems are inter-related for one effect of the decline of state socialism has been the:

> excitement occasioned by the collapse of the Soviet and Yugoslav federations and the fear that nationalism might also inspire popular radicalism in western Europe. Most academic radicals, having abandoned Marxism for the politics of difference, consider their enemy not to be capitalism but the imposition of identities such as that which might, they fear, be effected by nationalists. It is now academics, not capitalists, who fear the plebeian 'Other'. What they fear in nationalism is (to borrow Marx's phraseology) that it is the populist sentiment of a heartless world, the communitarian soul of soulless conditions. Fortunately for them, western European nationalisms are proving to be far more ineffectual than was, and is, capitalism. Politically effective nationalisms are almost always more vicious than virtuous but there is good reason to think that they will pose less danger in the future than they have in the past. Perhaps, then nationalism will not sell for much longer. (Knight, 1997: 179)

Within the European Union, the demand for participatory democracy characterizes the Corsican, Flemish and Celtic cases, while the tension between violent action and pragmatic legitimacy is best illustrated by the Basque case. The Basques enjoyed long periods of relative autonomy prior to their incorporation into the Spanish state. Their language and culture were deemed unique for they were unrelated to any Indo-European language group and were among the oldest surviving elements of civilization in Europe. Elements of their institutional distinctiveness, the *fueros*, survived until fairly recently and were used as evidence of a prior claim to legitimate statehood. Under the impress of state-building, non-Spanish elements were ruthlessly eradicated, producing a deep resentment within the Basque community. State oppression was confirmed during the Spanish Civil War and sustained thereafter when Spanish forces were regarded as constituting an internal colonial military occupation.

The creation of *Euskadi ta Askatasuna* (ETA, Basque Homeland and Liberty) in 1957 reinvigorated Basque nationalism. A particular combination of the defence of traditional cultural values, rapid industrialization and opposition to Spanish internal colonialism nurtured violence as a movement strategy (Clark, 1979, 1984). Issues animating the drive for nationalist autonomy, whether separatist or not, included:

1   A concern for the survival and promotion of a unique culture, its language and attendant social institutions.
2   A concern to influence the direction of economic change, so as to cope with either de-industrialization or regional economic growth.
3   A concern with political representation which maximizes democratic accountability.

Nationalism in this context represents the discourse of resistance to an unjust state. Having undergone a structural transition from Francoism to representative democracy (Diaz, 1993), Spain typifies the evolving European state form in that it is neither a regionalized unitary state nor a federal state but a hybrid form incorporating features of both.

A crucial influence in political transition was the development of the European Union which provided an additional context for relations between Madrid and the historic nations and regions (Loughlin, 1996d). Successive reforms at all levels in the political territorial hierarchy have given an international platform for nationalist and regionalist leaders who have been adept at promoting their version of the 'Europe of the Regions' programme. Even though outright separatism has been temporarily contained through political accommodation, there remains the significant issue of nationality in both Euskadi and Catalonia. Post-Franco reforms designed to introduce Eusquerra and Catalan to new social domains, for example education, commerce, the media and the law, have been relatively effective in assimilating Spaniards and North Africans into the new 'nationalist' social order, but there is resistance to the abandonment of a Spanish state identity in favour of a nationalist variant described as a 'trading nation with a European vocation' (Keating, 1997: 26). Both Catalonia and Euskadi recognize the value of pactism, a medieval tradition which it is argued sustains the convention of corporatist bargaining within the nation and strategic bargaining and deal-making in the Spanish and European contexts (Keating, 1997: 26). This multi-level analysis of nationalism is critical as the discourse of resistance has given way to the practice of negotiation and subtle accommodation following the establishment of national and regional assemblies.

Both nationalism and the territorial bureaucratic state are changing apace, and though the latter is fiercely criticized, it is a near-permanent political feature setting variable limits to the degree of regional autonomy available to sub-state nationalities (see Chapter 5). Separatist pressures were curbed by the ability of central states to respond to regionally based demands for tariff protection. In consequence, separatism is no longer necessarily the most effective option for radical nationalist movements. Keating has identified this important shift in ideological and political economic reasoning in that 'peripheral nationalists have moved from demanding protectionism to support for free trade' (1992: 54). Many regions have created an alternative agenda instilling a renewed sense of purpose and economic direction which focuses on promoting regional identity as the basic building block of European politics. A 'Europe of the Regions' vision is still a long way removed from regional economic reality but remains one of the most powerful visions of a refashioned Europe.

Nationalist movements derive their character either from ideological imperatives or from their group's historical and political geographical context. A significant issue concerns the timing of the so-called ethnic revival which gained popular awareness in the 'decolonizing' period of the 1960s and early 1970s and secured legitimacy for the nationalist cause in the form of national and regional assemblies instituted in the late 1980s and mid-1990s. The reforming political programme was articulated by the nationalist intelligentsia who sought to explain the plight of the masses by reference to the absence of indigenous and determining political institutions. The professionalization of ethnic activism was often accompanied by a turning towards the left, which in honouring the traditions of the national culture also sought to promote radical socio-economic intervention. Social justice within civil society replaced the protection of ethnic homelands and the conservative imagery of rural communitarian politics as the dominant consideration. The key issue in Corsica, Euskadi and Northern Ireland was whether violence could be deployed as a successful tactic for ethno-linguistic recognition and political reform. In other areas, such as Bavaria and Friuli, it was the turn towards an increasingly federal Europe which animated activists seeking to harness Europe's overlapping identities in the forging of political alliances with various green movements, anti-fascist organizations, advocates of a non-nuclear economy, women's groups and represen-tatives of the rainbow alliance. In several states, such as Spain and Belgium, such pressure has forced established political parties to advo-cate decentralist reforms, while increasing governmental action to manage tension and conflict through regional development programmes, by establishing sub-state assemblies and formulating more plural and inclusive socio-economic policies.

Nationalist activists insist on the defence of their unique homeland. Historically, nationalist ideology is suffused with the iconography of nationalist territory. It is evident that popular resistance and the politics of collective defence over land and territory dominate the relationship between the nationalist elite and incursive, hegemonic state power. Ecologists and others have taken up this concern with valued environ-ments and in significant cases, such as Estonia and Slovakia, they have formed a common alliance with nationalists to protect threatened identities and environments. In the UK the nationalist politician Cynog Dafis became the first green sponsored MP when he was elected on a Plaid Cymru/Green Party platform for Ceredigion/Gogledd Penfro in the April election of 1994. This trend will continue as socially excluded groups demand that sectoral approaches to politics give way to more meaningful holistic appeals to individual participation and ownership of the process of community representation. Questions concerning demo-cratic involvement and the accountability of the 'hollow state' are now central, but we should not interpret their relative neglect as evidence of the decline of the nation-state (see also Chapters 1 and 5). In many respects the state is still maturing (Mann, 1996: 298) and we should be careful not to presume that it was ever merely concerned with 'national' affairs. The sovereign state has always been trans-national and geo-political. What is different today is the presumption that its functions can be better transacted by political agencies at other levels within the political hierarchy, as we shall examine below.

The challenge of the regions

For some 30 years or so the nation-state has been challenged by forces from above, such as global economic inter-dependence, and from below, such as under-represented nationalities and marginalized social groups (Williams, 1980). Two trends influence the capacity of ethno-linguistic minorities to challenge the character of the state and to re-negotiate their role in the European division of labour. The most far-reaching is the weakening of national economic sovereignty and the transfer of economic powers from state legislatures to the European Commission. Despite difficulties with the exchange rate mechanism. and post-Maastricht negotiations, the EU seeks to achieve an integrated management of its constituent economies with new policies on competition, trade, monetary exchange rate, science and technological research and, to a lesser extent, its foreign policy. At the regional level, agreements such as the Four Motors programme linking Baden-Württemberg, Rhônes Alpes, Lombardy, Catalonia, together with Wales and Ontario, help to sustain an element of additional political-regional dynamism. In the case of Catalonia and Wales, increased economic activity can reduce out-migration and consequent language shift, thereby easing one of the key issues of nationalist antagonism. Such moves represent a broader structural pattern which seeks to by-pass aspects of central state authority and build up the regional level of power. There may be little need to establish a *de jure* federal Europe if regions increasingly operate as if they were members of a *de facto* federal Europe.

As conventional political authority is increasingly shared among units within the political system, the absolute nature of the territorial nation-state can no longer be sustained as if it were a closed system. Pooled sovereignty, permeable borders, Community-wide socio-economic and environmental policy-making, freedom of movement and, to a lesser extent, shared foreign policy through inter-related agencies such as the Western European Union, NATO and the OSCE, characterize the state system and render it more inter-dependent, both with respect to member states and to subordinate constituent regions. However, increased integration and mutual dependence are not without their own structural strains which pose new challenges.

In such circumstances the relevance of nationalism takes on a contextual-diachronic dimension. New forms of economic production coupled with globalization have been institutionally recognized by the EU single market (see Chapter 2). Often it is the intervention of EU agencies, such as DGXV1 (Directorate-General for Regional Policies), which strengthen regional-level participation in international forums and networks. Parallel developments in the networking of key city-regions also threaten to by-pass the central state apparatus and may lead to increased functional integration while loosening formal inter-state relationships (see Chapter 5).

The implications for regional government are profound. Governments are now more prone to adopt a neo-liberal macro-economic strategy. Any enthusiasm to yield to Keynesian steering of the economy has been dampened by the application of new rules of the single market which has forbidden traditional forms of direct macro-economic interventionism (Jeffrey, 1996: 4).

The 'Europe of the Regions' programme was initially advanced by Bavaria, but is now being advocated by the Assembly of the European Regions as an instrument of cohesion.[1] The concept is also recognized, if in a rather weaker sense, as a Europe in which the regions have found a new self-consciousness and taken on new roles which had long been denied them (Loughlin, 1996d). The debate over the 'new regionalism' (Jeffrey, 1996) has been more acute as a result of the Delors reforms, the Single European Act and establishment of the structural funds, the passing of the Maastricht Treaty and the formulation of new rules for expansion and entry. The debate is characterized by an emphasis on partnership and subsidiarity which is enshrined in the structural funds. It is articulated most forcibly by the Committee of the Regions created as a result of the 'relaunch' of the Community by Jacques Delors in 1985 which strengthened the European Commission, Parliament and the Council of Ministers at the expense of the individual member states. It gave greater recognition to EU regional policy as a result of the measures detailed above. Such developments have made the 'regional dimension' more central to European policies and strengthened the participation and representation of regions and local authorities in European policy-making. EU regional policy is now incorporated under the more systematic 'structural action policy' and is second only to the common agricultural policy (CAP) in terms of its budget (Loughlin, 1996d). Additional changes include the incorporation of the terms 'subsidiarity' and 'partnership' into the official vocabulary of EU policy-making, and relate increasingly to sub-national levels of government as well as the relationship between national governments and the EU.

A second set of reasons for the emergence of the Committee of the Regions derives from the evolving political systems of the member states. New partners, hence reforms, have created new opportunities for the regions led by the Bavarian regional government; it is seen as a cohesive idea that regions of various kinds become important policy actors in their own right. If they succeed they may even strengthen the sense of civic as opposed to ethnic nationalism which is so sorely wanting in the long unsatisfied regions of Europe. This is particularly acute in the case of applicant states in central Europe.

Initially, several of the stronger states welcomed such developments. Germany, which is highly centralized functionally, despite being a federal system, encouraged the joint operations of the federal and the *Länder* levels and was delighted by their actual contributions to policy-making. However, these reforms increased the influence of the Federal government on policy so that the *Länders'* competence was undermined, leading them to adopt a position where 'the *Länder* Strike Back', in the second half of the 1980s. The *Länder* felt increasingly threatened by the way in which European measures, especially EU regional policy, were being implemented (Jeffrey, 1996) because of EU directives which made 'national' governments responsible for their implementation. To regain these prerogatives, the *Länder* governments insisted on representation in the Council of Ministers when this dealt with issues falling within their range of competence (Loughlin, 1996d).

By comparison, other countries, such as the UK, were highly suspicious of these developments and sought to dilute the Maastricht Treaty, as did Spain which was unhappy about the emerging role of Catalonia as a relatively influential European actor. The Defere Reforms of 1982, followed by the 1986 elections, saw an increasingly equivocal pose being struck by the French government. After an initial period of suspicion toward the Committee of the Regions, together with the massive increase in the structural funds of the EU budget, the German *Länder* were represented in the Council of Ministers, as were the regions of Belgium. Consequently, they were less persistent in lobbying for the 'Europe of the Regions' idea because they now had a seat at the table and direct access to decision-making procedures.

The 'Europe of the Regions' programme is undergoing a weaker transformation because the drive towards a federal Europe gives a new policy dimension to the whole regional question. Issues of regional policy are no longer defined solely in relation to the nation-state. Accelerated European integration following the single market programme and the Maastricht Treaty revisions have established new conditions for regional self-assertion which in many ways have undermined the classical nationalist case.

Experts argue that 'although it is unlikely that there will be a federal Europe resembling present-day federal states or a Europe of the Regions where nation-states will have disappeared, it is certain that there will be a regionalised Europe where decentralised and regionalised states will be at an advantage' (Loughlin, 1996a: 90). This advantage lies in the ability of the regions to extract more from the central coffers of Brussels. It also derives from the new model of the state where many tasks previously carried out by national levels of administrations have been handed over to supra-national levels of government at the EU level and to sub-national levels of government at the regional level (see Chapter 1). The latter's competence is increased by the ability of regional economies to compete for scarce resources and attract inward investment.

grounds that it did not secure the twin goals of redistributive justice and the protection of the welfare state. However, the current Labour government's justification for assemblies is in terms of their contribution to reducing the democratic deficit and increasing participative involvement in European-wide trends. It is also argued that such institution-building will help Scotland and Wales to develop a local response to general issues of education, health, welfare and the like which recognize the salience of regional cultural awareness so as to empower citizens as they become 'stakeholders' in society.

Questioning the nature of the state

Political commentators have called for a fresh analysis of the relationship between state transformation and deregulation or de-bureaucratization. They warn of the dangers of the 'hollow state', which is a deceit, a misnomer, for the privatization of state functions has a real impact on civil society–state relations. In turn, nationalists are required to re-think their interpretation of the nature of the state by examining questions such as the connection between state transformation and the principle of subsidiarity and the rationale for decentralization.

The French experience is instructive in this regard for, despite several moves to counter the dominance of Paris within the political system, nationalists and regionalists including the FNLC (Front National de la Liberation du Corse), HB (Herri Batasuna) and the provos (The Provisional Irish Republican Army) continue to struggle against a mini-Jacobin position. Some modification of the unitary state was occasioned in the early 1980s by the Deferre reforms which enabled the decentralization laws, 'the *"grand chantier"* of François Mitterrand's first term, to be adopted, without any major problems and led to legal acknowledgement of the region as a fully-fledged local authority' (Jouve, 1997: 347). After 1983 the different rationale led to a change of method by which the French state devolved onerous burdens to lower levels of political authority. This entailed the creation of a schema which relies on flexibility, with lower levels taking on the initiative of decision-making and having to live with the consequences. Jouve (1997) avers that the emergence of a meso-level of government will be determined by the ability of regional councils to mobilize private and public interests to create alliances within networks focused on the regional executive bodies. However, they will be challenged not only by the persistent Jacobin political philosophy, but also by the *communes* and the *départements*.

Should direct contact between lower levels of administration and the EU develop in fields such as technological transfer, education and research, the promotion of European citizenship and culture, then it is quite possible that regional agencies will forge a new dynamic level

Currently, at least two types of nationalism are in contention within the EU. Primary nationalism representing the majoritarian nation-state, such as the UK and France, is preoccupied with defending the state's national interests within an evolving world order. Secondary nationalism representing long-marginalized peoples, such as the Basques, Bretons, Corsicans and Welsh, is preoccupied with gaining access to European networks but is having to re-define itself in the light of the new regionalism. Catalonia provides a good illustration of this tendency. During the latter part of the Franco regime the Opus Dei technocrats, who directed the drive towards modernization, sought to combine mass industrialization in Catalonia with continued political authoritarianism and centralization in Madrid (Keating, 1997: 18). But because of fear about recognizing the territorial and cultural solidarity of Catalonia the regime did not engage in coherent regional planning nor in building up Catalonia's external links. Catalonia's political culture is ideologically diverse though framed within a decidedly nationalist orientation. Such pluralism should be emphasized for too often all varieties of nationalist opinion are categorized together in contradistinction to other ideological groupings of civil society.

The range of opinion within Catalonia is impressive and a source of great energy to press the general case for Barcelona and the nation (Balcells, 1996). Following the creation of national assemblies, the Catalan leader Jordi Pujol's party, the *Convergence i Unió* (CIU), which is very pro-European in orientation, has promoted the case of city-regions through the Committee of the Regions and through Pujol's Presidency of the Assembly of the European Regions. Moves towards greater European federalism are supported by the Catholic Christian Democrats and the PNV which boost the pro-European, nationalist and federalist image of the Catalan political establishment. Catalonia's pragmatic politicians are not particularly concerned with the classical separatist line of the old-guard nationalist movement. Rather, they wish to take advantage of developments within the European Union. There remains the Catalan Republican Left (ERC) whose hard-line nationalist sentiment is largely separatist in orientation as it has not given up the fight for an independent Catalonia. The ERC has downplayed its nationalist credentials and need to re-define itself within the European arena so as to escape from the old argument between bourgeoisie and radical opponents.

Other European states look set to play a more influential role within the Committee of the Regions as, for example, in the UK a Parliament for Scotland and a National Assembly for Wales were endorsed by referenda in September 1997. Elements within the electorate, unlike in 1979, have come to recognize that in an increasingly plural and bureaucratic Europe, they need their own institutions so as to modify the nation-state project and control some aspects of the local economy and public services. In 1979, the Labour Party could not accept this argument on the

of radical, co-operative ventures and that the meso-level will re-emerge as the antidote to concerns for the 'hollow state' and the 'democratic deficit'.

In the new European order regional minorities are faced with the prospect of sharing responsibility for many state functions. Until some 15 years ago they were articulating a discourse of opposition rooted in ideal-type aspirations, unsullied by the compromises forced by political responsibility. The discourse incorporated elements from across the political spectrum including traditional classical nationalists, Christian democracy, pacifism, Marxist anti-colonial activists and ecological conservatives. The style of argument and opposition was largely idealist, defensive and reactive, rather than realist in terms of the economic and political difficulties of the day. Today nationalists share in the general failing of political movements to appeal to a largely unresponsive public.

A crisis of politics?

The general crisis of politics is reflected in the increasing difficulty of mobilizing young people on the conventional issues of national injustice, regional politico-economic decline, relative poverty, language decline and territorial loss. Of course, the crisis of politics has its parallels in other institutions, such as the Church, the trade union movement and in educational circles. But many conventional political parties have lost their radical allure and nationalist parties seem unable to compete with the single-issue movements of post-modern political culture. This is rather surprising as many radical nationalists who were active in student and counter-cultural movements some 25 years ago now occupy positions of leadership and influence but still fail to inspire younger members to engage in similar types of direct action campaigning in the name of the nation. Has the decline of resistance to a centralist enemy been encouraged by the construction of a benign state and the success of European integration?

The paradigm shift in the approach to the European state since the mid-1980s revolves around changes in the super-structural events of the Cold War. It is argued that, as the tensions between East and West began to ease, more room and energy were left for the focus to switch from global, international events to internal EU events. By itself this argument is unconvincing for it seems more in keeping with the facts to argue that the EU was anxious to promote itself *vis à vis* the perceived threat posed by other economic blocs notably North America and Japan. Individual European states, witnessing their own relative decline in the world economic system, banded together within a reinvigorated and strengthened EU framework, pushing it to become a strong trading and political bloc.

Further east, the cockpit of politics witnessed the emergence of the EU as a more significant player. The one central effect of the collapse of the Soviet Union was to re-direct European Union funds from the Mediterranean to the eastern European borderlands, especially following the re-integration of Germany (but see Chapter 6). It is proper that the newness of the EU's impact should be emphasized, but it needs to be interpreted within a historical geographical context. As Rex (1997) has cautioned, among the problems which we discuss as problems of nationalism are the machinations of multi-national states, but these problems should be set within a larger context of empire, whether that be overseas European empires or the former Soviet empire itself.

Resurgent colonial nationalism has to find the means to assert itself. Rex (1997) argues that if nationalism's traditional institutions have been destroyed it may have to reassert itself through the agency of dissident members of the former imperial bureaucracies. In the former Soviet Union, ironically, the cause of nationalism is often taken up by former *apparatchiks*, while the colonial subjects often lack the politicians and political skill to pursue this cause themselves (Rex, 1997: 458).

The uses and abuses of nationalism in eastern Europe

The leaders of 'unsatisfied nationalisms' make great play of the fact that the state system was effectively sealed during the late nineteenth century by the internationalism of imperialism. However, the twentieth century has witnessed the triumph of nationalism as the fulfilment of history, for it has defeated both imperialism and the new internationalisms of communism and Nazism (Anderson, 1991; Safran, 1995; Sugar, 1995). While significant alternatives to nationalism have certainly waned, they have by no means departed the political stage. In western Europe, non-violent nationalism is viewed as a component in the creation of citizenship and democracy, while in eastern Europe any sort of nationalism is interpreted by many as a threat to the construction of democratic citizenship, and this despite the emergence of a plethora of new states. If the Wilsonian principles of the post-Versailles period were applied today, there would be an unqualified celebration of the re-emergence of Estonia, Latvia and Lithuania, Armenia and Georgia as 'independent' actors. However, several other post-Soviet states such as Belarus, Moldava and Kazakhstan face a more acute form of the general crisis of confidence and of national identity as they struggle to assert a distinctly non-Soviet impress within civil society.

Why have ethnic and/or national issues gained such prominence in the post-Soviet competition for 'who gets what' in almost all the states of the region (Walker, 1996: 9)? Nationalism provides a primary channel for political mobilization in almost all cases but need it always be associated with open conflict? That depends on the construction of an integrative society as future expressions of nationalism are unlikely to repeat 'the often destructive interwar conflicts' (Bugajski, 1995: 234). However, traditional animosities will spill over into overt conflict as sources of instability relate to political rifts, economic disparity, social divisions, ethnic enmities and outright armed conflict. Nationalism is all too capable of being used as a mobilizing issue by chauvinistic leaders anxious to exacerbate socio-economic unrest and exploit minorities as scapegoats (Bugajski, 1995: 234). The nationalist claim to confer dignity and security on previously disadvantaged groups is a powerful political weapon for a number of reasons. Both socialism and nationalism have been pervasive European ideologies, but when the first was discredited and de-legitimized by *glasnost*, the apotheosis of the second was unavoidable (Bugajski, 1995; Walker, 1996). In the post-1989 climate, the international community provided unprecedented legitimacy to central and eastern European group claims articulated in terms of national sovereignty and self-determination. Finally, the 'demonstration effect' played an important role, as political actors throughout the region witnessed the cohesiveness and effectiveness of nationally based political movements in Poland and the Baltic States. By contrast, state socialism was incapable of responding effectively to the cumulative economic and political collapse of the old order.[2]

Snyder (1993) argues that ethnic nationalism appears spontaneously when an institutional vacuum occurs and that it constitutes a 'default option' when political institutions fail. However, Walker (1996: 9) counters that eastern Europe experienced less an institutional vacuum and more the collapse of a certain set of formal institutions:

> This in turn was related to the organisation of the Soviet-style regimes which were ethnically-based federations – with a hierarchy of units – from republics or provinces for large, territorially compact groups to autonomous regions and/or areas for less numerous, but territorially based groups. Educational and cultural institutions mimicked this hierarchy, with greater to lesser privileges accorded to groups at different levels. This federal system constituted a ready-made institutional structure, from the political to the cultural sphere, for the development of nationally-based political élites. (Walker, 1996: 10)

As bearers of nationalism, the elite sought to re-define the goals of their movement, trying not to repeat the mistakes of their nineteenth-century predecessors by limiting the practical programme of reconstruction to particular class-fractions. National elites nurtured within by Soviet policies confirm Smith's (1976, 1982) observations on the leading role of the intelligentsia as the social bearers of nationalism. As the control of Moscow waned, these elites were well placed to use the framework provided by ethnically based federalism for the promotion of 'national interests'. Walker (1996) and Levesque (1997) demonstrate how Gorbachev's reform programme enabled national elites to re-define and channel popular grievances and ethnic affiliations into nationalist movements during a period when alternative political organizations were still illegal.

Did the Soviet reform policy contain the seeds of its own failure? Levesque (1997) argues that more interventionist action on behalf of Gorbachev's reforming European policy would only have accelerated the collapse.

> A somewhat slower transition in Prague and Berlin, which a more activist Soviet policy could have supported, would have allowed Gorbachev to better push ahead with his European policy. His prospects were excellent in the summer of 1989. What he did miss was time, and above all the credibility and a minimal viability of the alliance on which he relied to advance it. (Levesque, 1997: 257)

Levesque doubts that such an ambitious and messianic project could have been successful, even if it drew on political realism either side of the East–West divide.

> It was only after the fact – after the failure – that the illusions turned out to have been just that. Therefore was it illusory to expect Mazozwiecki's Poland to remain in the Warsaw Pact? . . . it was essentially the precipitancy of German reunification which led to its dissolution, without the Pact being replaced by anything else. Leninism's departure from the European stage could have left a more solid European order in its wake, one that would have been worthy of the historic occasion which 1989 represented. The question of a new order remains on the agenda and still awaits its solution . . . Undoubtedly, however, the most important fact remains that Leninism left so easily and peacefully. This remains the most remarkable event of 1989 and a legacy of hope for the future of mankind. (Levesque, 1997: 258)

Within the post-Soviet states the domestic consequences of over-stimulated nationalism are a new round of discrimination against both socially marginalized groups and the former occupiers, ethnic Russians, who are seeking to protect their interests by insisting on individual civil rights over and above the collective group rights of the now 'dominant' 'official' nationality. Bugajski (1995) predicts that more aggressive attempts at regenerating a powerful pan-Russian presence or recon-stituting a Russian-dominated union will have an immense impact upon surrounding countries. Nationalism in the region is so closely linked to questions of economic and political security that it is inevitable that selected states are seeking out Western patrons and protectors. The eastern expansion of NATO and the EU will herald a new era of trans-European co-operation and exchange, but it will also complicate issues of division, rivalry and competition. In order to reduce ethnic conflict, there is an urgent need for investment in the identity-formation pro-cesses of the new states. This is a most difficult task facing the leadership of the new republics. It is made doubly difficult because of multiple identities and the doubtful legitimacy of asserting one of these variants as the national culture, together with all of its implications in terms of national iconography, national curriculum planning and developing inter-cultural competence within the new state, the wider geo-political region and the global economy. For Byelorussia, the Ukraine and Georgia to achieve this, without inviting further dissolution or lack of confidence in the formal structures of the state, is an almost impossible if necessary task. 'If these tensions are allowed to grow, it is not only the nascent states that are at risk; the security of the region as a whole will be threatened as well, and the patterns of hostility will require generations of rebuilding and reconciliation to overcome' (Kelleher, 1996: 351).

The politics of equal respect in multi-cultural societies

Catherine Kelleher's advice may be applied to all multi-cultural states for the foundation of any modern democracy is the ability of its citizens to derive maximum security and satisfaction from contributing to the common wealth of society. Habermas (1996) argues that democratic citizenship is under severe stress. His suspicion is that a liberal political culture can hold together multi-cultural societies only if democratic citizenship becomes an inclusive basis for participation. Too often within modern states the promise of liberty, equality and fraternity is premissed upon one's membership of the majoritarian culture:

> Hidden behind such a facade of cultural homogeneity, there would at best appear the oppressive maintenance of a hegemonic majority culture. If, however, different cultural, ethnic and religious subcultures are to co-exist and interact on equal terms within the same political community, the majority culture must give up its historical prerogative to define the official terms of that *generalized* political culture, which is to be shared by all citizens, regardless of where they come from and how they live. The majority culture must be decoupled from a political culture all can be expected to join. (Habermas, 1996: 289)

This is a powerful challenge to the historical legitimacy of the unitary nation-state and implies a complete reformulation of the liberal theory of democracy to suit a plural and inter-dependent world order. Within civil society, basic human rights have been expanded to include elements which earlier theorists would have considered to have lain outside the remit of the citizen–state relationship. This relationship is central since democracy avers that citizens are entitled to certain minimum rights, chiefly those of participation in and protection by the state. However, the changing nature of the state's ideology and praxis has demanded a more pluralist view of its responsibilities.

The clash between state nationalism and ethnic nationalism has highlighted the management of inter-group relations as the chief security issue of contemporary Europe. How do minority interest groups, often under the umbrella of a nationalist movement, influence the state structure so that it concedes certain rights which are not requested by the majority? Such concessions may be in terms of bilingual education, the legal system, differentiated media and communication systems, or the use of a previously disallowed language within public administration. They may be predicated on the basis of a personality or a territoriality principle or some expedient admixture of both (Nelde et al., 1992).

The conventional view was that the state should not discriminate against or in favour of particular sub-groups, however they may be defined. This individual rights approach is often justified by majoritarian principles of equality of all before the law, and is implemented through policies of equal opportunity for socio-economic advancement based upon merit and application. The fact that many states have persistently discriminated, by law, against Jews, Catholics, Protestants or Romanies in most multi-faith societies should never be forgotten, for so often the state has exercised a malignant effect upon minorities, thereby blighting their historical development. In principle, the partial improvement in the treatment of minorities, including constructive dialogue between various interest groups and governmental agencies at all levels in Europe, bodes well for the future enactment of minority rights.

This improvement presupposes that the state is in some way responsive to the legitimacy of minority demands. Historically, the recognition of linguistic minority demands is a very recent phenomenon. In accordance with the resolutions proposed by European parliamentarians, such as Arfé in 1981, Kuijpers in 1987 and Killilea in 1994, the European Commission since 1983 has supported action to protect and promote regional and minority languages and cultures within the European Union. In 1996, some 4 million ECUs were expended on European socio-cultural schemes (budget line B3-1006 of DGXXII). Equally significant is a raft of recent legislation and declarations upholding the rights of minorities to use their languages in several domains (Minority Rights Group, 1991; Plichtova, 1992; Declaració de Barcelona, 1996). However, ethno-linguistic minorities still face many structural barriers to their full participation within the EU system (Williams, 1992, 1993b).

An alternative view, the group rights approach, has found increasing favour as it recognizes permanent entities within society whose expectations cannot be met by reference to the recognition of individual rights alone. In the main, such recognition is offered grudgingly, and reflects a minimalist stance which seeks to extend the individual rights tradition into a multi-cultural context. Such extensions tend to obscure the key issue of group tension, namely the ability of the minority to preserve and, if possible, develop its own group characteristics and desires, in the face of state-inspired assimilation (Williams, 1986).

Reconstructing societies in central and eastern Europe face two opposing notions of justice: one which views justice as the apportioning of rewards to groups on the basis of proportionality; the other which suggests that justice should consider the established rights of individuals, regardless of national origin, language, religion or other cultural marker (Glazer, 1977). In an individualistic society the majority would favour merit as a guiding principle of selection (Havel, 1991). The beleaguered minorities would argue that this merely reproduces their marginalized and pejorative position as a permanent dependency.

Some European states and international bodies claim that multi-culturalism can serve as a popular, inclusive, official ideology whereby ethnic, racial and regional variation can be managed. And yet, unlike earlier ideologies, multi-culturalism has no implicit economic mandate, at least not one which is necessarily threatening. At its simplest, it is an expression of good will and democratic intent, devoid of particular fiscal or regional development implications. Critics argue that multi-culturalism also lacks a political mandate, excepting one which subtly reinforces the status quo division of power (Taylor, 1991, 1992; Bissoondath, 1994).

In a major review of Canadian multi-culturalism, Fleras and Elliott (1992) address four types of criticism which portray multi-culturalism as socially divisive, regressive, decorative, and impractical. They conclude that this criticism is outdated and misinterprets what multi-culturalism seeks to achieve. Those who mischievously construe it as a policy that 'pays people to maintain their culture and divide' a particular country have failed to grasp its role as an instrument to enable immigrants to come to terms with a new environment, to combat racism and to promote civil liberty and social justice (see Chapter 9).

It is for these reasons that opponents of right-wing nationalism, being advocates of an open, plural European social order, seek to reduce the power of the state by boosting the power of the region and that of the EU. Might an increasingly federal Europe witness a deepening of the idea of multi-culturalism so that it becomes the first post-modern, multi-cultural political system of the twenty-first century? For the optimist, a political emphasis on fluidity, flexibility, accommodation, openness and diversity is an expression of a highly developed pluralist society which demands *mutual respect and tolerance of its constituent cultures as a structural norm.* To the pessimist, such openness is a recipe for continued strife, inter-regional dislocation, inefficient federalism and the artificial reproduction of aggregate cultural identities which deflect attention away from more pertinent social categories. Rather than being a springboard for action, such a conception is seen as an open prison which will hamper the *unfettered development of the individual in a free and burgeoning society.* This is because we are still left with the basic political constituents of primary state nationalism, minority nationalist self-government, regional alienation and European federalism, which reflect different sources of political ambition. Any attempt to reconcile such disparate challenges requires herculean stamina let alone constitutional finesse in a future Europe. However, others see hope in a transformed and more thoroughgoing conception of institutional multi-culturalism.

Tensions between multi-culturalism and the nation-state idea

Commentators are acutely aware that the

> chaotic unregulated internationalisation of capital, communications and culture without popular participation in the European public sphere is producing disorientation, unpredictable and irrational reactions; a reversion to old, even buried ethnic and religious identities and their fundamentalist assertion, sometimes leading to violently imposed segregation, racist assaults and annihilation. (Bloomfield, 1993: 278–9)

Consequently, there is an urgent need to establish a framework which addresses inequalities and guarantees cultural autonomy, self-recognition, the recognition of others, and minority rights. Without a fresh appreciation of the need to build such structural pre-conditions, it is unlikely that ethnic and cultural pluralism and democratic, civil politics will flourish.

A major consideration is how to guarantee the relative autonomy of constituent groups within any multi-cultural framework. Principles which European federalists such as Denis de Rougemont and Alexandre Marc formulated in an attempt to celebrate a common western heritage now have to adjust to the reality of managing divergence in a multiracial society. Originally, both federalists and nationalists such as Yann Fouéré (1984), Saunders Lewis (1926) and Gwynfor Evans (1996) argued that the rights of smaller nations and stateless peoples could only be guaranteed within a common European home. A generation on we may ask how realistic is the idea of guaranteeing a permanent role for Europe's minority cultures? Are they self-sustaining or are many perpetuating a generation-specific conception of a sub-culture, subsidized by public coffers and made daily more dependent on official patronage? Such questions focus attention on the relationship between manifestations of national culture and the territorial–bureaucratic state. Most minority cultures and languages are increasingly dependent upon the state for legitimizing their access to the media, for granting them permission to establish bilingual or religious-based schools, for upholding in law several of their key fundamental values and principles. Understandably minorities appeal to international organizations such as the EU, the Council of Europe and UNESCO to act on their behalf. However, it is debatable whether such threatened cultures are better respected at an all-European level, rather than within the preserve of established nation-states.

Acceptable answers depend upon a recognition that individuals *per se* exercising private choice or public rights are incapable of sustaining the infrastructure necessary to support contemporary cultures. Until recently, voluntary collective action together with autonomous social institutions, such as the Church or an ethnic press, were the cornerstone of any widespread cultural reproduction. However, when language and the culture it represents become institutionalized in new domains via new agencies, such reforms by their very nature change the relationship between the individual and the state, at European, national and local levels. Thus new answers to old questions must focus on the partnership between citizens and government agencies. Power for enfranchisement is the key to how successful such a partnership will be in serving the needs and expectations of a multi-cultural society.

When we move from rhetoric to detailed application it is claimed that the multi-cultural framework is being overworked. It may flatter only to deceive. In discussing the relevance of policy and action, McAll (1990) has queried whether social inequality rather than ethnic diversity should be the focus of interest. McAll underestimates the extent to which a multi-cultural reality can be socially constructed. Multi-culturalism in Europe may come to be fully embraced for more mundane and instrumental reasons than critics allow as a result of an ever-increasing global commercial and trade orientation and a switch from a collectivist to an individualist rationale. However, such processes are far from being irreversible and their impact on social life may be overstated.

Some supporters of a federal Europe see multi-culturalism as a renewable resource. They argue that in the globalization–localism debate, Europe can use its heritage as a bridge to influence global markets further. Drawing on its colonial and imperial past, Europe is inextricably linked with most of the world's leading economic markets and will penetrate them more effectively only if it can further harness its constituent, if diverse, financial and institutional infrastructure, language skills, scientific/academic networks and technological advantages. Thus conceived, multi-culturalism is not only about preserving cultures and improving race relations, it is also about extending trade relations, employment, science and technology, global linkage and, yes, even the maintenance of peace and security.

Such sentiments are hard to gainsay. The difficulty is in convincing conservative nationalists that cultural pluralism is not necessarily a threat at the European level and in determining what proportion of the public purse is to be expended upon satisfying the legitimate demands of such a policy. I recognize that issues of principle, ideology and policy are frequently thinly disguised disputes over levels of resource expenditure. One arm of government is involved in extending the remit of pluralism, while another is reigning in the fiscal obligations to so act. Either way dependent cultures are tied inexorably to the largesse of the state. It is the exception rather than the rule for subordinated groups to be able to benefit from a large measure of private finance, and hence private control, so as to further their interests in domains such as education and language promotion. Hence groups which are at the bottom of the pile will be further disadvantaged by this switch to private enterprise. Government programmes to aid recent refugees notwithstanding, there are severe difficulties in assuming that the private sector will take up the slack of state largesse. Currently some EU governments are obliged to maintain their support for multi-cultural projects while simultaneously signalling their intent to withdraw public finances and welcome private sector funding. Either way the language and culture of marginalized minorities are in danger of being expropriated by external forces and cultural dependency is increased.

In contrast, when state-derived nationalism advocates cultural hegemony as the instrument for social integration within multi-cultural societies, it exercises a very powerful form of institutional legitimation and persuasion through social communication networks. Contemporary insistence on the inviolability of English/French/Russian 'national community identity' in response to alternative identities is a re-working of the nineteenth-century preoccupation with state and nation-building. However, this variety of nationalism is increasingly emasculated by processes of social transformation whereby non-essential functions, subordinate social groups and devalued territories are divorced from the new conceptions of space and time as realized in the development of global meta-networks (Castells, 1996: 477). Consequently, the closure implied by state nationalism appears discriminatory and dysfunctional. The synergy and collective vitality which it mobilized in previous generations now appears to threaten the spontaneity, creativity and human dignity of individuals seeking to construct a more open society, which is less tied to place, kin, ascribed group membership, custom and tradition: in short, an identity which is not necessarily predicated upon membership of a nation.

Gender, race and alienation: nationalism's unwitting contribution to social exclusion

Europe is witnessing an increase in social diversity and norms, in different familial arrangements, in social and residential mobility and in the opting out of 'given' cultures and the creation of 'new' social formations: part of what Roseneau describes as triumphant sub-groups which are 'founded on a goal – greater autonomy – that may not be readily met by the accomplishment of a legal status and, thus, needs to be continuously serviced' (Roseneau, 1993: 77). This drift towards single-issue politics and fragmentation may not necessarily be reflected in the conventional theorizing and systems based around the grand ideas of the nineteenth century.

There are signs that Europe is moving from a collectivist to an individualist conception of social order. This is reflected in a dilution of the commitment to the welfare state (see Chapter 1) and implies greater disparity between 'haves' and 'have nots'. The individualist emphasis favours the erosion of state intervention in many aspects of socio-economic life. The keywords of this democratic order are initiative, venture, partnership and flexible accommodation to the global–local nexus. From this perspective, identity is not a given, it cannot be taken for granted, it is to be negotiated and reconstructed within each genera-tion. It never *is*, it is always in a state of *becoming*, to paraphrase Heidegger. Culture has become a commodity, to be assessed, priced, served and re-packaged to suit the exigencies of each situation. Group claims to public recognition and resourcing are increasingly being heard and adjudicated by appointed, non-elected political servants.

One of the great fears sparked by such trends is that the pace of change and subsequent alienation will occasion systemic violence, inter-nal unrest and inter-group discord. This is especially true of societies straddling the fault-line between the EU and central Europe. From within the EU questions are raised as to how permeable will be the new frontiers of an expanded EU and what effect will EU enlargement have on the internal management of its constituent ethno-linguistic groups and nations, let alone the extent to which border tensions will exacerbate historical rivalries. Attention currently focuses on what role intractable ethnic conflicts play in triggering major regional clashes and how will Europe's security architecture react to such conflagrations?

Strategists and macro-economic theorists are preoccupied by such questions as they assess enlargement plans and advise Russia and its former satellite dependencies on their position *vis à vis* the EU. The crucial question is to what extent systemic change will be able to cope with the effects of the grand design of 'opening up' the frontiers of Europe?

There is a general tendency among western political commentators to play down the initial prospects for growth and stability, even if quite a different prospect is envisaged in trade journals and among consumer sales specialists, minerals and commodities experts. Bugajski (1995: 234) has cautioned that:

> all of the countries of eastern Europe have entered a post-communist era with fragile democratic systems and serious economic problems. The potential for domestic political turmoil coupled with eruptions of numerous long

suppressed and unresolved ethnic and international tensions could engender various forms of conflict within and between several states in the region. The more ethnically homogenous countries, such as Poland, Albania, and Hungary, have fewer internal rifts stemming from cultural, religious or regional cleavages. But in multiethnic countries with large and territorially compact minorities, cultural, linguistic, religious and regional differences will continue to fan friction and conflicts. These disputes could turn confrontational if economic conditions markedly deteriorate or if political reforms and administrative decentralisation fail to satisfy rising minority aspiration for cultural and political self-determination.

An equally serious problem for states, such as Byelorussia, is the re-configuration of their national identity. In making themselves distinct from Russia they face profound problems concerning legitimacy and authenticity. Within state socialization processes, such as the development of a national curriculum, strategic decisions have to be taken not just in terms of what is to be taught, but also in terms of which foreign languages are to be encouraged and for what reasons. Other former Soviet successor states, such as Estonia or Latvia, having embraced English as a symbol of economic modernization and democratic ideals, are re-evaluating the role of Russian as a link language, stripped of many of its former connotations but carrying still the deep meanings of a historic geo-political association. Thus both the role of Russia and the value of Russian within eastern Europe remain crucial to the overcoming of divisions in Europe. Despite the activities of commercial enterprises, western European governments are currently exercising caution in both the economic integrative and strategic spheres, for although they have been hesitant to expand NATO membership in the past, 'western states now have significant economic and political levers at their disposal, both to apply pressure on Moscow and to reassure Russia and all the post-Soviet republics that NATO does not constitute a threat to their territory, only a shield against neoimperialist ambitions' (Bugajski, 1995: 240).

A more positive illustration of inter-regional trends which have accelerated since the demise of centrally planned economies in central and eastern Europe is trans-border co-operation. New connections have transformed previously suspect or fragile strategic regions into pivotal nodes in an expanded European network of communication and trade, emphasizing how geography and place are periodically reinterpreted and transformed. Vulnerable, strategic minorities, such as German-speakers in the Alto Adige/South Tyrol, are now in a stronger position to re-build their relationship with geographically contiguous majorities to the north. Once again they can serve as a bridge between the Romance and Germanic culture areas and trade regions. Similarly, the Friulian–Slovene corridor now offers a strategic gateway to central Europe as it did in the days of the Austro-Hungarian empire. Whether economic development will invigorate the Friulian language in that borderland region or more powerful neighbouring languages will re-establish themselves is an open question.

Globalization and national economies

There is a widening gap between the theory of national economic management and the reality of globalization processes which involve an increased inter-dependence at the world level, in which widening circles of domination and dependence are accelerating the effects of uneven development, both internationally and within long-established states (Hirst and Thompson, 1992). Allied to this is the need to overcome temporal and spatial discontinuities in 'real time' communication and economic transactions which are increasingly independent of the limitations of specific locations (Brunn and Leinbach, 1991). However, the penetration of the globalization process is not merely the sum of its constituent parts; it possesses a simultaneity of *both* increased uniformity and increased diversity which allows scope for originality and the harnessing of new opportunities and technologies for relatively disadvantaged groups. This is despite the superficial current of homogenization throughout the world and the apparent inexorable development of a uniform global culture, for there is undoubtedly a counter-current of increased religious and/or ethnic identification and confrontation within and across national frontiers often manifested in violent and emotional forms (Mlinar, 1992; Williams, 1993b).

Globalization is an imprecise and developing process, an ideology and programme which challenges the current order. Together with European integration, it changes the context within which civil society is mediated, posing a threat to conventional territorial relationships and simultaneously opening up new forms of inter-regional interaction such as cable television and global multi-service networks. Ethno-linguistic minorities have reacted to these twin impulses by searching for European-wide economies of scale in broadcasting, information networking, education and public administration. They have established EU institutions and bureaux and entered new alliances to influence EU decision-making bodies (O'Riagain, 1989). They believe that by appealing to the super-structural organizations for legitimacy and equality of group rights, they will force the state to recognize their claims for varying degrees of political/social autonomy within clearly identifiable territorial/social domains. Globalization and its implications are thus a major factor for those ethno-linguistic movements which are seeking to engage in the process of European inter-dependence on more favourable terms than those which obtained until recently. Processes of social transformation, including the erosion of conventional networks and the disintegration of traditional homelands have forced minority movements to be less concerned with preserving than with transforming lesser used languages. Their cultures are becoming 'de-coupled' from territory and place and 'coupled' to new agencies and domains in predominantly urban environments, thereby tying them closer to a new state dependency through institutionalized education, public administration and legal reforms. Attempts to resist the 'folklorization' of places, the 'museumization' of communities, and the commodification and gentrification of ethno-linguistic regions through tourism and the heritage industry focus on the opportunities provided by new telecommunication networks, the mass media and the empowerment of selected communities through strategic intervention in the form of language enterprise agencies, linguistic animateurs and local authority resource centres and sustainable local economic development agencies. All of the latter provide the infrastructural support to encourage language reproduction and greater use of the threatened language within the community (Williams and Evas, 1997).

A more sociological perspective is actor-oriented and asserts that social change is the result of a conscious set of decisions reflecting the power positions of strategic political actors, with some input from social elites and elite-led social movements. One fragment of the elite has determined that multi-culturalism is the best approach to managing diversity. As multi-culturalism is integral to all aspects of society, one wonders whether it is specific enough to inform public policy. Does it absorb all other identities into its own? Is it in danger of becoming a hegemonic paradigm thereby losing its purchase as a guide to action? At present the answer is probably 'no' but multi-culturalism may push things in this direction, especially when the validity of more flexible and temporary groups (or constituencies) is denied. From this perspective the ideology serves the interests of government itself for it was born at a time when the integrity of that political system was under threat from political challenges to the future of Europe. However, it is clear that other powerful interests and new sets of actors will emerge to challenge the salience of multi-culturalism. The critical features determining how vibrant this period will be is a combination of structural reform, demographic sustainability and democratic accountability in an enlarged Europe.

All these issues influence the development of nationalism, regionalism and multi-culturalism. Cumulatively, these trends will enhance the productive capacity of European economies, but they will also challenge the conventional integrity of civil society and strain the finances of responsible local and regional government. During economic downturns the refusal to honour any of these initiatives is likely to antagonize immigrant groups who consider themselves to be an integral part of the realization of a multi-lingual Europe (Castles et al., 1984; Walsh, 1992; Miles, 1993; Held, 1995).

The social forces which emphasized the training in official languages for immigrant adults so as to integrate them economically are now lessening because governments are seeking to withdraw financial support and transfer responsibility to non-governmental organizations. Any dispute about the presence of immigrants revolves around issues of domination and oppression, such as employment and racism (see also Chapter 9); the host language is not an issue, just access to it via training. Nationalism's insistence on the primacy of the ethnic/national basis of culture reveals a preoccupation with past social formations and misses the current opportunities for new partnerships and the construction of post-modern alliances based on a variety of hitherto subordinated identities.

How political policies based upon state nationalism cope with identities derived from new social movements incorporating gender, race, place, ecology, religion is a vital issue. If these alternative markers of group identity achieve salience, will ethnicity as a base for social mobilization increase or decrease in reaction to greater political–economic integration?

Conclusion

It is claimed that European integration, especially EU regional policy, the incorporation of regions within European decision-making and the emergence of the new regional actors constitute the dawn of a 'Europe of the Regions'. The necessary structural pre-conditions may be in place but it is premature to believe that Europe has embraced the regional perspective as a framework for action. Nevertheless, there are hopeful signs that a political agenda is being constructed which reflects a more holistic, post-disciplinary stance on the following issues.

Power struggles are inherent in competitive ethnic and language contact situations. But we need to know far more about the influence of the international state system on constituent language groups, including the effects of periodization; that is, the analysis of temporal rhythms of opening and closing when fresh initiatives are launched, usually following the aftermath of mass destruction through warfare. We need to know how the imitation and diffusion of phenomena such as romantic and chauvinistic nationalism, linguistic revivals, language-planning schemes, anti-colonial sentiments, dissident revisionism and interest-group movements have each challenged the hegemony of class-based political activity in recent decades. We also need to know whether they can offer any prescriptive basis for social and economic development.

We need greater understanding of the iconography of nationalistic landscapes and the meaning which nationalist movements have for supporters and opponents alike. The pioneering work of Zelinsky (1984, 1988) and the detailed analysis of, for example, Wales (Gruffydd, 1994), Ireland (Johnston, 1995) and Catalonia (Folch-Serra and Nogué-Font, 1996) could be repeated in a systematic manner throughout Europe.

When self-determination and minority rights principles are revisited we need to know how effective is their application within all socio-economic domains not just within the strategic, constitutional and legislative aspects of particular states. We need constant monitoring of the role of super-structural agencies, such as the EU, the Council of Europe, NATO and the OSCE, in influencing the definition, legitimization and resolution of persistent ethnic/nationalist grievances.

We need incisive scrutinies of the tensions inherent in the contrast between dynamic trends, such as language switching, population mobility, telematic advances in information technology and mass entertainment/sport and the rather conservative fixed nature of ethno-linguistic claims rooted inexorably in a historic homeland. The contrast between openness at the international level and closure at the local regional level is a major structural feature of Europe's political economy. The defence of minority interests is riddled with ambiguity and duality in respect of the appropriate role for both 'majority' and so-called threatened languages in multi-lingual societies. We need to differentiate between purely linguistic and broader cultural features which are subject to erosion by hegemonic ethno-linguistic groups.

All the above require intense analysis of the complex relationship between globalism and localism in most spheres of social interaction within the constituent regions/nations of Europe. Comparative analysis between and within the so-called under-represented regions is insufficient, for such studies often compare typological constructs rather than shared realities. We thus need far more rigorous multi-level analyses of all European regions.

Finally, we need to be watchful of the ever-changing nature of political demands and of the mobilization of minority political movements. In interpreting nationalism, we need to ask in what ways are the political programmes of today's triumphal interest groups different from those which prevailed during the classic era of the late nineteenth century? In evaluating nationalism as a contemporary and historical version of a prescribed reality are we talking about the same phenomena?

The state-nation remains dominant and is likely to gain further prominence if it becomes the filter through which increased regional-level mobilization from below is mediated or, alternatively, the instrument by which top-down regional development and social equalization policies are enacted within an enlarged European framework. Such is the tenacity of the state apparatus that ironically both expanded superstructural organizations and regional-level actors are contributing to the renewal of a reformed, decentralist state-nation posing as the only truly representative political instrument in a multi-cultural world order. The challenge facing civil society is how well it can harness the material and political advantages of enlargement and globalization without sacrificing the identities and ambitions of constituent minorities on the altar of political–economic integration. Europe need not allow itself to be permanently divided, if it can transfer some of the energies it has diverted to preparing for war to be expended on a vigorous re-reading of the relationship between science, conscientious capitalism and democracy. The one virtue of nineteenth-century nationalism which has slightly dimmed today was its trust in the resilience of the human spirit to shape a better world through communitarian political action. The one all-too-tragic vice which nationalism has retained is that untold violence may be justified because it is employed on behalf, and in defence, of the nation – that most enigmatic of human devices.

ACKNOWLEDGEMENTS

I wish to thank my colleagues J. Loughlin and J. Mathias, Cardiff University, for their encouraging discussions on several aspects of this chapter.

NOTES

1 The term 'Europe of the Regions' is generally understood as signifying a federal Europe in which nation-states are gradually replaced by regions as the appropriate sub-unit of the federation. Its intellectual origins may be traced to a variety of sources, notably European federalists, such as Denis de Rougement and Alexandre Marc, who envisioned the atrophying of the nation-state, and early nationalists, such as Saunders Lewis, Yann Fouéré and Gwynfor Evans who wanted to reassert the role of the historical nations in a pan-European confederation.
2 I have resisted the temptation to discuss the Yugoslav situation in this chapter. For a comprehensive account of the tragedy unleashed by competing nationalisms, see Glenny (1993), Magosci (1993), Graham (1998), and Sugar (1995).

5

New Geographies of Democracy in Contemporary Europe

Joe Painter

For proponents and analysts of democracy alike, these are confusing times. Since the collapse of the Berlin Wall in 1989, authoritarian regimes in eastern Europe and the former Soviet Union have been replaced by self-styled liberal democracies (Smith, 1993; Pridham and Vanhanen, 1994; Lewis, 1997a, b; White, 1997). The 'victory' of liberal democracy or, more specifically, of liberal democratic capitalism has been declared by many, and most famously by Francis Fukuyama (1992) in his now well-criticized assertion that 1989 represented the triumph of the West and the 'end of history'. Among academics, a veritable democracy industry has sprung up dissecting and critiquing democratic theory and practice. Yet, amid this apparent landslide victory for democracy, doubts remain.

First, the victory is that of a certain form of democracy. For sure, it is one that Whiggishly proclaims itself the best, or occasionally the least worst (Churchill, 1947), form of democracy (and even of governance in general). But these are claims to be investigated, not taken on trust. Secondly, while the earlier authoritarian regimes were both unsustainable and founded on unequal and oppressive social relations, the records to date of their successors have been, to put it at its mildest, mixed. Few would argue that the governments of Milosevic in Serbia or Tudjman in Croatia are models of democratic politics, or that Yeltsin's Russia, with high levels of organized crime, large (and largely disaffected) armed forces, and continuing political instability, has reached the point where its democratic future is secure. On the other hand, in countries such as the Czech Republic, Poland and Hungary democratic institutions seem to function much more securely.

Thirdly, the sudden popularity of liberal democracy among formerly authoritarian elites is due, at least in part, to self-interest. In the process of transition there have been winners and losers, and existing elites (such as the members of the *nomenklatura*) often found themselves well placed to realize gains from the privatization of state-owned industries and property (see Chapter 7).

Fourthly, the upsurge of interest in liberal democracy in the former state socialist countries ironically coincides with significant challenges to the stability (perhaps even a crisis) of the liberal democratic nation-state in the West:

- In many countries mainstream parties have seen their support ebb away. In some cases, this is a result of increasing voter apathy; in others, support has moved to new political formations such as the Greens or to extremist and/or populist groups such as the *Front National* in France and the *Lega Nord* in Italy.
- Increasingly close links between political parties and major corporate interests have challenged simple notions that the conventional electoral system smoothly expresses 'the will of the people'.
- The very idea of a singular and homogeneous 'popular will' has been challenged and largely undermined by both the increasing complexity of modern societies and new ways of understanding the social world which stress multiplicity, decentredness and fragmentation.
- The legitimacy, authority and effectiveness of conventional electoral politics has been undermined by corruption scandals (seen most dramatically, but not only, in Italy); by 'steering constraints' generated by processes of globalization (economic, social, political and cultural); and by a decline in active participation in mainstream political activity.
- New information and communication technologies, most recently the Internet, have raised new questions for traditional theories of democracy and themselves have new and ambiguous relationships to democracy. Is the Internet a major threat to democracy with its ability to transfer financial and informational resources rapidly across political frontiers in ways which it is difficult or impossible for elected governments to control? Or is it itself a new model of democracy: decentralized, participatory, self-regulating and open?
- The internationalization of cultural media (especially newspapers and broadcasting) and their increasing concentration in the hands of a small number of private companies and transnational corporations, raise further problems for the conventional account of liberal democracy, with its central reliance on freedom of information and expression. At the same time, a growing awareness of the socially constructed nature of all information and knowledge, however produced, challenges rationalist theories of democratic practice (Benhabib, 1996).
- Where new spatial scales of governance are emerging, they often lack democratic legitimacy. The best known case is that of the so-called democratic deficit of the European Union itself (of which more below).

Fifthly, the term 'democracy' itself is problematic. Like 'development' and 'nationalism' it appears powerful and widely understood and yet like them it has become over-extended and attached to a huge range of often contradictory meanings. Like so many political terms, the word has become imprecise. Tempting though it is to add an adjective ('true' democracy, perhaps, or 'pure' democracy), the specificity offered is illusory since the adjective simply begs further questions or conceals a particular political position. More positively, however, the imprecision of the term can be seen as revealing: the range of uses to which the word is put speaks of the political power of the ideal of democracy and the complexity and contradictory quality of the discourses of democracy can, if carefully analysed, disclose much about the political processes surrounding the claims of the 'end of history' debate and the struggle to promote different democratic projects.

Models of democracy

Any analysis of democracy, therefore, must begin from the position that it is an essentially contested concept. No ultimate definition of democracy can be provided because the process of definition is itself a matter of political conflict. A simple indication of the range of meaning conveyed by the term at various times is given by the second edition of David Held's *Models of Democracy* (Held, 1996), in which he discusses *nine* developed models (of which, furthermore, three have two variants apiece).

As Raymond Williams (1983: 93–8) notes in his sketch of the changing meanings of the term in *Keywords*, for much of its history (largely until the late nineteenth century) the dominant connotations of the term democracy have been negative: it implied rule by the mass of the people and, by extension, mob rule or the tyranny of the crowd. Only in the second half of the twentieth century has democracy come to be seen largely as a positive thing, something which almost all parties and rulers have claimed to support, regardless of their actual political practices. However, even if we limit our consideration only to the more recent positive meanings of the term we are still faced with bewildering variety, such that it is sometimes difficult to understand how apparently firmly opposed beliefs can be attached to the same label.

In Britain, for example, Conservative legislation over the past 18 years to restrict the activities of trade unions is claimed by the political right to represent a substantial extension of democracy by protecting the rights of individual trade union members against arbitrary actions by their organizations and of the wider public against disproportionate interference by unaccountable special interest groups in the workings of the economy. For the labour movement and its supporters, by contrast, most of the reforms constitute a wholesale attack on democracy by eroding freedom of association and placing restrictions on the ability of workers collectively to withdraw their labour.

A somewhat different example is provided by regionalism. Increased regional autonomy arguably enhances democracy by bringing decision-making closer to the people and making it more sensitive to the problems posed by uneven geographical development (Ehrlich, 1997). At the same time, however, it can make access to the political process much more unequal as different regions find themselves with different combinations and layers of political institutions. This situation is particularly marked in a country such as Spain, where only some regions (such as Catalonia and Galicia) have regional autonomy (Guibernau, 1995; see also Chapter 4). These examples highlight both the contradictions and the contested nature of the mixture of connotations attached to the concept of democracy.

As Held's account makes clear, however, these and similar complications arise partly from the application of differing models of democracy. In the above trade union example, these might be a particular conception of liberal democracy and a version of participatory or associational democracy. It is worth listing Held's models in full as a means of clarifying the impossibility of producing a single, unitary definition of democracy (Table 5.1).

To some extent the variety of models arises from the gradual development of the concept of democracy over many centuries. In addition, though, there are differences in the status of the different models. Some have a distinctly normative inflection (including Held's own model of 'cosmopolitan democracy'). Others, such as competitive elitist democracy or legal democracy, are perhaps more descriptive of actual political practices; indeed, to some extent these models can be seen as *post hoc* rationalizations of actual practices. It should also be noted that the principles of justification shown in Table 5.1 are philosophical and abstract; the institutional arrangements and other features for each model are, however, spelled out in more detail by Held in his extended account of each model (see Held, 1996). Regardless of these details, though, it is clear that discussing democracy in the abstract, without specifying the form of democracy involved, is unlikely to be very fruitful.

Democracy and geography

Until recently, treatments of democracy within the mainstream literature of post-war geography were limited mainly to considerations of the geographies of specific practices, such as elections (Johnston, 1979; Taylor and Johnston, 1979; Johnston et al., 1990). Such studies were often focused narrowly on issues of spatial organization (for example, the areal definition of constituencies) and sometimes tended uncritically to accept dominant assumptions about the satisfactorily democratic nature

TABLE 5.1 *Nine models of democracy*

Model	Principle(s) of justification
(I) Classical Athenian	Citizens should enjoy political equality in order that they be free to rule and be ruled in turn
(II) (a) Protective republicanism	Political participation is an essential condition of personal liberty; if citizens do not rule themselves, they will be dominated by others
(b) Developmental republicanism	Citizens must enjoy political and economic equality in order that nobody can be master of another and all can enjoy equal freedom and development in the process of self-determination for the common good
(III) (a) Protective liberal democracy	Citizens require protection from the governors, as well as from each other, to ensure that those who govern pursue policies that are commensurate with citizens' interests as a whole
(b) Developmental liberal democracy	Participation in political life is necessary not only for the protection of individual interests, but also for the creation of an informed, committed and developing citizenry. Political involvement is essential to the 'highest and harmonious' expansion of individual capacities
(IV) Direct democracy	The 'free development of all' can only be achieved with the 'free development of each'; freedom requires the end of exploitation and ultimately complete political and economic equality; only equality can secure the conditions for the realization of the potentiality of all human beings so that 'each can give' according to his or her ability and 'receive what they need'
(V) Competitive elitist democracy	A method for the selection of a skilled and imaginative political elite capable of making necessary legislative and administrative decisions; an obstacle to the excesses of political leadership
(VI) Pluralism	Secures government by minorities and, hence, political liberty; crucial obstacle to the development of excessively powerful factions and an unresponsive state
(VII) Legal democracy	The majority principle is an effective and desirable way of protecting individuals from arbitrary government and of maintaining liberty; however, for political life, like economic life, to be a matter of individual freedom and initiative, majority rule must be circumscribed by the rule of law: only under these conditions can the majority principle function wisely and justly
(VIII) Participatory democracy	An equal right to liberty and self-development can only be achieved in a 'participatory society', a society which fosters a sense of political efficacy, nurtures a concern for collective problems and contributes to the formation of a knowledgeable citizenry capable of taking a sustained interest in the governing process

continued overleaf

TABLE 5.1 (*cont.*)

Model	Principle(s) of justification
(IX) (a) Democratic autonomy	Persons should enjoy equal rights and, accordingly, equal obligations in the specification of the political framework which generates and limits the opportunities available to them; that is, they should be free and equal in the determination of the conditions of their own lives, so long as they do not deploy this framework to negate the rights of others
(b) Cosmopolitan democracy	In a world of intensifying regional and global relations, with marked overlapping 'communities of fate', the principle of autonomy requires entrenchment in regional and global networks as well as in national and local polities

Source: summarized from Held (1996)

of western liberal capitalism. Any shortcomings in the democratic process were often perceived in technical, rather than political, terms and thus as amenable to technocratic problem-solving. Conventional electoral geography generated large amounts of useful data, and did focus on aspects of political practice (or at least practicalities), but did so without situating its analyses within a wider social theory. As a result, it relied on somewhat simplistic and mechanical explanations of political behaviour, such as the oft-quoted neighbourhood effect.

Besides elections, other elements of some of Held's models have been analysed by geographers. Work on the geographies of citizenship (Marston and Staeheli, 1994; Painter and Philo, 1995), for example, has stressed the constitutive role of space and place in the politics of citizen rights, and have drawn on contemporary political theories of the mismatch between abstract ideals of citizenship and the unequal power relations in which they are founded in practice. Social movements and the campaigns of pressure groups (central to some pluralist models of democracy) have also been the subject of geographic research (Castells, 1983; Lowe, 1986; Painter, 1995: 151–79; Raento, 1997). Overall, then, geographers have studied aspects of the structures and practices of democracy, but have rarely made democracy the central focus of research.

One exception to this has been recent interest among geographers in Chantal Mouffe's development of 'radical democracy', which was the subject of a special session at the Association of American Geographers Annual Meeting in San Francisco in 1994, with the debate published subsequently in *Society and Space* (Gibson-Graham, 1995; Massey, 1995; Mouffe, 1995a; Natter, 1995). Smith and Blanc (1997) have taken the debate in a slightly different direction by comparing Mouffe's 'radical democracy', as well as communitarian and associational ideas, with the French *transaction sociale* approach. Mouffe's project connects with contemporary work in social and cultural geography focusing on the constitution of identities (and the power relations between them) and on the role of anti-essentialist ways of thinking. Although the debate so far has largely been conducted at a fairly high level of abstraction, the potential of a critical engagement with Mouffe's ideas for informing substantive geographical research is clearly shown in Brown's ethnographic study of AIDS activism (M. Brown, 1997).

Mouffe's work is unusual in drawing together the interests of geographers and political philosophers. If geographers have shown little interest in the work of political philosophers, the reverse is also true. However, the limited extent of inter-disciplinary traffic does not mean that space and place are irrelevant to interpretations of democracy. In particular, if the analysis of democracy is extended to include embodied and institutionalized social practices, then it is likely that a geographical sensitivity will bring significant insights. To show what might be involved, this section will contrast the aspatiality of the dominant abstract models of democracy with a conception of democracy as social practice.

Democracy and Enlightenment universalism: aspatial democratic theory

As with much political theory and political philosophy, debates about the nature of democracy have been conventionally conducted in an implicitly aspatial framework. In part this stems from the Enlightenment origins of modern democratic thought. The 'Enlightenment project' was by no means homogeneous, not least because it had a complex geography. Of particular note in the present discussion is the mainly European character of the Enlightenment. The label 'Enlightenment' is commonly understood to refer to the work of the eighteenth-century French *philosophes* such as Diderot and Voltaire. However, a much wider geography (and history) can be traced, articulating (among other things) the Italian Renaissance (Leonardo, Galileo), English empiricism, pragmatism and utilitarianism (Bacon, Hobbes, Bentham), Scottish political philosophy and political economy (Hume, Smith) and German idealism (Kant, Hegel). A distinctive American tradition, involving thinkers such as Franklin and Paine, straddled the Atlantic, drawing on European ideas of progress, but prefiguring and contributing to the rupture with Europe that was the American Revolution. Politically, the Enlightenment reached its symbolic apogee 13 years later in the French Revolution, an event that expressed perfectly the contradictory currents in the Enlightenment project of liberation and authoritarianism that would exercise the Frankfurt School (and latterly post-modern writers) in the twentieth century. Notwithstanding this complexity, there are a number of themes common to much Enlightenment thinking. Central among these are the role of reason in the development of human knowledge and human societies, the right to liberty of the individual, and the formal equality of human beings. In each case a universal conception is promoted, such that rationality, liberty and equality are regarded as essential elements of the human condition regardless of the particularities of its concrete manifestations.

This universalizing impulse fed through to the development of modern theories of democracy. Many of these were developed in the eighteenth and nineteenth centuries as part and parcel of the Enlightenment itself, while most of the remainder find their philosophical underpinnings in Enlightenment thought. As a result, abstract democratic theory has lacked much explicit consideration of spatiality. By definition, universal models make no allowance for spatial variation. This is significant because, as we shall see shortly, geographical scale and spatially uneven development have important effects on the operation of democracy and on the practical applicability of competing models of democracy. One effect of the aspatiality of conventional models has been a persistent lack of correspondence between abstract models of democracy and 'actually existing' democratic practice. Two examples illustrate this. First, in majoritarian models of democratic decision-making, the use of territorial constituencies leads frequently to the election of governments with minority support. In Europe, this is most marked in 'first past the post' electoral systems such as Britain's. However, even in European countries with more proportional methods, the spatial structure of the electoral system limits the extent to which legislatures reflect the 'popular will'.

Secondly, geography affects the definition of the political community in which democratic governance is operating. In practice such definitions are usually territorially based (although functional representation is possible in theory). This raises fundamental questions about who is a member of the community and is thus eligible to participate in decision-making. The history of citizenship is fraught with conflicts and discriminatory practices (often on ethnic, religious or linguistic lines). In contemporary Europe, a host of ethno-territorial nationalist movements (Basque, Scottish, Irish, Breton, Walloon, 'Padanian' and many others) are seeking to redraw the borders (social and geographical) of political communities. Elsewhere, ethnic and national minorities struggle to acquire, maintain or enhance their rights to political participation within existing boundaries (examples include Russians in Latvia, Turks in Germany, Algerians in France, Muslims in Bosnia, and Albanians in Italy).

Some models of democracy do depend on implicit spatialities (frequently an assumption that the territorial unit of governance is the traditional nation-state). However, these geographies are rarely rendered explicit. For example, in Table 5.1, classical Athenian democracy was constituted around a city-state, while Held's own model of 'cosmopolitan democracy' is an explicit attempt to develop a conception of democracy fit for a globalized world economy increasingly marked by networks of social relations that fall beyond the scope of nation-states.

There have also been some attempts to consider what geographical arrangements are necessary to facilitate the operation of particular models. Electoral geographers, for example, have debated how the spatial organization of electoral systems can be developed to correspond with particular theoretical models of democracy. In the main, however, little explicit attention has been paid to the implications of spatiality for conceptions of democracy.

Spatializing democracy: democracy as social practice

While some of Held's models may be seen as in part descriptions of historical or contemporary political systems, they are all to a greater or lesser extent normative formulations. In some cases, the necessary conditions for their realization did not exist in full; in others, while the conditions may have existed, the key features detailed by Held for each model may not have been developed. More generally, political philosophy represents one strategy for approaching the history of democracy that involves establishing sets of underlying (universal) principles such as rights, duties, citizenship, authority and participation and examining the ways in which different thinkers and political traditions have mobilized and adapted them at different times. Empirical investigation can then be brought in to study how far particular regimes or political systems fulfil the normative requirements of different models. This kind of approach underlies the recent (extremely useful) 'democratic audit' of the United Kingdom published by Klug et al. (1996).

Valuable though such normative approaches are, I wish to adopt a slightly different perspective, and to consider democracy as a set of social practices and processes. This approach does not reject or contradict the normative model-based arguments; indeed, it depends upon them since the social and political actors and practices, which form its focus, draw heavily on conceptions of democracy in the prosecution of action. Understandings and theories of democracy are thus central to political activity. Moreover, as I shall suggest below, a critical analysis of knowledge(s), rationality(ies) and rationalization(s) should be seen as essential to a full understanding of democracy as social practice.

Democracy as a set of social practices thus involves what might be called 'democratic consciousness': some knowledge of one or more of the various models of democracy elaborated by Held that provides a set of resources through which participants in political practices make sense of what they are doing. Such knowledge is likely to be imperfect and may draw on more than one of the models in contradictory ways. It is also as likely to be at least partly tacit, rather than codified. In other words, it seems likely that while most citizens would be able to present some rationalization of their participation in democratic activities in terms which relate to one or more of the models, few would be able to elaborate any of the models in full (or see any need so to do).

In addition to the role of knowledge, democracy also involves a range of other social processes including embodied practices (the act of voting, for example), institutionalized practices (the various activities of governance itself, for example) and the mobilization of material resources. On this basis, the kind of historical and substantive account of democracy I have in mind might encompass, among many other things:

- the emergence and development of social movements in support of the extension of the franchise, such as the French revolution of 1848, Chartism in Britain and, more recently, the popular revolutions in eastern Europe in 1989;
- the symbolic significance of the ballot, demonstrated in a mundane way in every election in Europe when major political figures are photographed at the moment of posting their voting papers into the ballot boxes;
- the political struggle between competing conceptions of democracy;
- the transition from authoritarianism to liberal democracy, exemplified most strikingly by the transformations in the former state socialist countries of Europe in the 1990s;
- the discursive and material constitution of political communities, seen in the recent re-unification of Germany and the earlier unification of Italy, as well as in the demands (mentioned above) for the splitting off of minorities into autonomous political communities.

Conceptualizing democracy as a set of social practices challenges the aspatiality of universal models and reveals the necessity of conceptualizing democracy geographically. To show why this is so, we can consider the role of knowledge in theories of democracy in a little more detail.

Knowledge and democracy

Enlightenment rationality and democracy Although it is not always explicit, many accounts of democracy accord an important place to knowledge. It is often argued, for example, that a well-informed electorate is essential to effective democracy, or that representative democracy may be justified partly on the grounds that the knowledge possessed by full-time representatives is superior to that of the wider population and that better decision-making is thereby assured. The latter suggestion is also used as an argument against the use of referendums and plebiscites (seen, for example, in the decision of most European Union member states not to hold referendums on the Maastricht Treaty). Related to this, it should also be noted that the development of knowledge is not central to all models: one of the conditions specified by Held for the 'competitive elitist' model is the existence of an ill-informed and emotional electorate. In his classic, though now much criticized, account of citizenship, Marshall (1950) argues specifically for the development of mass education as a means to secure full citizenship involvement in the democratic process. A further testimony to the significance of knowledge is provided by one of the arguments advanced against the extension of the franchise to the working class in the nineteenth century, namely that 'the masses' possessed inadequate intelligence and/or wisdom to exercise political rights properly.

Although the importance of knowledge for the concept is clear, its analysis has not formed a major element of much of the writing on democracy. Despite this lack of explicit theorization, it is also apparent that within mainstream thinking on democracy an Enlightenment conception of reason and rationality underpins ideas of what constitutes knowledge. Democracy, in many models at least, becomes a mechanism for arriving through the application of reason, at decisions which give true expression to 'the general will'.

The kind of disembedded rationality of the Enlightenment is challenged by seeing knowledge formation as the outcome of geographically situated social practices, in line with the approach outlined above. In this perspective, reason cannot be a guarantor of knowledge which gives true expression to the general will because the idea of a singular universal reason and with it a universally true knowledge disappears. However, this does not imply that knowledge is of no longer any significance as one of the social practices of democracy.

Democracy as interlocution: deliberative, dialogic and agonistic models The relationship between a post-rationalist perspective on knowledge and democracy is illuminated by drawing on some models of democracy not discussed in detail by Held.[1] Giddens (1994), drawing on the work of David Miller (1992), distinguishes between epistemic and deliberative democracy:

The deliberative conception of democracy Miller distinguishes from an 'epistemic' one, sometimes attributed to Condorcet and Rousseau among others. The epistemic view of democracy asserts the existence of a general will and presumes that democratic procedures can realise it – that is to say, it holds that a correct or valid answer can be reached to questions facing the political community. Such a view, according to the proponents of deliberative democracy, sets an impossible standard for democratic institutions to meet. The deliberative approach accepts that there are many questions which have no single correct answer, or where solutions are thoroughly contested. In deliberative democracy, accord might be reached by various means. Those involved might agree on a norm or norms which guide the assessment of particular policy decisions; or they might agree on a procedure which can be applied to contentious cases . . . Democracy in this conception is not defined by whether or not everyone participates in it, but by public deliberation over policy issues. (Giddens, 1994: 113–14)

Epistemic democracy is effectively the Enlightenment view referred to above, while the deliberative approach is a version of what Giddens calls 'dialogic democracy'. What is important in Giddens' account is not that he charts a post-Enlightenment version of democracy (he cites Habermas's work as resonating with the deliberative conception, confirming that the latter too falls within the Enlightenment tradition of rational debate), but that he shifts the focus from the *outcome* of decision-making (whether a decision is right or wrong, good or bad, true or false) to the *process* of decision-making. The final decision, the ends, is in some senses less important than the path taken to reach it, the means. In stressing the process of dialogue ('dialogic democracy' is the term he uses to refer to the extension of the principles of deliberative democracy beyond the formal political sphere), Giddens' argument also raises the possibility of taking seriously multiple voices and hence competing perspectives and multiple knowledges and rationalities. There are clear implications here for democracy in Europe as European society becomes increasingly diverse and multi-cultural (see Chapter 4).

This line of thinking is taken qualitatively further in the work of Chantal Mouffe. The initial argument was set out in her joint work with Ernesto Laclau (Laclau and Mouffe, 1985) and has been further developed in her subsequent writings (Mouffe, 1992, 1995a, b, 1996). Mouffe's arguments are grounded in her conception of identity politics. For Mouffe, antagonism and conflicts between identity-based social groups are endemic and cannot be eliminated. Democracy, in her argument, is to be found in the processes of struggle, rather than in their eradication. However, in those processes identities are themselves constituted and modified. She refers to her approach as anti-essentialist inasmuch as there is no source of identity which lies outside of politics on which a rationalist politics might be founded. Mouffe makes use of the 'friend/enemy' or 'we/they' coupling to propose that political struggle should not involve attempts to annihilate the 'other'.

According to Mouffe (1995a: 263), the fundamental question for democratic politics is not 'how to arrive at a rational consensus reached without exclusion, or in other words how to establish an "us" which would not have a corresponding "them". This is impossible because there cannot exist an "us" without a "them". What is at stake is how to establish an "us–them" discrimination in a way that is compatible with pluralist democracy.' She goes on to suggest that:

In the realm of politics, this presupposes that the 'other' is no longer seen as an enemy to be destroyed, but as an adversary, somebody with whose ideas we are going to struggle, but whose right to defend those ideas we will not put into question. We could say that the aim of democratic politics is to transform an 'antagonism' into an 'agonism'. The prime task of democratic politics is not to eliminate passions, not to relegate them to the private sphere in order to render rational consensus possible, but to mobilise these passions in a way that promotes democratic designs. Far from jeopardising democracy, agonistic confrontation is in fact its very condition of existence. (Mouffe, 1995a: 263)

The implications of this approach for the role of knowledge and rationality are clear. There is no possibility of a universal rationality providing the basis of a democratic politics; indeed, in Mouffe's view such universalisms are anti-democratic. Her radically pluralist conception, however, does depend on the existence and interaction of a wide variety of multiple knowledges and rationalities. Democracy lies in their confrontation.

If Mouffe's argument that democratic politics is located in the agonistic struggle between social groups is accepted, then geography is not incidental to the operation of democracy, but lies at its heart. Moving away from a universalist conception of knowledge identifies knowledge formation as a socially – and spatially – embedded process. The social geographies of the identity-based groups through which democratic practice is produced affect both the constitution of the groups and the resulting democratic struggles.

Spatiality and democracy in contemporary Europe

In this part of the chapter I want to discuss what a geographically sensitive account of democracy as social practice might involve with reference to contemporary Europe. To do so, I want to consider two processes of political transformation which I take to be constitutive of contemporary Europe. The first of these is a re-shaping of what might be called the European spatial imaginary in the context of the aftermath of the Cold War. The second is more institutional and involves the restructuring of the European nation-state, referred to by Jessop (1994) as its 'hollowing out'. As we shall see, universal models of democracy cannot be straightforwardly applied to these changes. In particular, the relevance of different formal models depends crucially on the geographical scales at which these two sets of transformations are occurring.

The European spatial imaginary: Mitteleuropa comes 'home'

I use the term 'spatial imaginary' to refer to the ways in which the cultural and political geography of Europe is understood. Among the key elements are: the imagined location of the heart of Europe; the relationship between core(s) and peripheries; the relationships between north and south and east and west; the relationships between modern, metropolitan Europe and areas where traditional ways of life persist; and the cultural and political relationships between Europe and its multiple 'others'. During the Cold War, Europe's understanding of itself was simultaneously split (East and West) and tightly integrated around the dynamics of super-power conflict. The domination of the European worldview by the ebb and flow of that conflict is well known, but none the less remarkable for that. Of course, there were differences within each 'side'. In the West, the French state under de Gaulle withdrew from the military structure of NATO in 1966, and in general France pursued a much less Atlanticist policy than Britain, which stressed the so-called 'special relationship' with the USA. Other European countries were less strongly aligned (the neutrality of Switzerland, and to a lesser extent Sweden, for example) or for pragmatic reasons inclined to a working relationship with the USSR, as in the case of Finland. In the East, Romania developed a foreign policy somewhat independent of the USSR, while Yugoslavia was a pillar of the non-aligned movement. Therefore, to speak of a *dominant* spatial imaginary is by no means to imply that there was a singular European worldview. However, the fall of the Berlin Wall and the transition of the former state socialist regimes to various forms of liberal democratic capitalism has arguably led to greater diversity than before and the dominant spatial imaginary of an earlier era has disintegrated into a variety of competing worldviews. The Europe imagined by the European Commission, for example, is clearly markedly at odds with that envisaged by the warring elites of the former Yugoslavia, and different again from diverse popular conceptions in the West. In relating these changes to democracy, the reincorporation of central Europe into the West's image of the European project is clearly central.

Table 5.2 shows the pattern of transition of various central European countries towards something approximating to western conceptions of liberal democracy. Although they have been often remarked on before, two factors remain striking. The first is the hegemonic appeal of liberal democratic capitalism as the appropriate goal and aspiration for the transition countries and the second is the speed of the geo-political reconfiguration involved. With regard to the latter, Table 5.3 shows the list of central and east European Countries (CEECs) that have applied for membership of the European Union.

On 15 July 1997, the European Commission gave its opinion on the applications for membership by the CEECs (Commission of the European Communities, 1997c). In this opinion, each application is evaluated against a series of criteria laid down in 1993 by the European Council (see Figure 5.1). It is notable that in the criteria the western conception of liberal democracy plays a prominent role (indeed, 'stability of institutions guaranteeing democracy, the rule of law, human rights and respect for and protection of minorities' make up the first of the three conditions). Of the ten candidates, the Commission's opinion was that negotiations for accession to the EU should be opened with Estonia, Hungary, Poland, the Czech Republic and Slovenia. The remaining five countries each failed to meet one or more of the criteria listed in Figure 5.1. Bulgaria was described as making only limited progress towards a

TABLE 5.2 *Transitions to liberal democracy in central and eastern Europe*

	First multi-party elections	Adoption of liberal democratic constitution[1]
Albania	Apr. 1991	Apr. 1991
Bosnia-Hercegovina	Dec. 1990	Pending
Bulgaria	June 1990	July 1991
Czech Republic	June 1990[2]	Jan. 1993[3]
Former Yugoslav Republic of Macedonia	Nov. 1990	Nov. 1991
Hungary	Apr. 1990	Oct. 1989
Poland	June 1989	Dec. 1992
Romania	May 1990	Dec. 1991
Slovakia	June 1990[2]	Jan. 1993[3]
Slovenia	Apr. 1990	Dec. 1991
Yugoslavia (Serbia/Montenegro)	Nov. 1990	Apr. 1992

[1] In some cases, a series of constitutional amendments brought about transitions to liberal democracy somewhat before the formal adoption of a new constitution.
[2] As Czechoslovakia.
[3] Following partition of Czechoslovakia (liberal democratic constitution in force for Czechoslovakia at an earlier date).

Source: derived from Banks et al. (1996)

TABLE 5.3 *Applications for EU membership*

	Date of application
Bulgaria	16 December 1995
Czech Republic	23 January 1996
Estonia	28 November 1995
Hungary	31 March 1994
Latvia	27 October 1995
Lithuania	8 December 1995
Poland	5 April 1995
Romania	22 June 1995
Slovakia	27 June 1995
Slovenia	10 June 1996

Source: http://europa.eu.int/en/agenda/appmem.html

market economy and not being prepared for the obligations of membership. Latvia failed to meet any of the criteria in full, with particular reservations being expressed about the denial of Latvian citizenship to members of the minority Russian community. Lithuania was seen as having made good progress towards a market economy, and towards accepting the obligations of membership but not enough for accession at this stage. It did, though, meet the political criterion. Romania was regarded as having made good progress on the political and economic fronts, but not yet enough for accession; it was thought to be weaker on its ability to accept the obligations of membership with regard to the

found that, in November 1995, 55 per cent of respondents in the ten Europe Agreement Countries[3] were 'largely dissatisfied' with the way democracy is developing in their countries (Commission of the European Communities, 1996e). This finding, together with the comments of the Commission about the slow progress towards democracy in countries such as Slovakia, raises questions about the limits to the process of democratization. The Commission pointed to the lack of 'rootedness' of institutions and this point could be extended to political parties and movements, most of which have sprung up suddenly, and sometimes for very opportunist reasons. In addition, surveys suggest a lack of popular confidence in institutions such as the government, parliament, trade unions and the police (Plasser and Ulram, 1996) although there are significant variations between the CEECs. These problems are exacerbated by corruption within the state. There is considerable potential for future instability and this raises some doubts about medium-term prospects for the full integration of the CEECs.

Nevertheless, the geo-political transitions (realignment with the EU and with NATO) are accompanied, and to some extent also enabled, by a shift in the European spatial imaginary which re-casts and re-imagines the CEECs as 'really' European. The period of the Cold War is represented as an aberration and, with its end, the former state socialist countries can return to their imagined historic position at the heart of Europe.

The 'hollowing out' of the European nation-state

The second geographical transformation I want to consider is more marked in the West, and particularly in the member states of the EU. The nation-state as a modern political form is a product of long historical development that culminated in the nineteenth century. During the twentieth century, the nation-state has increasingly been challenged, and its status as the pre-eminent 'power container of modernity' (Giddens, 1985: 13) placed in question. The challenges are numerous and varied and include: the internationalization of economic activities and the growth of transnational companies, migration, new technologies, the development of new political movements, demands for regional and local autonomy, religious movements and a host of others. According to writers such as Ohmae (1990, 1995), we are seeing the 'end of the nation state'. Others such as Hirst and Thompson (1996b) and Hutton (1995) insist on the continuing significance of the national scale (but see Chapter 1). Perhaps it would be most accurate to say that the nation-state will remain as an important institution and scale of governance, but that it will no longer be the pre-eminent one. Rather, it is increasingly being equalled in importance by a variety of other scales of governance and institutional forms. The result is a *polycentric* pattern of governance in which economic and social life is steered by a diversity of organizations and processes.

In preparing its Opinion, the Commission has applied the **criteria established at the Copenhagen European Council** of June 1993. The Conclusions of this Council stated that those candidate countries of Central and Eastern Europe who wish to do so shall become members of the Union if they meet the following conditions:

- stability of institutions guaranteeing democracy, the rule of law, human rights and respect for and protection of minorities;
- the existence of a functioning market economy, as well as the ability to cope with competitive pressures and market forces within the Union;
- the ability to take on the obligations of membership, including adherence to the aims of political, economic and monetary union.

A judgement on these three groups of criteria – political, economic and the ability to take on the *acquis* – depends also on the capacity of a country's administrative and legal systems to put into effect the principles of democracy and the market economy and to apply and enforce the *acquis* in practice.

The **method** followed in preparing these Opinions has been to analyse the situation in each candidate country, looking forward to the medium-term prospects, and taking into account progress accomplished and reforms already under way. For the political criteria, the Commission has analysed the current situation, going beyond a formal account of the institutions to examine how democracy and the rule of law operate in practice.

FIGURE 5.1 *Criteria for admission to the European Union (Commission of the European Communities, 1997c)*

single market and economic and monetary union. Finally, Slovakia was seen as firmly committed to taking on the obligations of membership and likely to meet the economic criteria. However, the Commission noted 'the instability of Slovakia's institutions, their lack of rootedness in political life and the shortcomings in the functioning of its democracy'. In addition, all the countries listed, along with 16 other transition states (including Russia), are signatories to the NATO 'Partnership for Peace' agreement. According to NATO, 'active participation in the Partnership for Peace will play an important role in the evolutionary process of including new members in NATO' (NATO, 1996).

Among citizens of the CEECs the popular appeal of liberal democracy, and popular dissatisfaction with the rate of progress towards it, are clear from recent surveys from the Paul Lazarsfeld Society in Vienna and the European Commission's *Eurobarometer* respectively. Conducted in the last quarter of 1995, the fourth *New Democracies Barometer* of the Paul Lazarsfeld Society covered seven CEECs[2] and revealed a 65 per cent approval rating for the new regimes (compared with 59 per cent in 1991) and an endorsement by 80 per cent of respondents of the political system as they expected it to be in five years' time. (Centre for the Study of Public Policy, 1996). The sixth *Central and Eastern Eurobarometer* survey

One suggestive theorization of this polycentrism is provided by Bob Jessop's concept of 'hollowing out'. This terminology parallels the literature on the 'hollow corporation' which focuses on new forms of industrial organization, and specifically on the ways in which companies increasingly externalize many production and service functions while retaining a core command and control role. This is analogous to some changes in the contemporary nation-state. According to Jessop (1994), the nation-state is undergoing a threefold process of 'hollowing out' as its powers and responsibilities are being transferred 'up' to supra-national bodies (most notably the European Union), 'down' to revitalized local and regional governments (see Chapter 4) and to some extent 'out' to networks of institutions which bypass the nation-state altogether:

> some state capacities are transferred to a growing number of pan-regional, plurinational, or international bodies with a widening range of powers; others are devolved to restructured local or regional levels of governance in the national state; and yet others are being usurped by emerging horizontal networks of power – local and regional – which by-pass central states and connect localities or regions in several nations. (Jessop, 1994: 264)

To these shifts we might also add the transfer of state powers and responsibilities (particularly in the field of service provision) to private and voluntary sector organizations. Jessop notes, however, and this is significant here, that the state retains much of its political importance and sovereignty. There are at least three ways in which these geographical transformations connect to my comments on democracy.

The European Union First, with the growth in power of the European Union, the well-known 'democratic deficit' of its institutions becomes an increasingly pressing issue. Those opposed to European integration are able relatively easily to construct the nation-state as the site of democratic governance in contrast to a bureaucratized and unaccountable EU. The development of democratic practices within the institutions of the EU following the Maastricht Treaty constitute what might be called 'thin' democracy. In its report to the 1996 Intergovernmental Conference, the Commission of the European Communities (1995d) identifies a series of measures that have been taken as a result of Maastricht to enhance the democratic legitimacy of the EU. These include (a) the development of European citizenship involving rights to free movement, to vote in European elections, to diplomatic and consular protection and to petition the European Parliament; and (b) a series of institutional responses involving among other things limited extensions to the powers and responsibilities of the directly elected Parliament. Although these measures represent some shift in power from the Commission and Council to the Parliament, the Commission itself admits frankly 'there also has to be a reservation, concerning the weakness, not to say the absence, of democratic control at the Union level in the fields of activity where intergovernmental process still holds sway.' Moreover, there are some territories that are subject to EU influence but that are not able to exercise even 'thin' democratic control. These include those CEECs that have agreed to accept EU directives and practices prior to (and as preparation for) accession.

Much of the struggle to increase the democratic legitimacy of the EU focuses on the European Parliament (EP). The origins of the EP lie in the Common Assembly of the European Coal and Steel Community (ECSC), founded in 1952. Its purpose was to provide a democratic input into the decision-making of the ECSC, but it was initially very underdeveloped. The members were appointed by national parliaments, rather than directly elected, and its role was largely advisory. In 1957, the six members of the ECSC signed the Treaties of Rome setting up Euratom (the European Atomic Energy Authority) and the European Economic Community (EEC). The Treaties of Rome expanded the scope of the Assembly to cover all three communities and in 1962 it changed its name to 'European Parliament'. Since then its powers have been gradually extended by subsequent treaties, and its democratic legitimacy was strengthened by the introduction of direct elections in 1979. It now consists of 626 MEPs who sit in multi-national political groupings.

In 1975 the EP gained the rights to modify, amend and reject the annual budget. Following direct elections, in 1980 the EP flexed its muscles by testing how far these rights could be exercised, as Nugent (1989: 115–16) notes:

> Major confrontations with the Council, far from being avoided, seemed at times almost to be sought as the EP attempted to assert itself. Yet for all its efforts the Parliament could hardly be said to have re-shaped the budget in any fundamental way . . . The reason for this relative ineffectiveness was that although the EP enjoyed joint decision-making powers with the Council on the budget, the powers of the two bodies were not equally balanced. Parliament was still very much restricted in what it could do: restricted by the Treaty which gave it very little room for manoeuvre in the major budgetary sector, compulsory expenditure; restricted by the Council's attitude which tended to be one of wishing to limit Parliament's influence as much as possible; and restricted by its own inability – because of conflicting loyalties and pressures – to be wholly consistent and resolved in its approach.

Three key developments since this episode have, however, given the EP significantly more power in practice as well as in theory. In 1987, the Single European Act (SEA) came into force, legislating in each member state for the completion of the single market from 1 January 1993. The SEA gave the EP the right to a second reading on draft legislation dealing with the single market and to propose amendments (though the final decisions still lay with the Council). It also gave the EP the right to veto treaties concerning the accession of new member states. The second measure that enhanced the Parliament's power was the Maastricht Treaty on European Union (which introduced, among other things, economic and monetary union). Its impact is explained by the Commission of the European Communities as follows:

The Maastricht Treaty on European Union has marked another significant step in extending the authority of the European Parliament. The Treaty, which came into force in November 1993, extends the Parliament's right of co-decision with the Council of Ministers (now the Council of the European Union) to all legislation concerning the single market and to areas like the free movement of workers, research and development policy, the environment, health, education, and consumer protection. In these areas, Parliament now has the power to veto draft EU legislation it does not approve of. In addition, the co-operation procedure introduced in 1987 has been extended to new areas too. As a result of the Treaty on European Union, the Parliament exercises democratic control over the European Commission. The President of the Commission is now designated by EU Heads of State or Government after consultation with the Parliament before taking office. He [sic] and the entire Commission are also subject to a vote of investiture by the Parliament. In order to facilitate this function, the Maastricht Treaty extended the mandate of Commission members from four to five years to make it coincide with the five-year legislature period of the Parliament. Finally, the Treaty of European Union also enhanced the legitimacy of the European Parliament by enabling citizens of one EU country who are resident in another to vote and to stand as candidates in European and local elections in their host country. (Commission of the European Communities, 1994c: 2–4)

Thirdly, the Amsterdam Treaty of June 1997 extended the co-decision process further to make it more or less the general rule. The Amsterdam Treaty also strengthened certain other citizenship rights, such as allowing citizens of the EU to take any of the EU institutions to the European Court of Justice over any action that they consider breaches their fundamental rights (Commission of the European Communities, 1997e).

These various developments reflect a recognition that, as the powers of the EU increase and those of member nation-states decline in relative terms, the principles of liberal democracy, and the legitimacy that they are thought to provide, should be extended to the EU and, crucially, developed on a European-wide, and supra-national basis, reflecting the increasingly supra-national processes of governance operating in contemporary Europe.

Urban complexity and democracy The second area of 'hollowing-out', the shift of powers to the local or regional level, raises another set of issues for democracy. European cities and urban regions are central to debates about democracy for a number of reasons. As the site of everyday life they are the arenas in which democracy as social practice is lived out. For most people, participation in political activities is conducted at the urban, rather than the national or European level. In addition, European cities are sites of dramatic social and cultural change, especially that associated with ethnic pluralism and cultural mixing. As European urban regions grow in social complexity, democratizing the city requires a sophisticated attention to the inclusion of all social groups in the process of governance. Mouffe's stress on the necessity of accepting the differences between identity-based social groups without simultaneously seeking to destroy the 'other' takes on new force in the context of the combination of increasing urban diversity and continuing urban racism. Six aspects of the relationship between democracy and the urban and regional scales are worth highlighting.

(1) There are marked economic disparities between different urban regions in Europe (see Chapter 1). The richest regions of the EU (such as the urban regions of Hamburg, Brussels, Paris and Vienna) typically have levels of GDP per capita over one and a half times the EU average (even after geographical variations in the cost of living have been taken into account). By contrast, the poorest regions have levels of GDP per capita below half the average in the EU, and lower still in the CEECs (see Chapter 7). The promotion of measures to enhance 'social cohesion' is a largely economic response to these disparities within the EU. Relatively little attention is paid to the issues of democratic inclusion in this context.

(2) Much policy-making at the urban scale focuses on economic regeneration, including the competition for footloose international investment projects and national and European grants (Harvey, 1989b; Painter, 1998). The commercial and political sensitivity of many of these projects and the entrepreneurial styles of governance with which they are increasingly associated frequently seem to undermine openness and democratic accountability.

(3) The formal electoral accountability provided by conventional city council governance may be a necessary condition for urban democracy, but it is not a sufficient one. The powers of local councils are often restricted, even over matters concerning only the local area. According to Pickvance (1997), for example, the tendency towards decentralization of the formal political arena evident in eastern Europe cannot be taken to lead automatically to democratization without the simultaneous development of what he terms 'substantive democracy'. Substantive democracy depends not (only) on decentralization but also on the ability of citizens to exercise democratic rights in practice. Pickvance takes the activities of social movements to be an index of substantive democracy:

Many writers on democratisation in Eastern Europe adopt a definition which restricts democracy to formal features of the political system . . . However, 'formal' definitions . . . exclude the substantive issue of whether democratic rights and opportunities do give citizens an input into the political system, or whether, despite these rights, an elitist power structure exists. To address this question, I choose the example of social movements . . . The questions I ask are whether such groups find it easy to form and develop and whether they are successful in achieving their aims, or whether they encounter resistance and repression. The result will be taken as a substantive measure of democratisation as it concerns its use rather than the existence of the right of association. (Pickvance, 1997: 137)

The results of Pickvance's study of Budapest (more decentralized) and Moscow (less decentralized) suggest that in Hungary environmental movements are more numerous and successful than in Russia, but that there are more housing movements in Moscow (although they are largely unsuccessful). This suggests that formal measures of democratization, such as decentralization, are not the only, or necessarily the most important, influence on substantive democracy, and Pickvance cites a range of other factors, such as resource availability, state responsiveness and motivation that are also key.

(4) Pickvance's finding suggests that extensions of formal democracy (by adding an elected level of government at the regional scale, for example) will not necessarily lead to substantive democratization unless they are accompanied by the parallel development of civil society.

(5) Contemporary European cities are marked by significant levels of social diversity, notably, though not only, in their ethnic composition. Diverse ethnic groups bring to the urban political arena different, and often competing, social needs and wants, and the expression of this pluralism challenges unitary and formal models of democracy, particularly majoritarian conceptions. Stress is increasingly being placed on the formal protection of minority rights and interests, but there is also a need to couple this to the promotion of self-confidence, participation and the building of voice for marginalized communities.

(6) By way of summary, contemporary European cities are becoming increasingly complex. Economic circuits, levels and institutions of governance and social groupings all intersect and interact in the city. This poses significant challenges for governance of any kind as the steering constraints that arise from trying to manage a complex system are considerable. Urban complexity raises particular difficulties for *democratic* governance because it can confuse and obscure lines of accountability and undermines any simple notion of 'the' popular will.

The nation state and national identity As Jessop (1994) points out, 'hollowing out' does not result in the elimination of the nation state. Indeed, it remains politically powerful, and it also remains a crucial site of democratic practice. Nevertheless, some writers identify the nation-state as a major block to democratization. This argument has been made most strongly by Mary Kaldor. Writing about the conditions required for democratic governance, she says:

> I shall argue that the nation-state, at least in its nineteenth-century form, is no longer adequate as the dominant form of political institution. New institutions, at local, regional and international levels, will have to supplant the nation-state but not in the form of smaller or larger nation-states, but as new forms of governance. I do not want to suggest that there is no room for national identity or national self-determination, but rather these demands have to be channelled towards new types of political units that are less absolutist and less powerful than the nation-state. (Kaldor, 1996: 9)

As movement into and between European states grows, then the discursive link between the nation-state and national identity becomes increasingly problematic. One of the central questions for democracy as social practice is the definition of the borders of the political community. In a less mobile world, ethno-territorial definitions of political community were exclusionary, but perhaps sustainable. In the twentieth century, the ethno-cultural purification of political space leads ultimately to Auschwitz and Sarajevo. It is also evident in a host of smaller but potentially insidious ways, as Kymlicka (1995a: 23) shows in the case of German citizenship:

> Membership in the German nation is determined by descent, not culture. As a result, ethnic Germans who have lived their whole lives in Russia, and who do not speak a word of German, are automatically entitled to German citizenship, while ethnic Turks who have lived their whole lives in Germany and so are completely assimilated to German culture are not allowed to gain citizenship . . . Such descent-based approaches to national membership have obvious racist overtones, and are manifestly unjust. It is indeed one of the tests of a liberal conception of minority rights that it defines national membership in terms of integration into a cultural community, rather than descent. National membership should be open in principle to anyone, regardless of race or colour, who is willing to learn the language and history of the society and participate in its social and political institutions.

The need for a democratic practice which extends beyond narrowly defined ethnic-cultural communities, and indeed beyond national borders however defined, motivates Held's proposals for a cosmopolitan democracy. Held argues that in the context of social and economic globalization, the traditional realist conception of international relations, in which nation-states are seen as independent actors pursuing their own 'national interests' in an anarchic and unregulated international arena, is both undemocratic and dangerous. He proposes the development of a democratic international order. However, democratizing the international arena can only be part of the answer because ethnic and cultural pluralization is not only about getting along with others in other

6

The 'Divided' Mediterranean: Re-defining European Relationships

Russell King and Marco Donati

Despite Europe's 'opening to the east' and the new geo-political enviro-nment of the post-Cold War era, the Mediterranean remains in many respects the most problematic flank of Europe. The Basin itself expresses a contradictory character as a region of both unity and diversity, of centripetal and centrifugal forces. Recently, it has been cast principally as a divide, a kind of Rio Grande between Europe and Africa (King, 1998). More specifically, it has been regarded as an economic divide between the 'First' and 'Third' worlds; as a demographic divide between two very different population regimes; as a security divide between stable, democratic Europe and the less stable, less democratic regimes of North Africa and the Middle East; and as a cultural divide between Christian, 'western' Europe and the Islamic 'south' whose people are demonized as 'others'. Trans-Mediterranean migration flows are per-ceived as perhaps the biggest 'problem' from the European perspective, along with terrorist threats from countries such as Algeria and Libya. On the other hand, there are clear signs of a more co-operative relationship between Europe and many of the non-EU Mediterranean states, notably through trade and aid agreements discussed at the historic Barcelona meeting of November 1995.

This chapter examines the complex European relationships with the Mediterranean, a region variously defined as nodal, marginal, dicho-tomous, a theatre of both conflict and co-operation. Taking both a chronological and a thematic treatment, the analysis will first briefly record the early geographical roles of the Mediterranean, starting with the Roman empire when the Mediterranean Basin was the core and northern Europe was peripheral. During this historical review, particular attention will be paid to the period between the 1960s and the 1980s when Mediterranean Europe ceased to be the poor southern periphery of the continent and took on a more dynamic but regionally differentiated role. Migration will be used as a lens to reveal the changing position of southern Europe within the wider regional dynamics of Europe and the Mediterranean Basin. Next, economics, demography and geo-politics will be deployed to analyse the multiple nature of the Mediterranean divide between Europe and the southern 'beyond'. The final part of the chapter will trace the EU's evolving and intensifying relationship with the southern Mediterranean countries: a relationship which is asymmetrical and in the short term weighted very much in Europe's favour. In the longer term, Europe's interests will probably be best served by a policy which gives greater weight to the needs of the southern shore countries.

From core to periphery: the Mediterranean in history

How may we conceptualize the southern or Mediterranean flank of Europe? Following the lead given by Seers (1979), many geographers writing in the 1980s regarded the Mediterranean, including southern Europe, as a periphery or semi-periphery of Europe, whose core lay in the central part of north-west Europe (see, for example, King, 1982; Ilbery, 1984, 1986: 2–10; Williams, 1987: 82–5; Keeble, 1989). As we shall see presently, more recent work has reappraised the position of southern Europe within the wider Euro-Mediterranean space–economy. And even a cursory glance at history reveals that Mediterranean Europe was far from peripheral in the past. Under the Romans the whole of the Mediterranean Basin and southern Europe (plus much of central and northern Europe) was unified under one imperial power, operating from the centre of the Mediterranean. Clearly the Mediterranean was the core, and northern Europe part of the periphery (this is not to ignore the existence of smaller-scale colonized peripheries of Rome within the Mediterranean region). Subsequently, much of the western Mediterranean was incorporated under Moorish control, coming from the south but breaching the Pyrenees; and then the Ottoman empire spread from the east as far as the gates of Vienna. After the *reconquista* of Spain, east–west, not north–south, was the main axis of confrontation in the Mediterranean, as Christian Spain faced the Muslim Ottoman empire (Graham, 1997).

Fundamental to this brutally condensed historical discussion is the question of whether the Mediterranean should be considered a frontier region, a *limes*; or a unified, integrated region looking inwards (and outwards). Conventional geography (and contemporary geo-politics) regards the Mediterranean as the boundary between the continents, and different 'worlds', of Europe and Africa (and to some extent Asia). But there are powerful figures espousing a unified view of the Mediter-ranean. In the preface to the English edition of his great book on the Mediterranean, Fernand Braudel reaffirmed the 'major truth' of the unity and coherence of the Mediterranean region (Braudel, 1972: 14). For Braudel, the Mediterranean was a flexible concept, not to be defined in hydrographic terms as a sea, nor in ecological terms as the area of land that lies between the northernmost date palm and the northernmost olive, but as a vibrant (and plundering) culture that historically reached across the Atlantic via the Canaries and the Azores, down the Red Sea and across Asia in search of spices, up to the British Isles for grain and wool, and down the west coast of Africa for slaves and ivory.

Braudel was not alone. André Siegfried, a Parisian contemporary, was at pains to express the Mediterranean's unity: 'everywhere it is the same, for the shades of difference are less important than the resemblances . . . (even if) . . . this unity is the result of aggressive contrasts – sea, mountain, desert' (Siegfried, 1948: 28–9). With remarkable prescience, he also pointed out that it is a region 'where political tempests rise, gain momentum and burst forth'.

The Mediterranean's economic and cultural supremacy *vis à vis* most of Europe lasted throughout the Middle Ages, although politically this was a time of collapse of centralized power, seen most clearly in the case of Italy where city-republics such as Pisa, Lucca and Venice exercised autonomy over their own affairs. By the Renaissance period, however, despite the extraordinary cultural and economic revival of certain cities like Florence, commercial hegemony was already passing to north-west Europe where wealth was vested in industrial production and trade along the Baltic–Atlantic axis (Graham, 1997: 92). The subsequent industrial revolution, which created such marked disparities between European countries which industrialized and those which did not, largely by-passed the Mediterranean. Only small parts of northern Italy and northern Spain were positively affected. The rest of southern Europe saw its population locked into a subordinate role in the wider divisions of labour, leading inexorably to peripherality, dependency, rural overpopulation, poverty and pressures for emigration (Dunford, 1997: 127).

From the industrial revolution to the 1960s, then, southern Europe lagged behind northern Europe. Along with the rest of the Mediter-ranean Basin, it existed as a kind of semi-periphery positioned between the 'First World' of Europe and the 'Third World' of Africa and western Asia. Symbolic of the 'discovery' of southern Europe as almost a 'Third World' of Europe was the popularity of the region for social and anthropological studies of 'backwardness' and traditional 'peasant cultures' in the early post-war decades. Among the pioneering studies of this era were Pitt-Rivers' *People of the Sierra* (1954) and two contrasting books on Basilicata (Banfield, 1958; Davis, 1962). These, and the plethora of studies which followed in the 1960s and 1970s, created the impression that the Mediterranean (and especially Mediterranean Europe) was a specific anthropological region where key concepts and behaviours such as honour, shame, strong kinship and gender roles and the survival of peasant values distinguished it from 'the north'. The extent to which the Mediterranean is, or was, a meaningful regional construct in cultural terms, or whether it was merely an anthropological 'invention', has been keenly debated in recent decades. Goddard's (1994) review of this controversy tends towards the conclusion that the alleged uniqueness and coherence of 'Mediterranean society' is more an anthropological construction than a reality.

More relevant for our analysis here are the ways in which the inhabitants of the peripheralized regions of southern Europe and the Mediterranean reacted to underdevelopment and to the non-existence or failure of state development policies. One reaction was common to many countries of the Mediterranean Basin after 1950: emigration. In the 1950s and 1960s the 'migration frontier' was a clear expression of the divide between the prosperous migrant-receiving countries on the one hand (principally France, West Germany, Switzerland and the Benelux countries), and the less-developed countries of southern Europe and the Maghreb on the other. These latter countries – Portugal, Spain, Italy, Yugoslavia, Greece, Turkey, Tunisia, Algeria and Morocco – furnished labour migrants to stoke up the economic progress of the north. The migration divide thus ran through south-central Europe, roughly following the line of the Pyrenees, the south coast of France and the Alps. By the 1960s, selective development and modernization in Spain and Italy had led to an adjustment of the divide so that it included as net immigration regions (based on internal migration flows) Catalonia, the Basque Provinces, Lombardy and Piedmont (Figure 6.1). Rural–urban flows within southern European countries also led to the rapid growth of Lisbon, Madrid, Rome and Athens (King, 1984).

The renaissance of Mediterranean Europe since the 1960s

Within a neo-classical framework, emigration can be considered as a 'natural response' to the conditions of unemployment, poverty and overpopulation which existed in most rural and some urban areas of southern Europe in the early post-war period. This dovetails with, rather than contradicts, the Marxist interpretation of European labour migration as a movement stage-managed by industrial capital's perennial search for a 'reserve army' of cheap labour. Ideology apart, the structure of the south European economy in the 1950s was fairly clear: more than half the population engaged in mainly peasant agriculture, low per capita incomes, a poorly developed and technologically backward industrial sector, an inefficient and corrupt service sector, an almost non-existent welfare state (see Chapter 1), and a trade balance dominated by

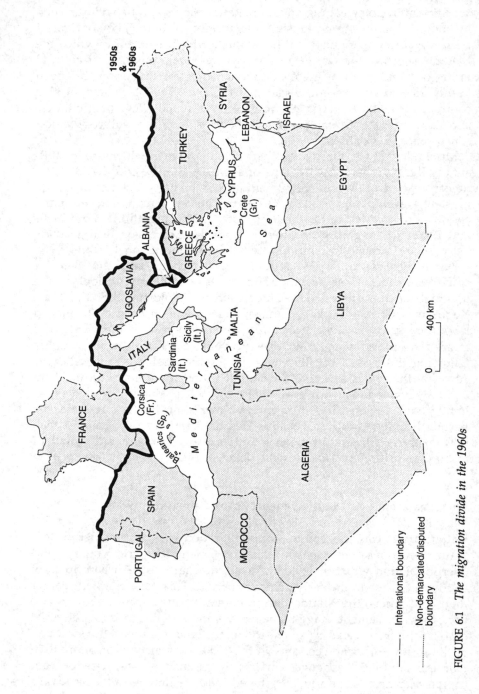

FIGURE 6.1 *The migration divide in the 1960s*

low-value raw material exports and high value-added manufactured imports (Williams, 1984: 1).

Yet already by the 1950s in Italy, and by the 1960s in Spain, Portugal and Greece, economic growth rates were accelerating sharply, at the same time as mass emigration was continuing. There is more than one way of interpreting the relationship (or perhaps non-relationship) between emigration and economic growth in southern Europe in the 1950s and 1960s. Under a functional equilibrium interpretation, emigration was seen as a causal factor behind the economic growth, siphoning off surplus population and unemployment leading to rising per capita incomes, and with remittances functioning as a valuable source both of foreign exchange and of private capital formation. Kindleberger (1965) argued this set of processes, paying particular attention to the case of Greece. Another view, linked to debates about the new international division of labour (Fröbel et al., 1980), stressed the role of cheap labour in southern Europe attracting inward investment, for instance in manufacturing enterprises. This leads to an interpretation of rapid economic growth based on domestic sources of abundant and under-rewarded labour which sustained a 'virtuous circle' of high rates of profit and reinvestment throughout the period up to about 1970.

But industrial development was by no means the only, nor the most important, economic process working to transform southern Europe at this time. Tourists flowed in the opposite direction to migrants and, from the early 1960s, became a mass phenomenon, making significant contributions to the invisible earnings of Italy, Spain, Portugal, Greece, Yugoslavia, Cyprus and Malta. By the early 1990s, other sectors played a role too: these included intensification and commercialization of agriculture in the irrigated plains, development of various infrastructural projects, such as the modernization of communication networks, and an expanding urban service economy.

An important political gloss was put on all this by the emergence of southern European countries from authoritarian rule, first in Italy at the end of the war, then in Greece, Spain and Portugal in the 1970s. The overthrow of military and paramilitary regimes with their fascist overtones, and the consequent removal of the 'democratic deficit', opened up European Community accession for the latter three countries in the 1980s, Italy having been a founder member of the Common Market. Membership of NATO had an equally symbolic meaning for these southern European countries' membership of the 'western democratic club' (Lesser, 1995: 14–15).

In short, all the indicators demonstrate the way in which southern Europe – principally Italy, Spain, Portugal, Greece, Cyprus and Malta – 'caught up' with (or moved very much closer to) western European standards of democracy, wealth and well-being during the period between the 1960s and the 1980s, increasing at the same time the scale of the 'development gap' across the Mediterranean, since the north African and eastern Mediterranean countries had recorded much more modest growth rates over the same period (Dunford, 1997: 134, 150). The dynamic profile of southern European countries (Portugal, Spain, Italy, Greece) over the period 1960–85 can be brought out by a comparison of their overall growth in GDP per capita (at 1980 US $ prices) of 165 per cent with both the countries of the southern Mediterranean (Morocco, Algeria, Tunisia, Egypt, Syria, Turkey: 118 per cent) and the northern EC countries (UK, Ireland, Netherlands, Denmark, West Germany, Belgium, France: 103 per cent): these calculations are based on data in Golini et al. (1991).

Three rather specific features of the transformation of southern Europe were important, however, and remain so today. The first concerns the trajectory of employment change. Twentieth-century economic development in northern Europe has seen the major share of employment move sequentially from agriculture to industry and then to the service sector. In the south of Europe, the shift has basically been from agriculture straight to the service sector, with industrial employment, although increasing for a time (notably during the 1960s and early 1970s), never reaching such a dominant position in the national structure of employment as it did in north-west Europe during either the first industrial revolution or during the Fordist era. By 1990, more than half of total employment in the four southern EU countries was in the tertiary sector, being highest in Italy (58.6 per cent) and lowest (47.4 per cent) in Portugal.

The second feature was the markedly uneven regional expression of southern European development. Hence 'uneven development' became a by-word for what went on in southern Europe between the 1960s and the 1980s (cf. Hudson and Lewis, 1985; Hadjimichalis, 1987). In Italy, despite the efforts of the Cassa per il Mezzogiorno, most of the industrial and advanced services development remained polarized in the economic core of Lombardy and Piedmont, although since about 1970 there has been considerable decentralization of industrial dynamism into adjacent regions like Veneto and Tuscany. Portugal and Greece saw increasing polarization of population and development in Lisbon and Athens, with subsidiary poles in Porto and Thessaloniki. Spain's pattern of regional development has been more complex, sometimes portrayed as an inverted core–periphery model where Catalonia, the Basque Country (until its economic crisis in the 1990s) and the touristically developed Mediterranean coast constitute an economic core which is spatially peripheral, complemented by a rather underdeveloped interior, within which the capital Madrid stands out as an isolated node of high development (Hebbert, 1990; see also Chapter 1).

Thirdly, while the nature of southern European development was based on a variety of processes and policies including foreign investment and state and EC regional aid, most important of all has been the endogenous expansion of the informal sector, also labelled the 'black' or 'parallel economy'. This local dynamism of the informal sector is perhaps the defining feature of the southern European 'model' of capitalist development (Mingione, 1995), other elements of which include a strong reliance on family-run firms, a carefully created balance of co-operation and competition, and an integrated blend of farming, small and medium industries, tourism and services. It is a model which, in many respects, is the antithesis of Fordism. But it is not uniformly developed over the territory of southern Europe: indeed, its successful expression in certain regions, such as the well-known 'Third Italy' or the central region of Portugal, and its virtual absence in others such as Calabria or the Alentejo, is one of the factors re-shaping the map of uneven development noted above (see also Chapter 2).

Hence, the 'rebirth' of Mediterranean Europe since the 1960s has occurred in particular ways. The changes have undoubtedly been great: the poverty, misery and malnourishment of the 1950s have largely been eradicated, industries and tourism have developed, agriculture has been modernized (or abandoned), cities have grown. But this progress has been achieved at a price. Unevenness of development has been a key problem; unemployment remains high over much of the region, especially southern Spain and southern Italy; and political corruption and organized criminal activities have not been eradicated – indeed, in some places they have flourished. Moreover, there is also a sense in which, despite all this 'progress' and 'modernization', the true nature of peripherality in southern Europe has never been grasped by economic planners and policy-makers. As Hadjimichalis (1994) has shown, this comment has special relevance in the new environment of the single market where integration has unconceptualized and unpredictable effects on southern Europe, both because of the failure to understand the historical processes of the reproduction of uneven development across the region, and because of the delicacy of the balance between the forces of integration and disintegration for the south within the 'New Europe'.

The Mediterranean – a divided space?

New migration trends

Returning to the theme of migration, the transformed economic status of southern Europe made it no longer a region of mass emigration to northern Europe and overseas. Return migration flows, from the emigrations of the 1950s and early 1960s, were already significant by the late 1960s. By the 1970s immigration was starting to become a recognizable phenomenon. The inflows, mainly from North Africa and a variety of other developing countries such as Senegal, Cape Verde and the Philippines, swelled during the 1980s, and were then given a further boost after 1989 when central and eastern European countries were 'opened up' as a source of emigrants.

The modernization and new economic prosperity of southern Europe was obviously an important 'pull factor' for these new migration flows, but it was by no means the only factor operating. Also important were the relative ease of entry to these countries (particularly given the much tighter controls implemented by the traditional migrant-receiving countries of north-west Europe after 1973) and the rising momentum of 'push factors' – population growth, poverty, unemployment, civil disturbances and so on – from the countries of origin. The openness of long sea borders and the geographical position of southern Europe as the 'vulnerable underbelly' of the European 'fortress', facing the direction of arrival of many migrants from North Africa and beyond, made entry very difficult to control: 'Fortress Europe' was not impregnable. As a result, a wide variety of immigrant communities quickly became established.

It is impossible to be precise about the scale of immigration into southern Europe since the 1970s. Annual flow data simply do not exist (apart from returning nationals and a small number of refugees), and the data on 'stocks' are largely based on enumerations of holders of residence and work permits; in other words, 'legal' migrants. But all authorities agree that there are also substantial numbers of 'illegal' or undocumented immigrants, especially in Greece where they constitute the overwhelming majority of the immigrant population. Periodic 'regularizations' of illegal immigrants have helped to produce more reliable statistics but even these data are questionable, for holders of permits may not remain in the country and there is some evidence of double-counting. What is clear is that there has been an upward trend in immigration, especially since the early 1980s. Simon (1987) estimated 2 million immigrants in the four southern EU countries; Werth and Körner (1991) estimated 3 million (based on 1989); in the late 1990s, the figure perhaps is in the region of 3.5 million, given the post-1989 influxes from central and eastern European countries like Poland, Romania and Albania.

The Yugoslav crisis generated a complex pattern of refugee movement both within the territory of the former Yugoslavia and abroad, especially to Germany where 325,000 were afforded temporary protection and are now being pressed to return. Of southern EU countries, Italy has a tradition of immigration from Yugoslavia which pre-dates the break-up of that country. Italian data on residence permits at 31 December 1995 showed 51,973 'Yugoslavians' (this is an ambiguous label since it includes both long-established migrants from the former Yugoslavia and recent arrivals from Yugoslavia–Serbia), 18,944 Croatians, 15,466 Macedonians (from former Yugoslav Republic of Macedonia), 10,224 Bosnians and 4,107 Slovenians (Caritas di Roma, 1996: 69).

Table 6.1 gives the stock of foreign population by main nationality for Italy, Spain and Portugal in 1995; note that these figures refer to legal immigrants only. There are no reliable figures for Greece where

TABLE 6.1 *Italy, Spain and Portugal: stock of foreign population by nationality (000s), 1995*

Italy		Spain		Portugal	
Morocco	94.2	Morocco	74.9	Cape Verde	38.7
USA	60.6	UK	65.3	Brazil	19.9
ex-Yugoslavia	52.0	Germany	41.9	Angola	15.8
Philippines	43.4	Portugal	37.0	Guinea-Bissau	12.3
Tunisia	40.5	France	30.8	UK	11.5
Germany	39.4	Italy	19.8	Spain	8.9
Albania	34.7	Argentina	18.4	USA	8.5
UK	27.7	Peru	15.1	Germany	7.4
France	27.3	USA	14.9	France	4.7
Romania	24.5	Dominican Rep.	14.5	Venezuela	4.5
Senegal	24.0	Netherlands	13.0	Mozambique	4.4
Brazil	22.1	Philippines	9.7	S. Tomé	4.1
Poland	22.0	China	9.2	Netherlands	2.7
Egypt	21.9	Belgium	8.9	Canada	2.4
China	21.5	Colombia	7.0	China	2.2
Sri Lanka	20.3	Venezuela	6.5	Italy	1.9
Switzerland	18.2	Switzerland	6.2	India	1.0
Spain	17.8	India	6.2	Pakistan	0.8
Somalia	17.4	Sweden	5.9	Iran	0.5
Greece	14.8	Chile	5.6		
India	14.6	Cuba	4.6		
Ghana	12.6	Denmark	4.5		
Argentina	10.5				
All countries	991.4	All countries	499.8	All countries	168.3

Source: SOPEMI (1997: 228, 233–4)

estimates of immigrant workers range from 250,000 to 470,000 (SOPEMI, 1997: 112). An examination of the individual nationalities reveals three major migration types. First, there are migrants from developing countries (note that Morocco and Cape Verde top the lists) who fulfil the function of cheap flexible labour in sectors such as construction work, agriculture and petty services, often working in the informal sector and contributing to the dynamism of the black economy (Pugliese, 1993). Secondly, there are immigrants from central and eastern Europe doing similar jobs. In Greece, Albanians and Poles are thought to be the two largest immigrant groups (Fakiolas and King, 1996). Thirdly, there are north Europeans who have settled in southern Europe for professional, recreational and climatic reasons. These are generally wealthy immigrants and include many retired people (A. Williams et al., 1997). Finally, there is evidence of considerable migration between southern EU countries: Spanish and Greeks in Italy, Portuguese in Spain and so on. Most of these are labour migrants, but there are also many professionals and students.

The above portrayal of immigration trends into southern Europe shows that this region now lies firmly north of the 'migration frontier'.

This frontier, described earlier as running through south-central Europe in the 1950s and 1960s, now lies positioned along the southern Mediterranean shoreline, south of Sicily, Malta and Crete (Figure 6.2). The intensity of the migration pressure across this divide, and the desperation of migrants to cross it, reveals the width of the 'development gap' which now runs like a fault-line through the Mediterranean.

Mediterranean economics and demographics

Table 6.2 presents some pertinent data to explore the economics and demographics of the Mediterranean Basin. The data confirm that the region's countries can be divided into two groups: the 'northern shore' countries (the southern EU states plus Malta, Cyprus and Israel) and the 'southern and eastern shore' countries. The 'northern shore' group has high levels of wealth, an employment structure in which between half and three-quarters of the population work in tertiary activities, a human development index (HDI) of around 0.9, low fertilities (especially the EU states in this group), and little or no difference between male and female literacy. The 'southern and eastern shore' countries have much lower incomes (real per capita GDP is only about one-quarter to one-third the northern level), higher proportions of the population working in agriculture (except Lebanon, Libya and Algeria), human development index values of around 0.6 or 0.7, much higher fertility rates, and (except for Lebanon) significant differences between male and female literacy, indicating a lack of gender equality. Under this scheme, only Albania is difficult to classify: geographically part of Europe, economically more like the North African and eastern Mediterranean states, and demographically in between.

While figures in Table 6.2 do indeed confirm that there is an economic and demographic fault-line running along the southern shore and then around the eastern Mediterranean (this line would be easier to draw if Albania and Israel swapped places!), some of the data indicate more similarities than differences. The share of industrial employment, for instance, is remarkably uniform throughout the region. More important, however, is the fact that Table 6.2 offers only a static picture; it is vital to look at economic, demographic and development *trends*.

Tables 6.3 and 6.4 provide a variety of trend indicators covering the evolution of Mediterranean Basin countries between the 1960s and the 1990s. Dividing the countries once again into two groups, the following observations can be made. Both groups of countries appear to be moving along roughly parallel trajectories but the gap between them remains significant. Table 6.3 suggests that on some criteria (for example, GNP per capita) it seems to be widening, while on others (for example, HDI) it is narrowing. For both groups of countries, both demographic and GNP growth were higher in the past than they are now. In the 1990s, population growth in the southern EU countries is

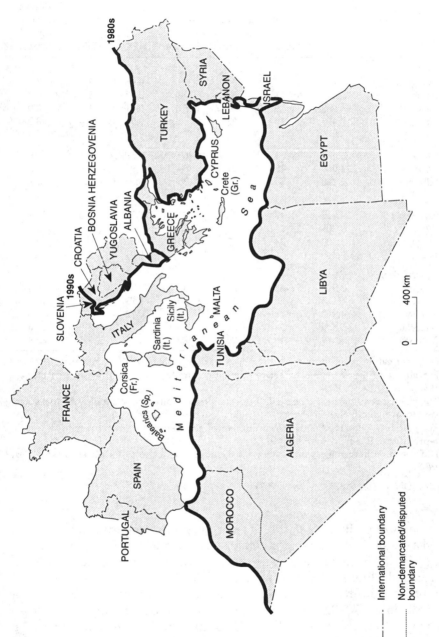

FIGURE 6.2 *The migration divide in the 1980s and 1990s*

TABLE 6.2 *Economic and demographic indicators for Mediterranean countries, 1992–3*

	Real GDP/ cap. (US $)[1]	Employment structure (%)[2]			HDI[3]	TFR[4]	Difference between M and F literacy[5]
		Agric.	Indust.	Service			
Southern EU							
Portugal	10,720	17	34	49	0.88	1.6	8.0
Spain	13,660	11	33	56	0.93	1.2	0.0
Italy	18,160	9	32	59	0.91	1.3	0.0
Greece	8,950	23	27	50	0.90	1.4	8.0
Other Europe							
Malta	11,570	3	28	69	0.89	2.1	n.d.
Cyprus	14,060	15	21	64	0.91	2.5	n.d.
Albania	2,200	56	19	25	0.63	2.9	n.d.
Eastern shore							
Turkey	4,210	47	20	33	0.71	3.4	21.1
Syria	4,200	23	29	48	0.69	5.9	31.3
Lebanon	2,500	14	27	59	0.66	3.1	4.9
Israel	15,130	4	22	74	0.91	2.9	n.d.
Southern shore							
Egypt	3,800	42	21	37	0.61	3.9	25.4
Libya	6,125	20	30	50	0.79	6.4	17.0
Tunisia	4,950	26	34	40	0.73	3.2	24.8
Algeria	5,570	18	33	49	0.75	3.9	26.8
Morocco	3,270	46	25	29	0.53	3.8	25.9

[1] GDP per capita based on purchasing power parities of the country's currency, 1993.
[2] Percentage of the economically active population employed in the three main sectors, 1992.
[3] The human development index (HDI) is a composite index composed of three measures, equally weighted: real GDP per capita, life expectancy and education (data for 1993).
[4] Total fertility rate (TFR) is the mean number of children per woman over the reproductive life-cycle (data for 1992).
[5] Difference, in percentage points, between the adult literacy rates for males and females, 1993.

Source: UNDP (1996: 135–9, 178–9, 203)

virtually zero. In fact, if the present very low fertilities continue, an increasingly ageing total population will start to decline quite sharply early in the next century, with Italy leading the trend (King, 1993a). Fertility decline is also well established in the southern Mediterranean countries – in the Maghreb states the total fertility rate fell by about a half during 1970–92 – but, as shown in Table 6.2, it still remains well above replacement level (TFR = 2.1). Other aspects of development – the evolution of the HDI and the rate of urbanization – also show more rapid progress in the southern countries. This 'catching up' is displayed in a different way in Table 6.4 where selected southern Mediterranean countries are compared with the average for the 'global North' across four

TABLE 6.3 Trend indicators for Mediterranean countries

	GNP/cap. annual growth rate (%)		HDI		TFR (1992)	Annual pop. growth rate (%)		Urban pop. as % total	
	1965–80	1980–93	1960	1993	(1970 = 100)	1960–93	1993–2000	1960	1993
Southern EU									
Portugal	4.6	3.3	0.46	0.88	57	0.3	0.0	22	35
Spain	4.1	2.7	0.64	0.93	41	0.8	0.1	57	76
Italy	3.2	2.1	0.76	0.91	55	0.4	0.0	59	67
Greece	4.8	0.9	0.57	0.90	60	0.7	0.3	43	64
Other Europe									
Malta	n.d.	3.2	0.52	0.89	99	0.4	0.6	70	89
Cyprus	n.d.	4.9	0.58	0.91	95	0.7	1.0	36	53
Albania	n.d.	-3.2	n.d.	0.63	59	2.3	1.0	31	37
Eastern shore									
Turkey	3.6	2.4	0.33	0.71	64	2.4	1.8	30	66
Syria	5.1	n.d.	0.32	0.69	76	3.4	3.4	37	51
Lebanon	0.6	n.d.	n.d.	0.66	57	1.3	2.3	40	86
Israel	3.7	2.0	0.72	0.91	77	2.8	2.1	77	91
Southern shore									
Egypt	2.8	2.8	0.21	0.61	64	2.4	2.0	38	44
Libya	0.6	n.d.	n.d.	0.79	85	4.1	3.4	23	85
Tunisia	4.7	1.2	0.26	0.73	49	2.2	1.8	36	56
Algeria	4.2	0.8	0.26	0.75	52	2.8	2.2	30	54
Morocco	2.7	1.2	0.20	0.53	54	2.5	1.9	29	47

For explanations of HDI (human development index) and TFR (total fertility rate) see notes to Table 6.2.

Source: UNDP (1996: 154–5, 176–7, 186–7, 196, 202–3, 208)

TABLE 6.4 *Progress of southern and eastern Mediterranean countries vis à vis the 'North' (= 100)*

	Life expectancy		Child mortality		Daily calories		Adult literacy	
	1960	1993	1960	1994	1965	1992	1970	1993
Turkey	73	90	19	33	85	100+	56	85
Syria	72	90	20	47	72	100+	43	72
Lebanon	86	92	48	45	80	100+	75	96
Egypt	67	86	16	35	78	100+	38	52
Libya	68	85	15	19	67	100+	40	77
Tunisia	70	91	17	53	76	100+	33	67
Algeria	68	90	17	28	58	83	27	62
Morocco	68	85	19	32	74	96	24	44

The 'North' refers to the 47 countries classified as 'industrial' (for a list see UNDP, 1996: 28). All figures are expressed in relation to the North's average, which is indexed to 100. The smaller the figure, the bigger the gap; a figure of 100+ indicates a performance better than the North's average. Indices for child (under 5) mortality are particularly low because of the way this is calculated as an inverse index (low = 'good', high = 'bad').

Source: UNDP (1996: 150–1)

development indicators: life expectancy, child mortality, food intake and literacy. Again, almost without exception, progress is registered.

These statistics give something of a lie to the nightmare scenarios of demographic explosion, economic collapse and consequent political strife and mass emigration in the southern Mediterranean countries. The full story is by no means straightforward and each of these 'dimensions of doom' needs more careful examination.

Fertility decline

The extent to which population growth in the southern Mediterranean is an 'avoidable crisis' (King, 1997) is open to a variety of interpretations. Both northern and southern shore countries have been experiencing declining fertility and population growth rates. On the southern shore, the mean annual rate of population increase was 2.8 per cent during the period 1950–80, falling to 2.2 per cent for 1980–2000. The corresponding figures for the northern shore (always excluding Turkey) are 0.8 and 0.4 per cent.

Understanding the future depends partly on appreciating the nature of the relationship between fertility decline and population change. The total fertility rate (TFR) may fall below replacement level but population growth may well continue because of the proportionate number of reproductively active females in the overall population. This distinction is particularly relevant in the Maghreb countries where fertility is declining fast and yet population growth remains rapid because of the structural effect of a large cohort of fertile women in the total population.

This can be contrasted with the situation in southern Europe where a longer history of fertility decline to historically unprecedented levels of only 1.2–1.4 in the 1990s has produced an ageing population with a diminishing replacement of young people. Recent Italian data indicate that the low fertility has 'bottomed out': the 1996 TFR of 1.21 represents the first time for more than 30 years that an increase in fertility has been recorded over the previous year (1.19 in 1995). Italy has had sub-replacement fertility since the late 1970s, although the TFR was as high as 2.7 in 1965.

If Italy can be regarded as the pioneer for population trends in southern Europe, Tunisia is the demographic leader for North Africa. Data for 1995 show that Tunisian TFR is now below 3.0 and projections show a further decline to around 2.0 by 2010; however, because of the structural component mentioned above, the absolute number of women aged 15–49 will not start to decline until after 2020. Figure 6.3 shows demographic indicators for Tunisia until 2030.

Three brief conclusions can be drawn from this analysis. First, the pace of fertility change in North Africa has been underestimated in the past so that the 'nightmare scenarios' need to be attenuated. Hence, secondly, the problem of a disturbing excess of young people entering labour markets which cannot possibly offer them all employment is likely to be shorter-term than thought. And thirdly, the equally dramatic (but qualitatively different) demographic decline in southern Europe opens up the potential for cross-Mediterranean migration to fill selected labour market shortages and to 'repair' the ageing population structure on the northern shore. In this vein, and especially in the western Mediterranean Basin, it may be possible to envisage a regionally integrated labour market and migration system.

Economic trends: from subordination to structural dependency

Compared to the demographic situation, the economic picture of the southern Mediterranean countries gives perhaps less grounds for optimism. As Tables 6.3 and 6.4 show, a good deal of social modernization has taken place, but the underlying economic trends remain extremely fragile, despite the boost given by large-scale hydrocarbon production in Algeria, Libya and, to a lesser extent, Egypt and Tunisia.

During the colonial period the Maghreb economies, especially, were subordinated to the strategies and needs of the colonizing powers, chiefly France but also Spain and Italy. Political independence – attained by Libya in 1951, Algeria in 1962 and Morocco and Tunisia in 1965 – was not complemented by economic independence: indeed, economic subordination to the European powers continued. The socialist or 'planned economies' experiments introduced during the 1960s and 1970s did not provide successful solutions and in some cases led to a worsening of economic conditions (Lacoste and Lacoste, 1991). In the late 1970s most

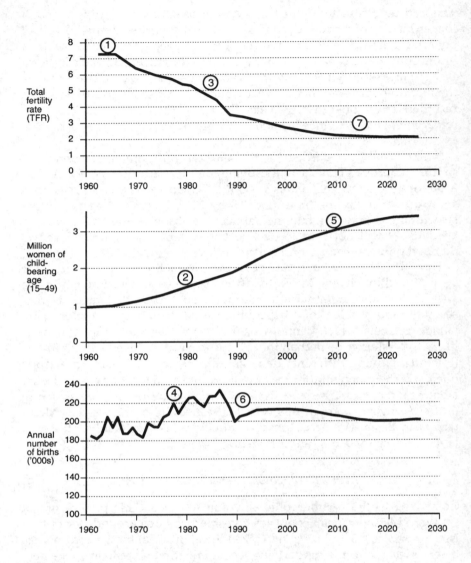

Key stages in demographic evolution (see figures on graphs):

1 High fertility until mid-1960s.
2 Growing number of women of child-bearing age as a result of past levels of high fertility.
3 Decline in fertility from late 1960s.
4 Rise in number of births continues into 1980s because of structural effect of large numbers of fertile women.
5 Stabilization in numbers of child-bearing women.
6 Decline in number of births when behavioural effect (declining TFR) outweighs structural effect (much slower rise in women of child-bearing age).
7 Replacement-level fertility early in next century.

FIGURE 6.3 *Tunisia's demographic development (Tapinos et al., 1994)*

TABLE 6.5 *Indebtedness in selected less-developed Mediterranean countries, 1993*

	Total external debt as % GNP	Debt service ratio[1]
Turkey	38	28
Egypt	109	15
Tunisia	58	21
Algeria	54	77
Morocco	81	32

[1] Debt servicing as % exports of goods and services.

Source: UNDP (1996: 172–3)

North African countries turned to foreign aid to help the modernization of their economies; however, most of these funds were dissipated on projects which failed to live up to expectations and to diversify the economic structure, leading to mounting trade deficits and external debts (Dunford, 1997: 147–8). Liberalization during the 1980s and 1990s has likewise given poor results so far. One of the major faults has been to underestimate the need to modernize agriculture in order to reach food self-sufficiency; instead, food imports constitute a major entry in the import balance sheet and, while farming still employs a major share of the labour force (Table 6.2), agriculture's contribution to GDP is minimal. The debt crisis is the major structural obstacle to the economic progress of many non-European Mediterranean countries. As Table 6.5 shows, the total external debt of Egypt exceeds the country's annual GNP, while servicing debt eats up one-third of Morocco's and three-quarters of Algeria's revenue from exports.

The problem of external debt and trade imbalance is a key issue in understanding not only the economic crisis of the southern Mediterranean countries but also the growing mistrust of these countries' peoples towards the western economic model and the political regimes that implement it. Worsening conditions at the macro-economic level have had severe repercussions on the life chances of whole sections of the population. Despite the progress registered on some measures of development, large numbers of people still live in poverty: 25 per cent of the total in Algeria, 30 per cent in Morocco, 34 per cent in Egypt and 54 per cent in Syria (UNDP, 1996: 170–1). Financial stringency imposed on many southern Mediterranean countries by the International Monetary Fund and the World Bank has led to declining spending on social and public services, generating widespread discontent while enhancing the importance of social assistance provided by religious organizations. As Azzam (1994: 95) explains, 'the appeal of political Islam has been aided by the apparent bankruptcy of secular alternatives. There is a search for the realisation of the unfulfilled promises by the regimes in power and the quest for prosperity merges with the search for cultural identity.'

political processes, on the other, are much harder to predict, and it is this unpredictability which Europe is especially fearful of. Hence the explicit move to a 'partnership' approach: the subject of the rest of this chapter.

Towards Euro-Mediterranean partnership

Given the strong historical links between the Mediterranean countries and Europe, it is surprising that the 'official' relationship took so long to develop. It was only in 1972 that the European Community established a so-called 'Global Mediterranean Policy' (GMP) with the rather limited aims of free trade in manufactured and agricultural products, and financial and technical co-operation. The reasons for setting up the GMP were much wider, however. Political instability within the region, notably between Greece and Turkey and between Israel and the Arab states, as well as the presence of the super-powers, led to a strategic policy by Europe to consolidate influence over the region in order to counter the bipolar power structure which split the Basin into conflicting East/West spheres. Economically, the Mediterranean had become an important trading and investment arena for Europe, a fact which reflected the growing regionalization of world trade.

Implementation of the GMP was delayed by a variety of obstacles both internal (disagreements among member countries, especially between the newly incorporated United Kingdom and France and Italy) and external (notably the Middle East War of 1973), so that the first initiatives were not taken until later in the 1970s (A. Jones, 1997). These agreements – with Israel in 1975, the Maghreb states in 1976, other Arab states in 1977, plus the customs union with Cyprus, Malta and Turkey – were then complicated by the applications for EC membership by post-dictatorship Greece, Portugal and Spain in 1977. So the GMP became a complex juxtaposition of *ad hoc* measures rather than 'global' in the Community sense of an overall co-ordinated approach (De la Serre, 1981).

The southern enlargement of the EC during the first half of the 1980s certainly enhanced the political relevance of the Mediterranean Basin in the foreign policy of the Community. In 1990, the Community launched the 'Renewed Mediterranean Policy', which provided 4.4 billion ECU over the period 1991–6 to support not only economic but also social development in the southern shore states. Roughly half of this total aid package was for country-specific financial assistance to Morocco, Algeria, Tunisia, Egypt, Israel, Jordan, Lebanon and Syria, and the rest was dedicated to joint programmes such as Med-Urbs, Med-Invest and Med-Campus to enhance co-operation between Europe and the Mediterranean countries in specific fields such as urban life, small and medium enterprises and research into energy, water and coastal management.

However, the 'Renewed Mediterranean Policy' proved insufficient to cope with the new international environment that emerged at the end of the Cold War. While the dismantling of the Iron Curtain turned western Europe's attention firmly towards the East, the widening economic and demographic divide to the south, and the potential implication of this for security, led the southern member states of the Community to emphasize the importance of the Mediterranean, at least as much as the central and eastern European dimension. This concern was greatly sharpened by the Algerian crisis and by Islamic movements in other southern Mediterranean states. Multiple preoccupations meshed worries over migratory pressure with a more traditional fear about the alleged proliferation of weapons of mass destruction and with paranoia about terrorist outbursts. The consecutive presidencies of the EU held by France, Spain and Italy in the early 1990s enhanced this southern European impetus and helped the process of defining a general framework for co-operation and security in the Mediterranean. While at the EU level this led, as we shall see later, to the historic Barcelona Conference of November 1995 and to the establishment of the Euro-Mediterranean partnership, at the level of regional co-operation *within* the Mediterranean several other initiatives materialized in the early 1990s. Three of these deserve mention: the Conference on Security and Co-operation in the Mediterranean, the Forum of Ten ('Five plus Five'), and the Mediterranean Forum (Tsardanidis, 1996).

In 1990, Italy and Spain presented a proposal for the creation of a Conference on Security and Co-operation in the Mediterranean (CSCM), on the model of the already existing Conference on Security and Co-operation in Europe. The CSCM was broadly based in scope and also geographically, features which help to explain its lack of progress. Differences of opinion soon emerged between France, Spain and Italy over their respective Mediterranean priorities. More fundamental was the sheer heterogeneity and complexity of the region from the security point of view, with obvious areas of tension in the Balkans, Aegean, Cyprus, Middle East and Maghreb. In the end, both the Italian and Spanish foreign ministers admitted that there was 'an element of utopia' in the CSCM scheme (Tsardanidis, 1996: 56).

Also in 1990, France proposed a more regionally circumscribed initiative, the Western Mediterranean Forum or the 'Forum of Ten'. Widely known as the 'Five plus Five', the forum included France, Italy, Spain, Portugal and Malta on the one side, and the five countries of the Arab Maghreb Union (Morocco, Algeria, Tunisia, Libya, Mauritania), on the other. This initiative, too, reached deadlock, principally because of internal problems within the AMU such as the tension between Algeria and Libya, and the Algerian crisis which escalated towards the end of 1991.

A third trans-Mediterranean initiative was the 'Forum for Dialogue and Co-operation in the Mediterranean' launched in Alexandria in 1994 as a counterpoint to the Western Mediterranean Forum, from which Egypt was excluded. This exclusion was seen by Cairo as a threat to

Egypt's traditional leadership role in Mediterranean affairs. The new forum reflected the ongoing successes of the Arab–Israel peace process. Comprising eleven countries (France, Spain, Portugal, Italy, Greece, Turkey, Egypt, Tunisia, Algeria, Morocco and Malta), it emphasized above all a pragmatic and flexible approach. Although the major sponsors of the Mediterranean Forum (Egypt and France) have tended to downplay its importance as compared to the Euro-Mediterranean strategy that has developed since 1995, there remains the belief that it can play a useful practical role in organizing regular consultations on key issues among a large group of Mediterranean countries (Tsardanidis, 1996: 60–2).

The Barcelona spirit

All of the initiatives outlined above proved insufficient and too narrowly based to make a real difference to the security environment of the divided Mediterranean region. Already in 1992, during the Lisbon European Council, the idea of a 'global partnership' for the Mediterranean was put forward. Further steps were taken in 1994, after the Corfu Council, when it was decided to establish a 'Euro-Mediterranean Economic Area' as a framework for co-operation and development. The Commission proposed that a free-trade area encompassing the EU and the Mediterranean be created by 2010; this would represent the largest free-trade bloc in the world. The Commission's belief was that a renewed emphasis on Europe's Mediterranean flank was necessary in order to formalize already-existing economic interdependencies and in recognition of the fact that the Community could not be isolated from the political instability around its southern and south-eastern margins. Others put the issue more bluntly, linking security, stability, development and migration into a set of interwoven concerns. According to Gillespie (1994: 1), if Europe did not concern itself directly with its southern neighbours, then these problems would be exported northwards, posing severe threats to stability and security.

On the eve of the Barcelona conference it was clear that political and security problems had to be linked to economic development and social reform. At the same time, confidence had to be built between the two shores in order to overcome mistrust and misrepresentation between the two sides. To a large extent, the Barcelona Declaration and the Euro-Mediterranean partnership are a major breakthrough, not only as far as the relations between Europe and the Mediterranean are concerned but also in terms of a broader understanding of security. Twenty-seven countries participated at the conference held in Barcelona in November 1995: the 15 EU countries plus all the other Mediterranean states except Albania, Libya and the republics of the former Yugoslavia.

The main assumption of the partnership rests in the belief that economic growth in the southern Mediterranean countries will enhance prosperity, thus creating the basis for the political reforms that should lead to increased political (and economic) stability. If these countries are stable, both among themselves and in their domestic affairs, then it is assumed that it will be easier to contain those phenomena that are seen in the EU as sources of insecurity and instability, such as migration waves and terrorist attacks. Stability on the southern shore is thus understood as security in Europe. This kind of rationale risks a repetition of the previous traditional approach to security which is based on 'hegemonic corporation' and emphasizes European security in terms of preserving western superiority. On the other hand, a closer examination of the Declaration reveals that the 'Euro-Med' initiative presents some original characteristics that fulfil the need for a broader understanding of security based on a dynamic model, so that stability does not merely equate to preservation of the status quo. A parallel approach has been set up for the implementation of the partnership: the conclusion of new Association Agreements and a multi-lateral dialogue.

The Association Agreements, which will replace existing ones, will be concluded between the EU and each single partner: the traditional bilateral approach. By mid-1997, agreements had been signed with Morocco, Tunisia, Israel and the Palestine Liberation Organization, while negotiations were under way with Algeria, Egypt, Jordan and Lebanon, and exploratory talks with Syria had started. Since December 1995 a customs union exists between the EU and Turkey. Apart from this latter initiative, none of the agreements has yet been ratified by national parliaments, however.

The multi-lateral approach is particularly important in order to cope with problems which are not contained by national boundaries and whose situation also depends on a 'global' understanding of their causes. This multi-national dialogue constitutes one of the most innovative aspects of the partnership. Of course, this is not to say that the institutional framework is strong enough to create the stable and balanced pattern of relationships that is implicit in a truly dynamic model; nevertheless, the approach is different from anything attempted before, and sets the basis for further networking even outside the institutional frame of the partnership.

The issues at stake in this open dialogue are indicated in the Barcelona Declaration and can be divided into three pillars: (a) political and security partnership; (b) economic and financial partnership; and (c) partnership in social and cultural affairs.

The first pillar: security, democracy and Islam

According to the Barcelona Declaration (1995: 137), 'the participants express their conviction that peace, stability and security are common assets which they pledge to promote and strengthen by all means at their disposal.' The 'means' are basically twofold: building confidence and seeking solutions through regular dialogue; and exerting pressure by imposing political conditions to the release of economic aid.

The Declaration does not linger over the meaning of security and stability but gives a clear idea as to the main threats to them: terrorism, organized crime and the proliferation of weapons of mass destruction. This perspective is clearly Eurocentric, for the threats listed are those thought to emanate from Arab countries. It was not possible to speak openly of 'Islamic fundamentalism' as a threat without jeopardizing the whole partnership process, but the plea for 'democracy, tolerance, pluralism and diversity' is an implicit condemnation of any fundamentalism. Nor was it possible, under the security pillar, to name explicitly the second major fear of Europeans, migration from the south. Only under the social pillar is the statement found that partners 'acknowledge the importance of the role played by migration in their relationships; they agree to strengthen their co-operation to reduce migratory pressure . . . [and] . . . in the case of illegal immigration they decide to establish closer co-operation' (Barcelona Declaration, 1995: 140).

Progress faces unavoidable contradictions between the assumption that prosperity needs a free market and western-style democracy and, on the other hand, the cultural legacy of the Arab world and the weakness of many Arab political regimes. For instance, there are basic differences in the assumptions that lie behind the idea of the state: 'if it can be argued that the concept of state in Europe cannot be understood in isolation from the concepts of individualism, liberty and law . . . the Islamic concept of the body-politic cannot be understood in isolation from concepts of the group, justice and leadership' (Ayubi, 1991: 7). Of course, there are different interpretations of political Islam among Muslims, but it is a matter of fact that the historical development of the Islamic state rests on different assumptions about the meanings of legitimacy, democracy and freedom than those that prevail in Europe. As Fatima Mernissi (1993: 49) reminds us, 'the importation of a parliament and a constitution, without their essence being carefully explained, does not allow the masses to reflect calmly about a concept that until now has been tainted with sin: the concept of freedom. In our time, freedom in the Arab world is synonymous with disorder.'

The economic pillar: money for stability?

The Barcelona Declaration (1995: 138) states that 'the participants emphasise the importance they attach to sustainable and balanced economic and social development with a view to achieving their objective of creating an area of shared prosperity.' In this case it is even more obvious that the initiative rests mainly with the EU, whose actions are based on (a) the establishment of a free-trade area by 2010; (b) economic co-operation; and (c) financial assistance.

The most ambitious goal is undoubtedly the free-trade area to be created through the Association Agreements and already foreshadowed in the EU's pre-Barcelona 'Renewed Mediterranean Policy'. Considering that, so far, none of the four agreements signed has entered into force, and the other five are yet to be concluded, the free-trade initiative seems far from being underway. The free-trade area will include all manufactured goods (to Europe's advantage), while for services and agricultural products it provides only for a progressive liberalization, which will take account of 'the results achieved within the GATT negotiations' and 'the various agricultural policies' of the EU (Barcelona Declaration, 1995: 138), a clear defence of the common agricultural policy (CAP).

Economic co-operation aims at enhancing the process of privatization, supporting small and medium enterprises, and introducing international and European standards in order to lay the basis for the creation of an information society. In the field of agriculture, intervention should enhance the process of productive diversification, enabling the Mediterranean countries to achieve a greater degree of food self-sufficiency and stable rural employment.

Thirdly, the EU has boosted financial assistance to the Mediterranean partners by making available nearly 4.7 billion ECU for the period 1995–9. The European Investment Bank will provide loans for a similar amount of money. Although the programme of EU financial aid to the Mediterranean is not trivial, on a per capita basis the level of support received by the populations of the Mediterranean partners is approximately one-third the aid received by the central and eastern European countries, and one-sixtieth the per capita value of structural funds to Ireland, Greece or Portugal (King, 1998).

The social and cultural pillar: state and society

The social and cultural pillar is potentially the most revolutionary product of 'Euro-Med'. Recognizing the common link between economic, political, social and cultural issues is a major breakthrough in the traditional approach to regional security. Promoting 'dialogue between cultures and civilizations' appears to be essential for bridging the gaps between the northern and southern shores of the Mediterranean, not only 'with the aim of breaking down prejudice, ignorance and fanaticism and fostering co-operation at grass-roots level' (Barcelona Declaration, 1995: 144), but also with the aim of *understanding* the radical differences between cultures, and their social and political institutions, which cannot always be reduced to a common ground of 'universal' values. Partnership requires the reciprocal acceptance and respect of existing different identities, and lays the way open to a questioning of the assumption that western civilization is the only bearer of 'truth'.

In this third pillar there are at least four main issues at stake:

- the promotion of dialogue between cultures and the involvement of civil societies;
- the development of human resources;
- the recognition of social rights; and
- co-operation in domestic affairs.

On the role of cultures and civil societies, it has proved difficult, in the early years of implementation of the partnership, to overcome the misunderstandings and mistrust that often divide the people of the northern Mediterranean from those of the south. The increasingly high profile given to migration in domestic politics and the media has certainly not helped, and as a result there has been an increase in the phenomena of racism and xenophobia. The Bologna Declaration agreed by the 27 ministers of culture in 1996 has enhanced the possibilities for cultural dialogue but this is mainly based on 'non-controversial' areas such as information dissemination and heritage management. To go further, this kind of dialogue meets enormous obstacles and is not always aware, paradoxically, of the major cultural differences that exist between the two sides of the Mediterranean. Some words and concepts are differently understood by western and Islamic culture. For instance, 'Islamic liberty, unlike the Western notion of liberty, is defined by Islamic laws and consists of the liberties that have been provided by God.' Similarly, 'Islamic equality . . . is also defined by Islamic laws and primarily refers to the social rather than the politico-economic equality of the believers' (Tamadonfar, 1989: 41).

The question of human resources is twofold: on the one hand, it relates to the need for the training of workers to improve their skills and hence productivity; on the other hand, it refers to the effort to contain the migration flow. The partnership provides for the promotion of cultural exchanges and knowledge of other languages, as well as 'strengthening co-operation to reduce migratory pressure, among other things through vocational training programmes and programmes of assistance for job creation' (Barcelona Declaration, 1995: 140). However, these programmes are still well behind in the matter of implementation and the allocation of funds is negligible considering the ambitious task of increasing competitiveness and employability in the southern partners' economies and populations.

Thirdly, the Barcelona Declaration stressed respect for fundamental social rights, 'including the right to development' (1995: 140). The programme specifies that living and working conditions should be improved, particularly in the case of women and of 'the neediest strata of the population' (Barcelona Declaration, 1995: 145). However, the Declaration contains only generic reference to the 'key role of women in development' and says little about trade unions. These are obviously sensitive topics in the dialogue with Muslim countries where current political regimes are being urged to make social reforms that are not necessarily consistent with the Islamic tradition and which, if successful, would probably undermine their power.

Finally, domestic policies comprise co-operation in the fields of illegal migration, terrorism, drug trafficking and organized crime, as well as measures to combat racism, xenophobia and intolerance. Many of these are also of concern to the security debate (first pillar). Yet progress has been minimal. The terrorism paragraph remains blank and the record of the latest meetings held in 1997 does not include debate on this or the migration issue. A first meeting on drugs and organized crime was held in Taormina in 1996 but the partners failed even to agree on the contents of a draft agenda.

Critique of 'Euro-Med'

It can be argued that the rationale behind the Euro-Med partnership is, first of all, the preservation and enhancement of Mediterranean, and hence European, security. Hence, when the EU speaks of economic development in the southern Mediterranean it is mainly thinking of its own political and strategic security. Furthermore, the measures proposed follow the assumption that free market economics and free trade promote economic growth, which is seen as the key element in creating political stability and facilitating the trend towards full democratization (Pace, 1996).

Whatever the merits of a free-trade philosophy in general terms, under Euro-Med it is unbalanced and weighted towards European interests. In practice the EU approach to trans-Mediterranean economic integration is 'Janus-faced' for it 'combines liberalism within and mercantilism without' (Wolf, 1995: 333). On the one hand, the partnership endorses the belief that the economic benefits of the liberal economic system will automatically spill over to all those who adopt it; on the other hand, the EU is very concerned about the domestic implications of the impact of free trade on 'sensitive sectors', and the creation of unemployment. This means that, in practice, the EU has a tendency to adopt a protectionist stance in certain products. In fact, most analysts agree that in economic terms the Euro-Med partnership is far from being a good deal for the southern Mediterranean states, even though, in *realpolitik* terms, it probably represents their only option. Joffe, for instance, points out that 'it is difficult to avoid the conclusion . . . that without significant amounts of additional development assistance . . . the damage done in the medium term to the economies of the South Mediterranean region may well outweigh the longer term benefits, if any do accrue' (1996: 186). The Barcelona Declaration does not provide for a specific structural fund for the southern Mediterranean countries; it does little to promote intra-south trade; and there is no direct provision to tackle the fundamental problem of indebtedness – only a 'dialogue in order to achieve progress in competent fora' (Barcelona Declaration, 1995: 138).

Finally, we have to stress that the traditional European (and American) conception of the problem of security as consisting of threats – terrorists, unstable governments, fundamentalist ideologues, even migrants – tends to produce a defensive approach, likely in turn to trigger power dynamics and therefore further insecurity. Instead, vulnerabilities, rather than threats, characterize today's security environment; vulnerabilities which are enhanced by globalization and have a transnational nature. As we have seen, migrants are among the most vulnerable actors, or victims, moving across the Mediterranean's divided space: marginalized by unemployment and perhaps alienation in their home countries, and by social exclusion and xenophobia in the destination countries (see also Chapter 11). States, too, feel more vulnerable because they are less able to control, on their own, the environment that produces insecurity. In this perspective, co-operation is more effective in producing security than a defensive approach.

The Euro-Med partnership acknowledges, to some extent, the need to rethink security along more co-operative lines, broadening its realm from the political arena to embrace economic and cultural aspects and including different actors within a stable interactionist framework of dialogue. But it has a much more traditional posture as far as stabilization policies are concerned. A top-down, western-dominated hegemonic view of co-operation prevails, privileging not only a western model of development but also revealing sharp inconsistencies between EU interests, on the one hand, and those of the wider Mediterranean region, especially its southern shore countries, on the other. Similar conflicts are present in the approach to political reform and socio-cultural dialogue because of the need to reconcile long-term ideals and goals with the contingencies of everyday politics in a fragmented and volatile region (Kliot, 1997).

Conclusion: Europe and the future of the Mediterranean

At first glance it would seem that a supportive European Union is coming to help the southern Mediterranean countries in their hour of crisis and to bridge the divide within the Mediterranean. However, the problem is far more complex, as we have tried to show in this chapter. The crisis in the Mediterranean countries, seen at its clearest in the Maghreb states, is characterized by an economic failure, which is undermining the existing political establishment, and by an identity crisis which stems both from the colonial past and from the current 'western globalization' process that makes a return to an Islamic identity appealing. While this might seem, to western observers, merely a reaction to the economic crisis which could easily be neutralized by promoting prosperity, it has to be understood in its deeper roots.

The Islamic religion, as well as its representation of society and politics, is embedded in an ancient tradition and history, and underestimating the importance of the human quest for identity, whether religious, ethnic or nationalistic, could prove to be a serious error. The conviction that, at some time or other, all societies will march along the same path of western capitalism and its set of values risks pushing the West into a paranoia about threats to its own identity and stability from any other different or alternative civilization.

In this perspective, there are three paradoxes that Europe tends to ignore. First of all, while it declares that the solution to all problems would be for other countries to adopt its economic model (and the political and cultural 'package' that surrounds the model), in practice it does little really to promote such development, fearing that this might interfere with its own interests and showing therefore little confidence (or perhaps too much) in the supposed self-regulatory capacity of the market.

In second place, to regard the core values of western civilization – such as the notions of individual freedom, human rights, democracy and so on – as universal and necessary to economic development, ignoring different cultural interpretations of society and politics, can lead to further dilemmas. This is the case with the partnership agreements between the EU and the associated Maghreb countries, which require a commitment to democracy and a respect for human rights as the premiss for financial aid, without taking into account the possibility that:

> the transformation of the present regimes into democratic ones may, if it is allowed to happen before the economic reforms have been given the chance to bear their fruits, lead to the transfer of power from governments which are not democratic (according to western yardsticks) to governments which actually do not believe in the western notion of democracy. (Pace, 1996: 112)

The rise of the Welfare Party in Turkey is a relevant example here. In these cases, the EU will need to develop the capacity to accept different political and social regimes; to act against them would only confirm and exacerbate the mistrust that already exists on both sides. Now, several policy-makers in Brussels, as well as analysts elsewhere in Europe, are convinced that political reforms are possible only after the successful development of the economy and the achievement of a measure of well-distributed prosperity. Prosperity reduces social tensions and provides the conditions for internal stability and a fuller democratization of societies. This is not necessarily a proven argument but it seems to many to be a more effective strategy than acting the other way around, opening the door to radical political change and the possible establishment of regimes that are anti-democratic in principle, as the case of Algeria showed.

Thirdly, notwithstanding all the debate over sustainable development and environmental damage, a debate which has been sharply focused on the Mediterranean through the Blue Plan (Grenon and Batisse, 1989), the West still refuses to believe that its model of economic development is part of the problem rather than the solution (Dalby, 1992: 112). Only slowly are signs emerging that, in an era of globalization, Europe is willing to question its self-confidence in the capitalist model and the universalist ambitions of its civilization.

7

Transformation and Division in Central Europe

Allan M. Williams and Vladimir Balaz

The remaking of relationships between eastern and western Europe inevitably involves re-shaping social and territorial inequalities. While these were relatively muted in central Europe prior to 1989, the attempt to import neo-liberal models of economic reform has created new and intensified forms of inequality within the region, and between central and western Europe (see also Chapter 2). Old forms of regulation have been dismantled and market mechanisms have been introduced without the creation of a new coherent regulatory framework. Political capitalism has emerged in the ensuing vacuum whereby the *nomenklatura* have been able to use their networks, traditional practices and their privileged positions with respect to both state and private companies to accumulate vast amounts of wealth (Staniszkis, 1991). In turn, these inequalities are shaping the processes of global reintegration and transition. The production and reproduction of inequalities in central Europe therefore need to be seen in the context of globalization which, following Held (1995), involves a nexus of social interdependencies between state and civil society and across spatial scales.

The Cold War political division of Europe resulted in the creation of two divided economic spaces, with distinctive institutions and practices. There was disruption and re-orientation of international capital, trade and production links, and the two economies diverged further as they evolved largely in isolation from each other (Vintrová, 1993). By the 1980s, there were complex layers of divisions between central and western Europe. Not only were the patterns of inter- and intra-industry trade different, but so were the macro and micro forms of state and corporate governance. There were also differences within central Europe, with Hungary and Poland moving towards more decentralized management systems than Czechoslovakia, thereby creating the conditions for *nomenklatura* privatization.

There was no inevitability about the pace and form of the transition process that began after 1989; rather, there was a series of contested arenas. Essentially, there were two main competing transition models (Gowan, 1995): a relatively slow transition process or a more rapid 'sharp shock' transition model. The sharp shock transition model, particularly associated with Sachs (1990), was adopted throughout central Europe. To some extent this was imposed by external agencies (Knell, 1996) and so provides crude support for the theories of 'dominant' globalization (see Chapter 1). However, more importantly, the neo-liberal reform agenda, and the failure to create a coherent regulatory framework for emergent markets, coincided with the interests of some dominant political groups within these countries. We therefore contend that Held's (1995) 'nexus of relationships' theory of globalization, with an emphasis on the interplay between the global and the national (and between the state and civil society in the latter) provides a useful theoretical starting point for the analysis of inequalities (see also Chapter 1).

Privatization served the interests of some sections of civil society and the state (Nielsen et al., 1995). The model of privatization was another contested area, the outcome of which was the emergence of new economic elites, often drawn from the old ruling elites who, in Poland and Hungary, had already benefited from *nomenklatura* privatization. For these new elites, privatization represented a once and for all opportunity to benefit from the redistribution of property rights as part of the privatization agenda. This can be seen as part of the emergence of 'political capitalism', whereby new relationships were created between political power and the control of capital by those able to utilize networks and practices partly inherited from the centrally planned economies (Staniszkis, 1991).

Informed by the objective of rapid disengagement of central and eastern Europe from the USSR, the Sachs model required a rapid shift to a *particular* model of capitalist institutional structures (discussed below). As such it has to be seen in the context of the neo-liberal agenda which was challenging existing forms of regulation in western Europe, in the 1980s and 1990s (see Chapter 1). In central Europe, however, market liberalization was proposed without comparable attention to the regulatory framework of economic relationships. Moreover, there was an assumption that political reforms should follow market reforms, for democracy was considered to be rooted in capitalist relationships. This was contested and Dahrendorf (1990), for example, argued that political reforms were a necessary pre-condition to economic reforms in order to prevent social upheaval; in other words, they were a pre-requisite for the creation of an appropriate regulatory framework. In contrast, adoption of the sharp shock model implied simultaneous co-evolution of political and economic reforms, which has meant a lack of checks and balances in the emerging social and economic formations, and the conditions for what has been termed 'Wild East' capitalism (Balaz, 1996);

that is, speculative capitalism characterized by deep social cleavages. An important element in this transition has been the concerted strategy to weaken the welfare states of the socialist economies, often without creating alternative institutions (Göting, 1996), hence deepening both poverty and polarization.

We have examined the processes of re-internationalization and privatization elsewhere (Williams and Balaz, 1998; Williams et al., 1998). Here we focus on how these interact with social and territorial divisions in central Europe, and the consequences for re-integration into the larger European economic space. The chapter is divided into three main sections which consider: (a) the nature of the economic transition; (b) the role of privatization; and (c) changing social and regional inequalities. The conclusion considers how these divisions are influencing the remaking of economic relationships. Given the diversity of experiences within central and eastern Europe, the discussion focuses only on the Visegrad Four (V4) countries – the Czech Republic, Hungary, Poland and Slovakia – where there are some similarities in their economic transitions.

The neo-liberal model of transition: a framework for the re-working of inequalities

The legacy of central state planning

Among the Soviet/central and eastern European economies, even the V4 countries were not a uniform economic system, and there were important differences in the extent to which they embraced economic reforms both before and after 1989. Hungary and Poland – compared to Czechoslovakia – had more experience of market signals from western countries and had proceeded further in dismantling central planning and decentralizing enterprise management (Fogel and Etcheverry, 1994). Yet there were similarities in the construction of social and economic divisions in the (then) three countries, and between them and western Europe.

First, there were significant differences between the economies of western and eastern Europe. In the two decades before 1989, the gap between the western and eastern economic systems widened in a number of respects. The gap was particularly wide in the case of Czechoslovakia in terms of the structure, modernity and the marketability (in global markets) of goods produced (Landesmann and Székely, 1995). The increasing problems of economic stagnation were evident in faltering GDP growth rates despite high investment rates. In other words, high levels of investment, with associated reductions in consumption and living standards, were failing to produce either high growth rates or the basis for long-term sustained economic expansion.

The re-integration of central Europe into the global economy took place against this background of a widening trade and production divide between eastern and western Europe.

Secondly, a particular pattern of social inequality evolved in central Europe prior to 1989. In many respects, social and territorial divisions were muted, at least in terms of income distribution (Atkinson and Micklewright, 1992). There were, however, sharp differences in the distribution of power in these societies. Enterprise managers were able to strengthen their powers to dispose of assets and benefit from these from the 1970s and, in Poland and Hungary in particular, were able to use *nomenklatura privatization* to translate this into *de facto* property rights. There were also significant territorial divisions of production and incomes even under central planning, as well as regional differences in cultural, educational and scientific networks which constitute the critical, non-traded externalities of production.

Reforms, restructuring and social division

The neo-liberal sharp shock transition model was implemented in the central European countries with scant regard to domestic and international economic realities. It did not take into account the realities of world trade (especially the intermediary role of the state in the global system), political and social structures in central Europe, or the need to create a new regulatory system for the emergent economic and social relationships. But the fundamental weakness in this neo-liberal strategy was that tough budget controls stemmed the flow of state resources to enterprises without creating new mechanisms (such as financial institutions) to allocate capital to them. The neo-liberal model had four main constituent elements: market liberalization, privatization, currency convertibility and trade liberalization. These are reviewed in detail elsewhere (Williams et al., 1998), and only some salient points are noted here in respect of stabilization; privatization is considered in the next section of the chapter.

Turning first to trade, a particular model developed within central Europe during the period of Soviet domination. The national economies were not only largely closed to global trade, but also to bilateral trade with each other (Dangerfield, 1995: 8); this stands in marked contrast to the post-war globalization of trade and investment in the capitalist economies. After 1989, there was re-internationalization of trade, following the introduction of convertible currencies and trade liberalization measures, particularly with the EU. The Europe Agreements, which came into force in 1994–5, phased establishment of free trade over ten years, but they were 'strongly asymmetrical' in the EU's favour in at least three critical respects (Gowan, 1995: 25): maintenance by the Union of non-tariff barriers, of stringent anti-dumping powers, and of protectionist barriers in sensitive sectors. The last point is critical for central

Europe had the greatest export potential in precisely those sectors which faced recurring crises of production in the EU. Therefore, exports of agricultural products, textiles, chemicals and steels to the EU were ring-fenced by protectionist barriers, even though these sectors accounted for 35–45 per cent of central European exports to the Union.

There were two major features of the re-internationalization of central Europe's trade in the 1990s. First, as expected, central European imports from and exports to the remainder of the ex-Soviet bloc collapsed. And, secondly, the EU's share of the imports of the central European countries expanded sharply to between 40 and 60 per cent, while there was a concomitant increase in its share of the exports of central European countries to 50–60 per cent. Adjustment was asymmetrical, however, and central Europe's pre-1989 trade surplus with the EU was sharply reversed to become a major trade deficit. The loss of traditional export markets, and the increased competition in domestic markets, coincided with the depressive effects of macro-economic stabilization programmes (including reduced social and territorial transfer payments). The impact on GDP and unemployment will be considered later in this section, but first we turn to foreign direct investment (FDI).

There had been FDI in central Europe before 1989 but thereafter the growth of foreign capital accelerated. Hungary attracted most FDI until the end of 1993, not least because of the prominent role allocated by the state to foreign investment in privatization, and thereafter it has competed with Poland for lead position. The Czech Republic started slowly and gradually closed the gap on the other two, but Slovakia has received a very low share of FDI. The impacts of FDI are conditioned by the particular strategies of trans-national capital. To some extent, FDI has been predatory, and has curtailed production. For example, part of the Hungarian cement industry was acquired by foreign capital which then stopped its new affiliate from exporting, and so competing with it in central Europe (Gowan, 1995). While there are a number of examples of such predatory behaviour, FDI has mostly been market seeking, being concentrated in food, drink and consumer durables. Despite the general collapse in purchasing power (and in production and employment), the polarization of incomes has created new markets attractive to FDI. In addition, central Europe features in the global production strategies of western companies, such as VW and Fiat, for the neo-liberal reforms have created attractive conditions of production for inward investment. The attraction is not simply low wage or labour costs, but is also related to the flexibility and skills of labour, as well as the requirements of 'just-in-time' production (Zon, 1996: 109) and corporate strategies for multiple sourcing. While this strategic FDI is boosting production and employment in the short term, its truncation effects are more questionable, given that it tends to be weakly embedded in regional economic structures. Moreover, in the short term, at least, the negative trade impacts of re-internationalization outweigh any positive gains from FDI.

None of these trade effects emerged in an institutional or regulatory vacuum. Instead, globalization and re-internationalization were shaped by the slow emergence of the institutionalized practices of market economies (Poznanski, 1995). A considerable part of international trade continued to be channelled through companies which remained in direct or indirect state ownership, or was mediated by what Smith (1994) has termed 'mercantile capitalism', whereby new trade companies emerged to market and distribute the production of (often state-owned) manufacturing companies. At the same time, the Europe Agreements symbolized the effective withdrawal of the state from the regulation of large areas of international trade, thereby confirming the powers of groups of, and individual, enterprises.

Growth, unemployment and the neo-liberal model

The sharp shock economic model had a major impact on GDP growth rates in central Europe; there was what Gowan (1995: 17) terms 'a savage double depressive shock' as a result of the disintegration of pre-1989 trade flows and the impacts of stabilization policies. The outcome is evident not only in GDP growth rates (Table 7.1) but also in unemployment and living standards throughout central Europe in the early 1990s. This was particularly marked in Slovakia which had an especially unfavourable economic structure, given its reliance on heavy industries and the defence sector, and greater dependence on depressed eastern European markets (Smith, 1994). The overall severity of the economic collapse in central Europe is indicated by the weighted average GDP decline of one-quarter between 1989 and 1993, a reduction without parallel in modern peacetime Europe. Poland, in 1992, was the first to experience renewed output (if not employment) growth, but recovery had spread to all four countries by 1994. By the mid-1990s, strong growth rates were prevalent throughout the region, exceeding 7 per cent in 1995 in both Slovakia and Poland, although more subdued (1.8 per cent) in Hungary as a result of measures to reduce its current account and budgetary deficits. However, the notion of recovery needs to be qualified with the knowledge that, with the exception of Poland, none of the central European economies had regained their 1989 GDP levels even as late as 1997. Recovery was also hampered because internationalization had made these economies increasingly sensitive to the recession in western Europe in the early and mid-1990s and globally in the late-1990s.

The effects of the double depressive shock are starkly evident in unemployment growth (Figure 7.1), although this increased far less than the decline in output (Nath and Jirásek, 1994: 76). Unemployment levels in central Europe had been close to zero before 1989, although the official statistics concealed considerable under-employment. However, unemployment has since risen sharply in all central European countries.

TABLE 7.1   *Selected international comparisons of changes in GDP and in fixed capital formation, 1975–89*

	Gross national fixed capital formation as % GDP		% change in GDP	
	1975–9	1985–9	1975–9	1985–9
Belgium	21.7	16.9	1.9	2.6
Netherlands	21.3	20.7	2.4	2.8
Czech Republic	28.0	28.1	2.8	2.2
Hungary	33.1	22.3	4.6	2.0
Poland	n.a.	20.9	−2.2	3.5
Slovakia	33.7	30.3	3.5	3.1

GDP estimates prior to 1992 were based on the material product system (MPS), which is different to the national output system of western economies which were based on market prices. A particular problem of the MPS is that most services, such as education and health care, were considered non-productive and had no prices and so were not included in official calculations. These estimates must therefore be approached cautiously.

Only national income rather than output estimates are available for the Czech and Slovak Republics for 1975–9.

Source: National statistical offices of the central European countries and authors' own computations to disaggregate Czechoslovakia

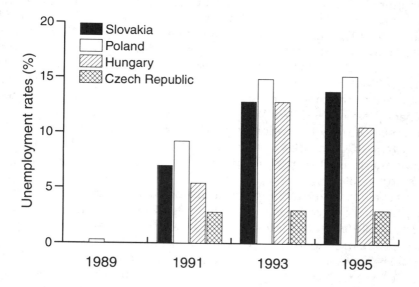

FIGURE 7.1   *Unemployment rates in central Europe, 1989–95 (data from the national statistical offices of the V4 countries)*

The one exception has been the Czech Republic due to its initially strong economic performance and limited restructuring in many large state-controlled enterprises, but by 1997 unemployment rates were also rising here. Elsewhere unemployment rates peaked at around 13 per cent in Hungary in 1993, and 17 and 15 per cent in Poland and Slovakia respectively in 1994, before subsequently falling; but they continue in double digits in most parts of central Europe. The rise in unemployment has constituted one of the major elements of the fiscal crisis of the state in central Europe, contributing to the attempt to redraw the boundaries between the public and private sector in respect of welfare provision (Göting, 1996).

Privatization: property rights and social relationships

Property rights are of fundamental importance in the distribution of both wealth and income. While the recasting of property rights, with attendant effects on social relationships, has been one of the hallmarks of the transition in central Europe, it is important to note that there were fundamentally different models of property ownership within central Europe before 1989. In Poland, even under central planning, private property ownership had survived in some sectors of the economy, especially agriculture. On the eve of transition, in 1989, about one-third of the labour force in Poland was employed in the private sector, producing one-quarter of GDP (Commission of the European Communities, 1997a: 16). Poland and Hungary had also experienced the so-called spontaneous or *nomenklatura* privatization. In Poland, under the 1981 State Enterprise Law, many managers of state enterprises had formed their own private companies and then entered into agreement with the parent state enterprises. In Hungary, spontaneous privatization was a form of commercialization with decentralized state institutions ('holdings', enterprises and banks) becoming the new owners (Voszka, 1995: 89). While this did not create private property, it did result in decentralization of property rights and *de facto* control by managers. In Czechoslovakia, in contrast, there was little experience of either quasi market reforms or of decentralized, let alone, private property rights.

In terms of understanding social inequalities, privatization has been one, if not the most important, of the internal reforms that constituted the post-1989 economic transition model. All four countries have experienced three overlapping phases (Williams and Balaz, 1998).

(1) The so-called 'small privatization', based on auction, was dominant in the first phase. Shops, restaurants and small service outlets were sold to domestic investors at close to book or market values. This phase usually lasted two or three years. In Slovakia, 9,757 units with a face value of $454 million, and in the Czech Republic 22,212 units sold for

$1,077 million, were included in the first phase of privatization up to the end of 1993. In Hungary, 9,800 small businesses with a total value of $157 million were sold in a so-called 'pre-privatization' programme, 1990–4.

(2) Medium and large state-owned enterprises (SOEs) were privatized by means of tenders and direct sales. Domestic and foreign capital participated in these tenders to varying degrees. The former usually relied on bank loans and on the large discounts from market values that were given to management-led employee collectives. In Hungary and Poland, the legacy of *nomenklatura* privatization gave managers a privileged position in such privatizations. More open tendering was used in the Czech and Slovak economies, although 'government-friendly' managers were increasingly favoured in the latter, where the average buying price was considered only 10–25 per cent of nominal enterprise prices (Williams and Balaz, 1998). Hungary was the only country to allow relatively unrestricted participation of foreign capital in such auctions: between 1990 and 1994, 385 Hungarian companies were acquired by foreign capital for an estimated $4 billion, equivalent to 10 per cent of GDP in 1994 at current prices. These forms of privatization are continuing but also overlapped with the introduction of mass privatization programmes in the Czech Republic, Slovakia and, latterly, Poland. The mass privatizations converted SOEs into joint stock companies and distributed their shares widely by means of (usually free or low cost) coupon issues. For example, in the Czech Republic coupons worth $11 billion were distributed to the Czech population.

(3) The third phase was not the direct outcome of state privatization programmes but was generated by major industrial and financial capitals. These accumulated diluted shareholdings from the mass privatization programmes in the Czech and Slovak Republics, and used the leverage effects of majority stakes in investment funds to secure control of larger volumes of capital in underlying companies (Balaz, 1996; see also Chapter 3). The best known of these is the Harvard Fund of Viktor Kozeny (McIntyre et al., 1994). In the third phase, in Slovakia, the five largest investment companies acquired 43 per cent of all investment coupons and the 20 largest companies acquired 58 per cent of these. While foreign capitals were largely excluded from the initial sales of state assets, they have subsequently made a number of acquisitions in the third phase; this has been facilitated by the lack of capital available to the original purchasers to complete their transfer payments.

The outcomes of privatization have not followed any simple pattern for, as Smith (1997: 331) argues, the 'breaking of the old and the construction of the new is complex, unevenly developed, multi-determined and "embedded" in the nature of the state socialist system, and its current transformations'. Nevertheless, some generalizations are possible. First, privatization has created particular versions of market economies in

TABLE 7.2 *Estimated percentage[1] contribution of the private sector to GDP in central Europe, 1990–95*

	1990	1992	1995
Czech Republic	5	28	75
Hungary	16	23	70
Poland	31	45	65
Slovakia	5	22	65

[1] These data may over-state the size of the private sector as an enterprise is classified as private if the public stake is less than 49%; in addition, indirect state ownership through state-owned banks is not taken into account. Only estimates are available for the micro-enterprise sector and the shadow economy.

Source: United Nations (1996) and National Statistical Offices of the CE countries (various years)

central Europe in a very short time; the proportion of GDP accounted for by the private sector in 1995, according to official statistics, ranged from 60 to 70 per cent of GDP in all four countries (Table 7.2). However, in some cases, private ownership is illusory; for example, the Czech coupon privatization scheme was controlled by 14 institutions, including nine banks, most of which were in at least partial state ownership. Secondly, while the impacts of privatization on output and efficiency are contested (Rondinelli and Yurkiewicz, 1996), there is evidence of a link between privatization (irrespective of model) and productivity (Pohl et al., 1997); this is partly due to the shedding of previously under-employed labour rather than to deeper changes in the labour process. A third function of privatization has been to boost state revenues and help balance the state budget by means of both income from sales and reduced subsidies, thereby contributing to the stabilization programme – but at a high social cost. For example, in Poland subsidies to SOEs had by 1993 been reduced to a mere $1.1 billion. At the same time, however, the restructuring of enterprises after privatization had increased unemployment, leading to additional budgetary pressures. A fourth contribution has been to shape the international re-integration of these economies either through direct sales to foreign investors (mainly in Hungary) or through subsequent transactions in what we have termed the third phase of privatization.

Finally, privatization has also had important implications for social equality. Most of the populations of central Europe lacked the means to participate in the bidding process, even where this was genuinely open, and they lacked the knowledge to benefit from the coupon privatization. Instead, limited numbers of the old and new elites, who had insider knowledge, power and access to finance, were able to control enormous assets in a very short time. In other words, there was the emergence of

TABLE 7.3 *Public opinion of the privatization process in Slovakia*

Response to the question 'Whom will privatization help'	%
Selected speculators only	40
Only individuals who have privatized companies	40
Only workers in 'promising' industries	13
All the people	2
Don't know and other responses	5
Total	100

Source: Úspešná privatizácia (1997)

what Staniszkis (1991) has termed 'political capitalism'. Privatization was also accompanied by frauds (Guoth, 1994). Given the opportunities for short-term personal gain, privatization often became the main objective of political and economic life in most of the republics, often at the expense of overall strategies of economic development. Popular cynicism and disaffection with the redistributive effects of privatization are reflected in public opinion polls, such as that published in Slovakia in October 1997 (Table 7.3).

Social and territorial inequalities

Economic reforms and inequalities

Re-internationalization, privatization, stabilization policies and the lack of a coherent regulatory framework have all shaped the emergent patterns of social and territorial inequalities in central Europe. There were, of course, inequalities before 1989, especially in terms of political power and associated privileges, but there were also high levels of state redistribution, so that material inequalities were relatively muted. In the 1990s, in contrast, the redistributive role of the state has been severely undermined by budgetary constraints and policy changes, linked to the stabilization programmes which are intrinsic to the neo-liberal transition model (see Chapter 1).

Many of the social costs of transition have arisen directly from cutbacks in state expenditure affecting levels of collective consumption, wages and jobs in the state sector, as well as general subsidies to prices and the standard of living. For example, in Hungary the proportion of GDP redistributed by the state budget fell from 61 to 48 per cent in only three years. The resulting costs were very real: domestic consumption fell by 8 per cent in two years; real wages fell by 12 per cent in 1995 and 4 per cent in 1996; and the real value of pensions was reduced by 25 per cent in two years, a decline without parallel even in central Europe.

Financial Times (16 December 1996) reports the leader of one Budapest think-tank as stating that: 'In France, Germany and Italy, far smaller adjustments have produced much greater protests. This is proof that Hungarians don't believe serious political or economic alternatives exist.' Other social costs were generated by the high rates of inflation that accompanied the transition. In the latter years of the Soviet system, most segments of society had accumulated substantial savings because of shortages of goods to purchase. This created a monetary overhang that it was feared would lead to inflation after liberalization of prices. In the event, the scale of inflation was such as to virtually wipe out these savings. This was particularly harsh for pensioners who tended to be left with nothing more than their (diminished) state pensions.

Internationalization has also impacted on social inequalities. There has been an uneven sectoral effect in terms of the relative growth and decline of particular sectors of the economy, resulting from re-internationalization. The largest production and employment declines have been in heavy goods and armaments, which lost their protected markets in the eastern bloc, and consumer goods, which experienced the full force of competition from western European imports. As these industries are territorially unevenly developed, and the regional conditions of production have been revalued as a result of re-internationalization, the outcome has been a deeply uneven regional pattern of employment, unemployment and income.

Privatization was supposed to create a strong middle class which would buttress democratic capitalist society, but the emergent social structures are strongly differentiated and the bases for potentially destabilizing social tensions. The transition has distributed both costs and benefits, and the particular neo-liberal model adopted, combined with weak regulation, have meant that these have been highly polarized. First and foremost, the redistribution of property rights – as part of the creation of political capitalism – has led to the concentration of wealth and control of enterprises in the hands of new elites. In addition, the associated rising production prices of privatized companies, downsizing and redundancies have widened the gap between social classes.

Income redistribution

There is a lack of reliable information on income inequalities in central Europe, particularly for the post-1989 period. However, Atkinson and Micklewright (1992) have compared the ratios of top to bottom income deciles in central Europe and the UK in 1982–5: these were 2.4 in Czechoslovakia, 2.6 in Hungary, 3.0 in Poland and 3.9 in the UK. Their sophisticated analysis subsequently makes allowance for undeclared incomes in both the UK and central Europe, for higher levels of self-consumption in agriculture in central Europe (which supplement incomes of the lowest decile) and for the unreported privileges of

TABLE 7.4   *Poverty index for central Europe,*[1] *1989–92*

	% pop. with income less than cost of living	
	1989	1992
Czech Republic	4	25
Hungary	14	20
Slovakia	5	34

[1] Figures for Poland not available.

Source: BIT Director General, September 1995, p. 54; quoted in Paparela (1997)

political elites in the Soviet bloc. But they still conclude that there is no reason to believe that the range of incomes was not far smaller in central Europe, especially in Czechoslovakia which had deviated least from central planning.

Despite the absence of reliable quantitative data, it is clear that a state socialist society with low levels of social inequality has been replaced by a 'new rich and new poor' structure. This remarkable redistribution of wealth is unlikely to be repeatable in the foreseeable future in Europe. Some information is available that allows us to trace this deepening of social cleavages. Paparela (1997) has shown that in the early stages of transition there was a rapid extension of previously low levels of poverty to encapsulate between one-fifth and one-third of the population; this was particularly marked in Slovakia (Table 7.4). Using different data, Milanovic (1994, quoted in Jackson, 1997) calculates the proportions living in poverty, defined as per capita incomes less than $120 per month at 1990 prices (hardly generous, being only about one-half the level used in German calculations). Between 1987–8 and 1992–4 the proportion living in poverty in Poland increased from 5 to 17 per cent, as a result of consumer goods price inflation, devaluation of real pensions and wages, and soaring unemployment. Héthy (1997) estimates that real wages declined throughout central Europe in the early 1990s; in one of the most difficult years, 1991, they fell by about one-quarter in the Czech Republic and Slovakia, and by 7 per cent in Hungary. Growing tax burdens and erosion of welfare provision added to the general deterioration of real incomes.

The decline in living standards, although initially almost universal, was accompanied by social polarization. Between 1988 and 1991–2, there was a consistent widening of the gap between the highest and lowest income earners in all four countries (Table 7.5). This was particularly marked in Hungary. The net result is that income differences in the 1990s have become comparable to those found in western countries (Héthy, 1997), but in the context of much lower income levels. The evidence for individual countries confirms the picture of social polarization. In

TABLE 7.5 *Comparative earnings in central Europe in 1988 and 1991–2*

| | Percentile (% of median earnings) | | | | | | |
	5th	10th	25th	75th	90th	95th	Ratio of 90th to 10th
1988							
Czech Republic	54	60	74	119	144	163	2.40
Slovakia	54	62	77	123	149	168	2.42
Hungary	50	59	74	135	183	226	3.14
Poland	55	63	78	126	163	192	2.60
1991–2							
Czech Republic	51	56	70	117	154	184	2.74
Slovakia	–	68	–	–	170	–	2.50
Hungary	48	56	73	146	204	257	3.64
Poland	–	62	77	132	180	219	2.92

Source: Vercernik (1995)

Hungary, for example, the ratio of the top and bottom deciles of incomes increased from 5.8 to 7.0 during 1987–94 (Ehrlich and Révész, 1997), even without taking into account the distortions of undeclared income which tend to favour the rich. An estimated 12 per cent of families live in chronic poverty, and suffered a loss of at least one-half of their real income in this period, primarily from lost employment earnings. In Poland, while the salaries of white-collar workers became more dispersed, the wages of blue-collar workers actually became more compressed (Jackson, 1997: 95). In contrast to the situation before 1989, earnings inequality is now higher among white-collar than blue-collar workers; this has been due to rapidly growing salaries among the highest-paid workers rather than from greater dispersion of wages in the lower range of earnings. Wage inequalities are, of course, higher in the private than the public sector due to the development of a class of highly paid white-collar jobs in the latter, which emphasizes the key role of privatization. In Slovakia, too, there have been major decreases over time in the incomes of all major groups (Table 7.6). In real terms, the incomes of employees fell from an index value of 100 in 1989 to 64.9 in 1991 and, after a brief recovery, plunged to a low of 55.8 in 1994, before climbing to 76.3 in 1996; that is, still little more than three-quarters of the 1989 base level. The same overall pattern applies to the Czech Republic, where the main differences are the less precipitous decline in incomes, and the fact that employees did marginally better than pensioners. Surprisingly, pensioners in Slovakia did not fare worse than employees according to these data, for the ratio of worker to pensioner incomes never fell below 69.9 per cent in this period. However, these data must be approached cautiously for they seriously under-report those workers with non-salaried incomes and multiple incomes; that is, the complex ways in which economic reproduction occurs in central Europe.

TABLE 7.6   *Real household gross money incomes in Slovakia and the Czech Republic, 1989–96 (1989 = 100)*

	1989	1990	1991	1992	1993	1994	1995	1996
Slovakia								
Employees	100.0	100.1	64.9	69.3	58.2	55.8	60.4	76.3
Pensioners	100.0	106.6	78.6	84.9	69.9	70.7	76.1	85.6
Czech Republic								
Employees	100.0	110.4	80.6	92.8	73.7	76.7	84.2	88.5
Pensioners	100.0	100.2	83.5	84.4	73.6	74.6	80.1	87.9

Source: Czechoslovakia Federal Statistical Office (1990); Slovak Statistical Office (1997); Czech Statistical Office (1997)

Moreover, there are far greater social cleavages constructed around wealth (especially property) than incomes.

Privatization and the redistribution of wealth

Privatization is essentially the redistribution of property rights, understood as the attribution of authority to dispose of goods or enjoy the income or other benefits from their use (Dallago, 1995: 234). In turn, property rights influence the distribution of income, wealth, power and life chances in a society. While it is difficult to obtain reliable, detailed information on the redistribution of property rights in central Europe, there is ample anecdotal evidence of the emergence of a new elite.

The composition of the new elite of political capitalism shows lines of continuity from the old elite. Many are former communist state officials, state enterprise managers or secret police members, who have been able to utilize existing networks and practices in the accumulation process. Privatization of production and circulation was particularly important as a source of wealth for the new elite. This took different forms in different countries. Ash et al. (1994: 223) contend that the free distribution of shares via coupon privatization could have brought about a redistribution of income from richer to poorer individuals, through subsequent sales of shares. However, in practice, a lack of information and of resources to buy shares (the distributions were not always free) has meant that the poor have largely failed to benefit. Instead, large institutions have monopolized control over shares and individual companies as part of the 'Third Wave'. Worker buy-outs or the preferential distribution of shares to workers also added to inequalities; they favoured workers in more profitable enterprises, and excluded large sections of these societies. The practices of privatization have had even more pronounced effects on inequalities, with managers using their insider knowledge to secure control of newly privatized enterprises at advantageous prices. It is true that the 'small privatization' process did help create a new *petit bourgeois* class (Mertlik, 1995: 223), but its effects were limited compared to the redistribution of assets as part of the 'large privatization' programmes.

Privatization was supposed to create a strong middle class as the backbone of new democratic capitalist societies; this class was to be instrumental in implementing large-scale capitalism, rapidly and irreversibly, by creating social groups with a self-interest in perpetuating the new political and economic order (Estrin, 1994: 19; Nuti, 1995: 107). In practice, however, managers have remained the most influential group in post-communist societies. This process started in the 1970s, when educated technocrats were fully incorporated into the communist party bureaucracy. This group ruled the state *de facto* from behind the administrative structures of the communist regimes. Any director of a large SOE had far more real political and economic influence than the local communist party governor. Since 1989 this group has begun to rule the state *de jure*. Batt (1994: 87) writes that: 'Many have seized the opportunity presented by sketchy legislation to convert their politically based positions into secure legal ownership of firms.'

Slovakia provides an example of the above process. The wealthiest individual in Slovakia is Alexander Rezes whose family privatized the major Slovak enterprise, the VSZ steel mill (Tyden, 1996). It was notable that, on 11 March 1994, when parliament dissolved Meciar's government, the Slovak government – on the same day – sold 1,568,159 VSZ shares to the Manager Plc, of which Rezes was one of the leading co-founders. Shares, with a nominal value of Sk1,568 million (US$50 million), were sold for only Sk314 million (US$10 million). Rezes also gained effective control of VSZ because the remaining 57 per cent of its shares were diluted among individuals. It was no surprise that Rezes subsequently proved to be one of Meciar's strongest backers. There are other examples of capital accumulation under questionable circumstances. For example, Mr Martinka has been accused of using his position as general director of the Devin Bank for personal ends. In any event, in 1996 his Vadium Plc company acquired Slovakia's largest and most modern spa, Piestany, with an estimated value of Sk1.6 billion (US$53 million) for a symbolic price. This transaction was subjected to several public inquiries, but these were later halted without due explanation (Leško, 1997). These may be extreme examples of the ability of individuals to benefit from privatization, but they are by no means unique. For example, the privatization of housing has widened social inequalities in Hungarian and Polish cities, although the evidence is more equivocal in the case of the Czech Republic (Duke and Grime, 1997: 886–8).

Uneven territorial development

Central planning had produced a relatively muted set of territorial inequalities compared to those pertaining in western Europe, but none the less there was an important legacy of inherited structures in terms of the spatial distribution of production, infrastructure and non-traded externalities such as educational and scientific facilities. During the course of the transition these were modified and deepened by privatization and re-internationalization, and by the weakening of state intervention in respect of uneven territorial development (see also Chapters 1 and 2). In practice, the lack of reliable data make it difficult to quantify the changes in such territorial inequalities.

European Commission reports (Commission of the European Communities, 1995a: 26) have identified two extreme forms of territorial structures in central Europe. On the one hand, innovative regions are 'the leaders' in the transformation process and create islands of relative prosperity and accelerated structural change. These contrast with lagging regions which are deeply dominated by the negative legacies of central planning: destruction of the natural environment and technical infrastructure; the existence of old industries that are unable to find new markets; and the role of social structures particularly related to the Stalinistic model of industrialization. During the transition period, these inequalities have widened (Commission of the European Communities, 1994b: 164–6). Against a background of general employment decline, job markets have collapsed in mono-industrial regions, notably those specializing in coal, iron, steel and armaments production; examples include Kladon in the Czech Republic, Katowice in Poland, Baranya in Hungary and much of Slovakia. In addition, there have been major decreases in agricultural employment around the major cities, but less so in remoter rural areas where there continues to be substantial under-employment. In contrast, while service employment growth has occurred in all regions, this has been especially pronounced in capital cities, and in other large urban areas and regions with strong tourism products. There has also been a shift in employment from the public to the private sector, especially in urban areas, reflecting the role of privatization and the growth of self-employment as well as the decline of state enterprises. Other regions facing particularly severe problems of economic adjustment include those with under-developed infrastructure such as north-east Poland and eastern Slovakia. There are also regions with poor accessibility to product markets, such as Zamosc in Poland and eastern Slovakia, and those with extreme environmental degradation such as the north-western Bohemia coal basin in the Czech Republic.

The extent of regional inequalities in the mid-1990s are somewhat easier to map than are changes during the transition period. Figure 7.2 uses GDP per caput in Slovakia and Hungary to illustrate the enormous gap between the poorest and richest regions in central Europe, and the EU mean. Taking GDP per caput at exchange rate prices, virtually all regions have levels which are less than half of the EU average. In Slovakia, the only exception is Bratislava at 57.9 per cent, while all other regions (excepting Kosice) had levels below 16 per cent, and these were

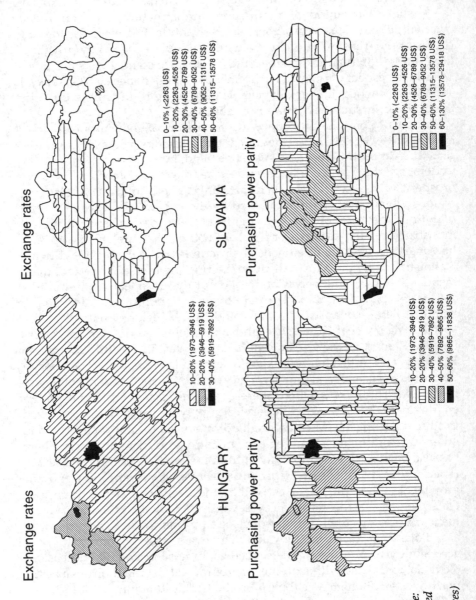

FIGURE 7.2 *Regional distribution of GDP per capita in Hungary and Slovakia, 1994–5 (Source: unpublished data provided by the Hungarian and Slovakian statistical offices)*

especially low in the east. In Hungary, Budapest is the most prosperous region but only achieves just over one-third of the EU level and, excepting Gyor and Vas, all other regions are below the 20 per cent mark, especially in the north-west. The use of purchasing power parity (PPP) data modifies the picture in some respects, and remarkably Bratislava is actually estimated to have a level which is greater than the EU average. However, the remaining regions (excepting Kosice) all record levels which are less than 36 per cent of the EU mean. The differences in Hungary are less striking. The comparative position of Budapest improves but it still only records a level of 56.8 per cent, while all other regions have less than one-third of the EU average. These data must be interpreted carefully given inaccuracies in the base data set, difficulties in making PPP adjustments, the tightly drawn boundaries of the capital cities and Kosice, and the impacts of individual investments such as the VW plant at Bratislava. However, they do underline the deep-rooted territorial inequalities within central Europe, and the vast gulf between most regions and the EU in terms of GDP per capita. There is similar evidence of territorial inequalities elsewhere in central Europe (for example, see Gorzelak, 1996).

There are a number of reasons for these extensive territorial inequalities; here we focus on the socio-cultural conditions for production, privatization and re-internationalization. First, there is an important legacy of uneven conditions of production. To some extent these were masked prior to 1989 by extensive state transfers and by the central direction of production, as is evident in the Czechoslovak government's programme to industrialize large swathes of previously rural Slovakia. However, these concealed real differences in the material and socio-cultural conditions of production. Thus, in Hungary, the lowest unemployment rates are in Budapest and the western regions, not only because of accessibility to 'new' western markets, but also because these 'are historically the areas whose inhabitants have the highest levels of culture, flexibility and multiple working skills' (Ehrlich and Révész, 1997: 291). In addition, there is the legacy of regional production structures (Commission of the European Communities, 1994b) for the previous economic system created a high degree of regional specialization and mono-industrial regions. In the face of market reforms and privatization, the result was that most agricultural and manufacturing regions declined. There were particularly sharp declines in regions with bases in energy production, heavy industry and defence as in central and northern Slovakia, for example. Capital cities and towns with strong service economies, especially tourism, are the regions that have experienced increases.

Secondly, privatization is inherently a deeply divisive process. Not only are the most efficient and competitive enterprises most likely to be privatized, but there is also evidence that, despite some reservations, privatization is followed eventually by improvements in productivity and profitability (Pohl et al., 1997). Territorial inequalities are therefore embedded in the process of privatization, which in turn serves to deepen these. For example, in Poland privatization is strongly territorially differentiated along two axes: between east and west, and according to the level of urbanization. The private business belts around Warsaw and Poznan developed even in the 1970s and have maintained their exceptional economic structures in the 1990s. Gorzelak (1996: 85) concludes that 'the localities which have long traditions of private business, which are well equipped with municipal and technical infrastructure and which are located close to the big markets of large urban centres, have far better chances for adapting to new conditions.' Thus in Lomianki, near Warsaw, there are 85 limited companies, while a further 550 companies are owned by individuals or groups of individuals. In contrast, he finds that in rural Krynk, there are only 79 private firms which employ only 157 people, and they are oriented to highly localized markets.

Thirdly, territorial inequalities have also been shaped by the reinternationalization of the central European economies, as can be seen in trade and capital flows. Regions dependent on heavy industrial production and armaments have suffered most from the loss of traditional markets and the difficulties of competing in new European markets. In contrast, other regions have proved attractive to FDI, or have been able to compete in international markets, as for example in the case of tourism in Prague or car production in Bratislava. Smith (1996a: 152) provides a detailed analysis of changes in Slovakia, identifying a set of emerging regional pathways, and the role of international linkages in these:

- Metropolitan regions which have been able to diversify into small and medium enterprise development, often through the tertiary sector, and the recycling of rents rather than industrial production.
- Regions of growth, with strong FDI, where export-led growth is important.
- Regions with lower levels of industrial decline and some export growth, often based on relatively low-value products and dependent on the availability of low-wage, highly skilled labour.
- Large group of marginalized and increasingly peripheral regional economies which underwent late industrialization under state socialism and have found restructuring difficult under pressure from the global capitalist economy.

Smith establishes a clear link between trade and regional production. The areas of greatest industrial decline have had negative export growth, while the areas where production decline has been least have had export growth and are largely located in the west. 'Export-led global integration therefore seems to emphasise the process of fragmentation, enabling dynamic enterprises and regional economies to begin to forge a pathway out of the general pattern of decline' (Smith, 1996a: 151). FDI is also regionally concentrated; for example, Prague had 49 per cent of all firms with FDI in the Czech Republic in 1991 and Budapest had 56 per cent of the total in Hungary (Commission of the European Communities, 1994b). The same pattern exists in Bratislava which has 54 per cent of all FDI in Slovakia (Smith, 1996a: 149–50) due to specialization in trade and finance, and the dominance of a few large projects such as the VW works. In contrast, most regions have little FDI, and the more peripheral ones have virtually none. Foreign capital therefore serves to reinforce the disparities brought about by trade impacts and restructuring.

Conclusion

Social and territorial inequalities are intrinsic to the processes of transition in central Europe and the reinsertion of the region into the global economy following the seismic political shifts of 1989. First, the ability of central Europe to compete in global markets has been sectorally uneven, and this has been embedded in the high degree of territorial specialization inherited from central planning. Industries and territories have also been differentially integrated into the strategies of transnational capital. Secondly, re-internationalization and the ability to compete in international markets has been linked to privatization. However, the latter has been controlled by particular groups (often termed the *nomenklatura*) able to straddle the state/civil society divide, who have been able to benefit by the redrawing of that division. Privatization, as a part of the construction of political capitalism, perhaps more than any other process, has contributed to the deepening of social inequalities, particularly in terms of wealth. While precise data are difficult to obtain on changes over time, there can be no doubt but that the transition has been accompanied (partly shaping, partly being shaped) by the deepening of social and territorial inequalities since 1989. Stabilization policies have also contributed to the deepening of social and territorial cleavages, particularly as the result of shifting the boundary between the public and private sectors – that is, declining state employment – and reduced levels of transfer payments.

While developments in central Europe have been influenced, via trade and capital flows (and to a certain extent labour migration), by the re-integration with western Europe, the latter has not been unaffected by changes in eastern Europe, which have contributed to the continuous recasting of the international division of labour. Not least, the growth of structural unemployment in the EU in the 1990s has made member states (see Tödtling, 1997) increasingly sensitive to the trade and jobs implications of further enlargement. For example, one German executive in Bavaria commented that 'People talk a lot about the competition we are facing from developing economies in Asia . . . The fact of the matter is that we have Hong Kong on our doorstep' (*Financial Times*, 8 November 1996). It is difficult, however, to assess the extent to which trade, investment and employment displacement would have occurred anyway because of global shifts in the conditions of production. At present, the collapse of production and living conditions in central Europe after 1989 (with only Poland having regained these levels even as late as 1997) means that the gap between eastern and western Europe has been deepening. However, such divergence is not inevitable. Some of the regions of central Europe with the most favourable conditions of production (the capital cities and those most accessible to EU markets) are likely in future to converge on the EU average, simultaneously increasing the international competition for some western European regions.

There is also no inevitability as to the form of future relationships between central Europe and the EU. These are contingent upon developments in both central Europe and the Union. They depend in part on developments in the central European countries themselves. In this respect, the key issues in the Commission's opinion on the applicants for membership include: how and at what pace neo-liberal economic reforms continue to be implemented; the extent to which high rates of unemployment and the social tensions associated with the transition are contained within emerging new forms of social regulation; and whether they meet the tests of functioning democracies and respecting human rights (see Chapter 5). There are, of course, a number of contradictions between the consequences of continued rapid implementation of neo-liberal reforms and the containment of social tensions, resulting from the social inequalities identified in this chapter. Even if these contradictions can be contained, then other uncertainties surround the future development of the EU itself, so that there is no certainty as to the type of economic union that the central European countries will eventually become members of, or its ability to provide assistance to ameliorate these tensions.

One of the key areas is the future of the EU structural funds. With a few exceptions, such as their capital cities, per capita incomes in most of central Europe are less than one-third of the EU mean, even in terms of purchasing power parities. Most, if not all, of central Europe would therefore qualify for Objective 1 status under the existing Community Support Framework, as well as for the cohesion fund. However, the EU has estimated that if existing structural fund policies were extended to central Europe, then this would cost about 14–38 billion ECU (compared to an existing total budget of only some 60 billion ECU). Under existing allocation rules, the scale of such transfers would be enormous, amounting to 7 per cent of GDP in Slovenia and rising to 34 per cent in Romania. It is highly questionable whether such economies could absorb this level of funding. In practice, this scale of transfers is unlikely to be politically acceptable (especially among the potential 'losers' within the member states) and, instead, there is likely to be significant prior reform of the structural funds (see Chapter 1).

The structural funds are only one of the key areas of reform in the EU. Equally important is the evolution of EMU (see Chapter 3) for it has already been agreed that the new members must be able to meet the obligations of full member states with respect to the *acquis communautaire*, and endorse the objectives of economic and monetary union. This means that the convergence criteria for EMU will have to be met but even the Czech Republic, which has made considerable progress towards setting up the institutions of a market economy, is likely to have difficulties in terms of meeting some of the criteria. Moreover, the real significance of the convergence criteria is that they have to be adhered to after accession which will severely restrict the range of economic strategies available to the new members, particularly where these involve state expenditures. This may, therefore, constrain the financing of essential infrastructural and environmental programmes. But the point should also not be missed that the ability of these economies to meet the convergence criteria, and to adhere to them, is also dependent on the trajectories of their constituent regional economies.

Finally, while it is evident that globalization and the upwards drift of political power to the EU from the state will mediate the future role of the latter, there is no reason at all to believe that the state is likely to wither away in the foreseeable future. The ability of the central European economies to compete depends in no small measure on how the nation-states develop an entrepreneurial role in support of their fledgling private sectors (and still significant SOE remnants). Moreover, the state also has a key role to play in dissipating the social tensions in 'Wild East' capitalism. Whether they will be able to do so depends not only on the renewal of economic growth (facilitating enhanced levels of state transfers) but also on the ability to construct effective regulatory regimes, which in part depends on the construction of wider and deeper democratic institutions and practices.

8

Deconstructing the Maastricht Myth? Economic and Social Cohesion in Europe: Regional and Gender Dimensions of Inequality

Diane Perrons

As Europe moves towards the millennium, many countries are converging, albeit haltingly, towards the nominal criteria for EMU set down in the Maastricht Treaty. However, less progress is being made towards economic and social cohesion. Regional disparities have widened in the period 1983–95 in all member states except the Netherlands and gender inequalities and inner-city poverty continue (Commission of the European Communities, 1996f). The central argument of this chapter is that the economic policies which are implied by the Maastricht convergence criteria, together with the related competitiveness and employment strategies set out in the White Paper *Growth, Competitiveness and Employment* (Commission of the European Communities, 1993b), map out a trajectory that is not only incapable of realizing economic and social cohesion, but is counter-productive to these goals and counter also to long-term growth and competitiveness. Moreover, aspects of the recommended employment strategies are implicitly based on the continuation of unequal gender roles despite greater gender equality being one of the key objectives of the EU, as set out in the Fourth Community Action Programme, and fundamental to the concept of European citizenship. Despite 'Blairism' and the success of the French Socialist Party in the 1997 elections, there is no evidence so far that the amendments made to the Maastricht Treaty at the Intergovernmental Conference in Amsterdam of 1997 will lead to a fundamental change of direction in economic policy-making.[1] Furthermore, problems of cohesion are likely to intensify following the widening of the EU proposed under Agenda 2000.

In the first section of the chapter, I set out some fundamental weaknesses of the dominant economic trajectory and indicate how it conflicts with economic and social cohesion as defined by the European Commission (Commission of the European Communities, 1996f). To illustrate and substantiate the argument, changing patterns of regional development and gender inequality in employment are explored in the following section. In the final section, I consider why the orthodox growth strategy has become dominant to the extent that it is impervious to clear evidence of economic distress and the failure to achieve growth or cohesion objectives, and yet remains as unchallenged as the 'emperor's new clothes' despite the existence of valid alternative economic strategies.

EU objectives – conflict or complementarity

Economic growth and competitiveness

The European Union proclaims the philosophy of the social market economy which endorses a 'system of economic organisation based on market forces, freedom of opportunity and enterprise with a commitment to the values of internal solidarity and mutual support' (Commission of the European Communities, 1996f: 14). At the same time, however, its overriding economic philosophy is one of macro-economic orthodoxy and economic stability. A further objective is to obtain levels of economic growth and employment similar to Japan and the USA, and to reduce the EU's comparatively high level of unemployment. Economic growth has averaged 2 per cent per annum since 1983 which represents a doubling of GDP every 30 years. But this rate of growth is slightly lower than the USA's and significantly lower than Japan's. Furthermore, the employment rate is lower and the unemployment rate is higher than either of these countries. Seven million jobs have been created in this period in the EU but the rate of employment growth (0.5 per cent per annum) lags behind the growth of the economy and insufficient jobs have been created for the number of people entering the labour market (Commission of the European Communities, 1996f; see also Chapter 2). Unemployment in the EU (11 per cent in 1994), especially among young people and women and even more so in the more peripheral regions, is significantly higher than in Japan (3 per cent) and the US (6 per cent). However, while the rate of female 'inactivity' has declined, male 'inactivity' has increased in all countries leading to a narrowing of gender differences (see Figure 8.1).

The European Union seeks to emulate American employment and growth rates, but overlooks other aspects of the US model: in particular, connections between employment growth and falling wages arising

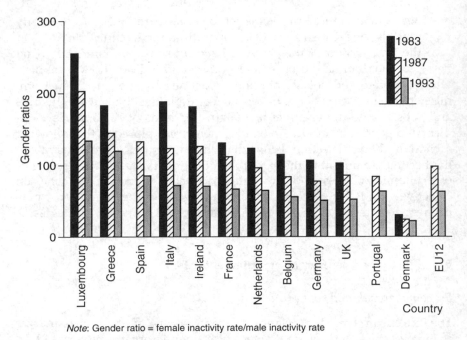

Note: Gender ratio = female inactivity rate/male inactivity rate

FIGURE 8.1   *The relative decline in the European Union, 1983–93 (calculated from Eurostat, 1997b)*

from deregulated labour markets (Reichs, 1996) and the corresponding increase in poverty, even within the working population. In contrast, the Maastricht Treaty aimed to increase the performance of the EU economy in terms of economic growth and cohesion while simultaneously protecting the rights of workers. More recently, the Commission has stated that equal opportunities are 'on a par with the struggle against unemployment' (Commission of the European Communities, 1997b: 15–16). However, the methods of achieving these more social goals have not been fully specified (Mayes, 1995) and the possible conflicts with the economic goals have not been fully recognized.

Economic and social cohesion

Greater economic and social cohesion is consistently referred to in the treaties of the European Union, including the Treaty of Rome (1957), the Single European Act (1986) and the Treaty on the European Union (1992) (Barnes and Barnes, 1995), while the Cohesion Report (Commission of the European Communities, 1996f) attempts to give this objective a clear and measurable meaning (see Chapter 1). The objective is broken into two parts: *economic cohesion*, which is measured in terms of reducing spatial, that is, regional disparities, and *social cohesion*, which is much less clearly defined and measured. Reference is made to the overall philosophy of the European Union in terms of its adherence to the principles of a social market economy and its solidarity objectives; namely, universal systems of social protection, regulation to correct market failure and systems of social dialogue. It is argued that 'promotion of social cohesion requires the reduction in disparities that come from unequal access to employment opportunities and the rewards in terms of income' (Commission of the European Communities, 1996f: 14). These objectives are, however, to be achieved without undermining the overall growth and competitiveness strategy, and without prejudicing or undermining the growth potential of the more prosperous regions: 'cohesion is not concerned with negative growth of the more prosperous regions' (Commission of the European Communities, 1996f: 15). Potential conflicts between the objectives are recognized in the designation of the cohesion fund, but are not fully addressed or resolved even in conceptual terms let alone in terms of policy instruments and resources.

Further conflicts have been identified. Under the Conservatives, the UK government maintained that economic efficiency and the protection of workers' rights were incompatible and advocated employment deregulation and increased flexibility (Commission of the European Communities, 1993b; see also Chapter 2). The importance of flexibility has been restated frequently by the new Labour government and endorsed by legislative changes in other member states. Moreover, the Cohesion Report recognized the continuing dilemmas of reconciling both efficiency and cohesion during the process of economic modernization (which could involve industrial restructuring and employment losses in some regions), and the effective development of research technology and development and its more even distribution (Commission of the European Communities, 1996f). However, while conflicts are identified, no clear pathway towards their resolution is specified. In my view, these objectives are probably irreconcilable within the context of an overall economic strategy which advocates macro-economic stability and, in particular, limitations on public expenditure. There are, however, alternative strategies even within a capitalist economic and social framework (see, for example, Boyer and Drache, 1996) within which economic growth and social well-being, including greater gender equality, are more compatible, but these are rarely voiced in official policy discourses.[2]

These conflicts are illustrated in Figure 8.2 which is, in part, based on a modification of Mayes' diagrammatic representation of the Maastricht objectives (Mayes, 1995). The conflicts within the existing economic paradigm of macro-economic stability are contrasted with the possibility of harmony in the context of a different growth trajectory based on a scenario where the gains from productivity increases are shared more evenly between those contributing towards their creation, and where the state plays a greater role in economic and social organization. The rationale underlying these figures is discussed in the next and final sections of the chapter.

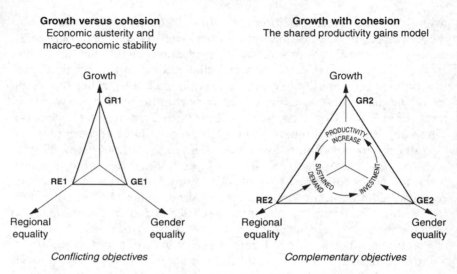

FIGURE 8.2 *Contrasting models of economic development: conflicts and complentarity between European Union policy objectives*

Economic orthodoxy, growth and social well-being

The essential elements of the macro-economic policies being followed by the European Union and the OECD states, and being pressed on the developing economies and eastern Europe, are macro-economic stability, economic orthodoxy and austerity. Within the EU this strategy can be summarized by the EMU convergence criteria which relate to nominal economic variables at the nation-state level. Ceilings have been set on the permissible levels of the following macro-economic indicators: inflation rate (3.5 per cent); interest rates (11 per cent); debt as a per cent of GDP (60 per cent) and the budget deficit as a percentage of GDP (3 per cent) (Commission of the European Communities, 1994b: 146). Member states have to be able to meet these criteria in order to enter economic and monetary union (EMU) which, in turn, will fix exchange rates irrevocably. In short, it is believed that, if achieved, these conditions in the context of the single market will create an environment conducive to competitiveness, growth and prosperity. Furthermore, the greater the area over which the market can operate in a free way the greater the gains are said to be.

The single European market now has 370 million consumers and it is believed that European firms will obtain efficiency gains by expanding to their minimum efficient scale. Nominal welfare gains are also predicted as a consequence of countries and regions specializing according to their comparative advantage (Cecchini, 1988; for a critique of this view, see Thompson, 1993). Thus, collectively, the EU member states will be in a stronger position to compete with Japan and the United States of

America both of which are thought to benefit from the relatively large size of their internal (single) markets. For this supply-side strategy to work effectively, however, in addition to stable monetary criteria and a large internal market, market rigidities have to be eliminated. As a consequence, labour market deregulation and flexibility in labour markets have been emphasized (Commission of the European Communities, 1993b, 1996f). However, this model of growth seems to depend more on widely accepted beliefs about the efficacy of stability policies in lowering interest rates and on the realization of nominal welfare gains arising from the increased economic efficiency of units operating within a single market than on empirical evidence which is rather mixed.[3]

Historically, during most periods of economic transition, either for individual nations or for groups of nations, new growth trajectories have depended on significant increases rather than reductions in state involvement in economic management (List, 1909; Freeman, 1987; Häussermann, 1993).[4] The efficacy of the orthodox neo-liberal strategy for central and eastern Europe has been challenged theoretically and by growing empirical evidence which highlights large-scale losses in GDP, rising inequalities, child poverty and abuse, and social dislocation associated with the short sharp shock therapy (see, for example, Gowan, 1995; World Bank, 1996; UNICEF, 1997; Dunford, 1998; see also Chapter 7). In relation to the West, problems with this form of growth are indicated by the change in the positions of Sweden and Finland relative to the other states during the period 1983–93. Both have experienced a sharp fall in GDP per capita (measured in terms of purchasing power standards): Sweden from 12 per cent above EU average to 2 per cent below; and Finland, equal to the EU average, to 9 per cent below (Commission of the European Communities, 1996f). In both cases the decline occurred largely in the early 1990s. While the loss of many markets in the former Soviet Union was important in the Finnish case, part of this relative decline, in Sweden especially, can be attributed to shifts in economic policy in the early 1990s (Gonas, 1998) in an attempt to harmonize with the EU, leading some writers to refer to the death of the Swedish model (Meidner, 1994).

One reason for the low growth levels is that the austere economic policies associated with the convergence criteria led to reductions in public expenditure and to depressed internal demand. As the strategy was adopted by all countries seeking to join the EMU, paradoxically it led to competitive deflation, unemployment and low growth throughout the EU member states (Dunford and Perrons, 1994). The effects of these measures are also gendered, with the cuts in public expenditure particularly affecting women's incomes and opportunities as a consequence of their over-representation in the public sector (see Escott and Whitfield, 1995). Furthermore, as the Cohesion Report indicates, there is little evidence that either regional or gender inequalities have narrowed within the EU except on the basis of very rudimentary measures such as the employment rate in the case of the latter (Commission of the European Communities, 1996f; Perrons, 1998).

Policies for increasing cohesion and solidarity within the EU, financed through the structural funds and equal opportunities policies, have now been 'mainstreamed'.[5] So, in principle, all EU policies have to be examined for their cohesive effects in terms of regional and gender equality; similarly, regional policies have to be examined for their equal opportunities implications and vice versa. However, there has been little discussion of how to establish policy priorities should conflicts arise. By examining the nature of regional and gender inequalities in the following two sections, I hope to indicate that the present macro-economic strategy, together with related employment policies, runs counter to the objectives of economic and social cohesion. In this discussion more attention is paid to the gender dimension to cohesion than is customary in official cohesion documents.

Regional disparities

Recent history suggests that when the overall growth rate is low then regional performance is weak. At the national level disparities have narrowed as a consequence of above-average levels of growth in the four poorest countries which, collectively, have increased their GDP per capita from 66 to 74 per cent of EU average. Ireland, with an average annual growth rate of 4.5 per cent per annum between 1983 and 1995 has moved from 80 to 90 per cent of the EU average GDP per capita. Even so, the gap between the EU four and the rest has narrowed by only one-quarter between 1983 and 1993. Furthermore, the rate of employment growth has been less impressive. In Ireland, employment has only grown at 0.2 per cent per annum and unemployment levels remain very high, being 11.9 per cent in 1996 (Central Statistical Office, 1997). Despite the above-average growth in GDP per capita in Spain (3 per cent per annum), one in five of the population is unemployed and the proportion of young people, especially young women, who are unemployed is even higher (Commission of the European Communities, 1996f).

The growth that has taken place has been uneven and has largely concentrated in the already developed regions meaning that there has been no narrowing of regional disparities (see also Chapter 1). The uneven pattern of growth is particularly noticeable in Portugal. GDP per capita in Lisbon increased from 81 per cent of the EU average to 96 per cent between 1983 and 1993, making it well over twice that of the Azores, but in neighbouring Alentejo it fell from 48 to 42 per cent. Consequently, wide disparities between regions remain and, indeed, are at least double those of comparable regions in the USA (Commission of the European Communities, 1996f). In terms of unemployment, this disparity is indicated by the 1:10 ratio in unemployment rates between the most (3.4 per cent) and least advantaged regions (34.7 per cent) (Commission of the European Communities, 1996f). Hamburg, the richest region in 1993, was four times as prosperous as the Azores or Alentejo, and the ten richest regions were 3.3 times as prosperous as the ten poorest regions. The gap between the top ten regions and the EU average widened between 1983 and 1993, while the gap between the bottom ten and the EU average narrowed slightly. There is also remarkable stability in terms of the composition of the top ten regions. One-half are (West) German, and the others are city regions: Brussels, Ile de France, Vienna, Luxembourg and Greater London. The bottom group was dominated by Greek and Portuguese regions, with the addition of one of the East German *lande*, Mecklenburg Vorpormmern.

When 25 rather than ten regions are taken for comparison, there is similar stability in the extent of the disparity but more variation in the composition of the most and least favoured regions. In the 25 richest regions income per head rose marginally from 140 per cent of the EU average to 142 per cent. In the 25 poorest regions, income per head also increased from 53 to 55 per cent (Commission of the European Communities, 1996f). Taking all regions, the average dispersion around the mean has remained stable across the period as a whole with a slight widening of disparities during the relatively unfavourable years of the early 1980s, a narrowing during the period of recovery in the late 1980s and then renewed widening during the years of crisis in the early 1990s. These fluctuations indicate that there is some degree of correlation between the performance of the peripheral regions and the performance of the economy as a whole (Dunford and Perrons, 1994; Commission of the European Communities, 1996f), although the direction of this relationship is contested. For example, one view is that during periods of boom greater assistance can be given to the less favoured regions, while an alternative view suggests that it is the maintenance of demand in the periphery that assists the more developed regions and so contributes to overall growth.

Regional policies associated with the structural funds are, at least theoretically, conducive to a narrowing of inequality, but there are many other EU policies which move in opposite directions, not least the agricultural policies which, despite changes, still absorbed almost 50 per cent of the EU budget in 1994 (Commission of the European Communities, 1996f). Furthermore, research, technology and development (RTD) policies tend to favour the already developed regions. Despite cohesion being 'mainstreamed', it is clear that the European Commission still prioritizes competitiveness. The less-developed regions are expected to adjust to the competitive conditions within the single market and forms of assistance have to be conducive to this end. The question remains, however, what are the appropriate mechanisms to achieve this, and whether any form of cohesion can be maintained if, at the same time, public expenditure and transfer payments are reduced?

The European Commission favours 'strategic economic services' in the form of 'regional policies research and development, trade and promotion . . . which address the long term competitiveness problems' of the regions and are thought both to resolve problems of cohesion and to prevent these emerging in the first place. The Commission argues that more reflection should be given to the appropriate balance between this 'productive expenditure' (too little) and expenditure on consumption or transfers to people which is regarded as being unsustainable in the long run (Commission of the European Communities, 1996f: 57). However, little evidence is provided to support this view and there are few studies which evaluate the effectiveness of regional incentives or training programmes for long-term regional competitiveness. In recent years more formal evaluation procedures have been introduced but few results have been reported so far (Bachtler and Michie, 1995).

The major source of redistribution between regions has, in fact, come from general taxation in most countries which, if even only mildly progressive, will tend to lead to significant shifts of resources to the less-developed regions. Cuts in public expenditure, one probable requirement of the Maastricht criteria, are therefore likely to reduce cohesion. A large part, estimated to be between 50 and 70 per cent of public spending on average in each member state, is spent on services to the community (education and health) and transfer payments to households and individuals (social security payments). Many of these expenditures are redistributive between individuals and between regions and, to the extent that taxation is progressive, it contributes towards cohesion.

Given that women are over-represented among the poorest sections of society and disproportionate recipients of welfare expenditure, as well as being over-represented in public-sector employment, they are likely to be particularly hard hit by public expenditure cuts (Meulders et al., 1993; Commission of the European Communities, 1996f; Gonas, 1998). Clearly, a situation of permanent dependency is not desirable (see Hudson, 1994b) but, in the absence of clearer evidence of the direct benefits of other kinds of policies to the population of peripheral regions as a whole, redistribution through taxation is very important. Direct aid helps to sustain demand in the peripheral regions, especially bearing in mind the higher marginal propensity to consume (and to consume local goods) of the lower income groups, and so indirectly it aids local firms and industries. Furthermore, by helping people in their daily lives, this form of aid facilitates the internal cohesiveness of regions. Some monetary income is necessary to sustain local social networks, such as child care, cultural and tenants' groups. These networks are run predominantly by women and play a vital role in maintaining the social fabric necessary for cohesion. In fact, in 1992, jobs meeting personal needs, such as hairdressing, cleaning and so on, outnumbered those in the iron and steel industries by 50 per cent (SEIES, 1995), and thus the indirect job creation effects of redistributive taxation and its role in sustaining regions should not be underestimated and denigrated in comparison to the more visible projects more clearly associated with regional policies. Ireland provides an illustration where there has been considerable modernization, illustrated most dramatically perhaps by the rise of the software industry in and around Dublin (Coe, 1997). GDP has been growing, but GNP and employment to a much lesser extent,[6] suggesting that in terms of cohesion the impact of such measures has been more limited, benefiting only some within the region and, in this particular case, many outside Ireland.

Economic orthodoxy, employment flexibility and equal opportunities

The convergence criteria have led to the retrenchment of the welfare state which has had adverse effects on female employment (see also Chapter 1). The public sector is both a large employer of women and, in some countries also subsidizes childcare, which facilitates women's involvement in the labour market given the persistently unequal division of domestic labour. These effects can be seen most clearly in Sweden where employment has fallen dramatically since 1989 and unemployment and inactivity have risen. Furthermore, there have been reductions in the level of social benefits, including unemployment pay and parental leave, from 90 to 75 per cent of the wage, while the cost of childcare facilities has increased and the quality of care has in some cases fallen (Gonas, 1998).[7]

Current EU and national economic policies and policy statements in relation to competitiveness and equal opportunities also emphasize the importance of employment flexibility (Commission of the European Communities, 1997b: 44). However, both the concept of flexibility and the actual operation of flexible working practices have to be carefully analysed (see also Chapter 2). Depending on the form of flexibility, there can be conflict rather than complementarity between flexibility and long-term competitiveness and between flexibility and equal opportunities and, as in the case of regional disparities, these conflicts raise questions about the practical meaning and status of 'mainstreaming'.

The Commission recognizes that the resources devoted to equal opportunities have been modest given the scale of the inequality. It also claims that its overall policies, including monetary stability and economic growth, indirectly contribute to equal opportunities, although somewhat contradictorily recognizing that 'the positive effects of these actions on the situation of women are often not very apparent and sometimes uncertain' (Commission of the European Communities, 1996g: 4). However, no rationale is given for this conclusion and no consideration is given to alternative policies. Before investigating these conflicts further, the concept of 'mainstreaming' needs to be elaborated.

'Mainstreaming' is said to involve not only the promotion of measures to assist women but also the application of a gender perspective and analysis to all policies, programmes and actions of the Commission (Commission of the European Communities, 1997b). More specifically, the objective is to 'introduce measures aimed at adapting the organisation of society to a fairer distribution of men's and women's roles' (Commission of the European Communities, 1997b: 15–16) 'by adapting the organisation of work to help women as well as men reconcile family and working life, and to *provide more flexible employment solutions*, again for both men and women' (Commission of the European Communities, 1996g: 5, emphasis added). Thus 'the promotion of equality must not be confused with the simple balancing of statistics . . . [but] . . . it is a question of promoting long lasting changes in parental roles, family structures, institutional practices and the organisation of work and time' (Commission of the European Communities, 1996g: 5).

Mainstreaming policies would thus seem to imply a profound break with the tradition of social policy in the EU which has previously subscribed to the public–private divide of policy space whereby there are discussions of remunerated work but not of the hidden reproductive and unremunerated work or of the organization of family life that has customarily belonged to the private domain, and therefore constituted 'forbidden territory for explicit state intervention' (Hantrais, 1995: 75). This departure would, at face value, seem to satisfy critics who have argued that issues such as childcare and the domestic division of labour cannot continue to be marginalized within official policy discourses if real change is to occur (Duncan, 1996). However, the actual policy measures deriving from this concern are extremely weak. The following section considers progress made towards gender equality in employment and, more specifically, whether existing employment strategies, particularly flexible working, contribute towards greater equality or simply reinforce existing divisions.

Gender inequalities in employment

Four dimensions are generally used to measure equality: (a) participation, which can be further disaggregated by status and extent; (b) earnings; (c) occupational segregation; and (d) unemployment. Although changes have taken place on each of these measures, inequality remains firmly entrenched and equal opportunities 'are not yet within reach' (Commission of the European Communities, 1997b: 57; see also Perrons, 1998). On one measure, employment participation, gender inequality has narrowed considerably. Women's employment rate, especially among prime-age workers, has increased dramatically and at a faster rate than men's in all countries and in the vast majority of regions in the EU in the past 15 years (see Figure 8.1). Parallel with this, women's unemployment rate has also expanded as employment growth has failed to keep pace with numbers of women seeking paid work. However, much of the employment growth is flexible. As flexibility is advocated as a means of increasing competitiveness and of moving towards more equal opportunities within the EU, the implications of flexible working are emphasized in this discussion. In relation to the other dimensions of employment inequality, neither the gender wage gap[8] nor segregation has changed significantly since the early 1980s (Plantenga, 1997). There has been some growth of female employment in professional occupations but there has been a greater growth of female employment in the casualized sector (see Chapter 2) and women are between two and three times more likely to be low paid than men (Commission of the European Communities, 1997b: 41). Because part-time work is associated with particular sectors and occupations, this expansion also tends to reinforce the tendency towards employment segregation.

Flexible working refers to part-time and temporary work as well as irregular hours and term-time only contracts and job shares (Dex and McCulloch, 1995). It has been increasing throughout the EU since the mid-1980s and in some periods has been the only form of employment increase.[9] Women are over-represented among the flexibly employed, especially part-time working: 83 per cent of part-time workers, 70 per cent of family workers and 50 per cent of temporary workers in the EU are women (Commission of the European Communities, 1997b),[10] but only 24 per cent of full-time employees were women in 1994 (estimated from Commission of the European Communities, 1995e).

Flexible work has been particularly prevalent in the service sector which has accounted for the major share of employment growth (see Chapter 2). On average 70 per cent of women's jobs are in the service sector, which ranges from 50 per cent in Greece to 90 per cent in Luxembourg, and one-half of the service jobs in the EU are carried out by women (33.3 per cent in Greece; over 50 per cent in Denmark). This sector, however, is characterized by higher levels of flexible and precarious working practices and, in fact, one-third of women employed in service jobs in the EU are part-timers. In the Netherlands, where the highest rates of part-time working are found (67 per cent of all women who work), 68 per cent of all service jobs are filled by part-time women workers (Commission of the European Communities, 1997b). Besides the Netherlands, part-time working is also common in the UK, Sweden and Denmark but is much less prevalent in the Mediterranean countries where family and informal work are more significant (Vaiou, 1995).

While flexibility has been argued to be a means of increasing the competitiveness of the economy and expanding equal opportunities, for example by providing opportunities for women returners, in reality this depends very much on specific terms and conditions. For example, the conditions of part-time work vary between countries (Perrons, 1995;

Plantenga, 1995). In the Netherlands, where part-time work can involve relatively long hours and pro rata pay and conditions, both feminists and trade unionists support this as a means of expanding employment and reconciling paid work with other activities for both women and men (Pfau-Effinger, 1995; Gregory and O'Reilly, 1996). In the UK, by contrast, although the proportion of women working part time is high (44 per cent), often only short hours are involved. In many cases flexible working has been introduced to minimize labour inputs and there is very little concern with equal opportunities. In one international retail company a new form of flexi contract has been introduced in which the employees have permanent contracts but their contracted hours vary between 3 hours and 25 hours a week according to seasonal sales fluctuations. 'What we have tried to do is introduce unique working patterns so that individuals truly do come in when the function and the operation requires them to' (interview reported in the Commission of the European Communities, 1998).[11] Besides gains arising from the increased intensity associated with part-time work (Neathey and Hurstfield, 1995; Barrientos and Perrons, 1996), there are also significant financial incentives for employers.[12]

Overall, part-time work is more common in female-dominated and low-paid sectors and thus tends to reinforce occupational segregation, but where employment is highly regulated, as for example in Spain, there has been an increase in fixed-term, temporary working. In Spain in 1996 15 per cent of women workers were on fixed-term contracts compared to 10 per cent in 1990 (Commission of the European Communities, 1997b). Part-time work can provide a means of reconciling work and family responsibilities. In fact, three-fifths of female part-time workers (compared to one-third of men) in the Community in 1994 expressed a preference for part-time work. However, the context in which this preference is expressed, and the longer-term implications in relation to equal opportunities, need to be explored. Women, for example, often make this choice as a consequence of the lack of affordable childcare (see below) and, furthermore, promotion opportunities for part-time workers are often limited (Commission of the European Communities, 1998). Flexible working practices must be differentiated before any statements can be made about their effects on both competitiveness and in terms of equal opportunities.

In many cases employers draw upon the assumed traditional division of labour between women and men in the household and women's financial dependency, and create jobs with limited prospects and partial wages. The hours worked and the pay received may not be sufficient to support even a single individual. Furthermore, in the UK, employers also rely upon state benefits to supplement wages, especially of lone parents.[13] Similar conditions for low-paid workers exist elsewhere in the EU. In a comparative study of flexible working in the retail sector in six EU countries, for example, in most countries, even if the employees

worked more hours, their wages would be insufficient to support themselves and their children (Commission of the European Communities,
1998). These employees are trapped for their earnings are insufficient
to sustain their households. As part-timers, their opportunities for
promotion are highly constrained and, even if they worked more hours,
their overall income, given childcare costs, would still be insufficient.
Yet these workers are contributing to the competitiveness and profitability of their organizations.[14]

In these circumstances flexible working does not provide an independent income (Orloff, 1996) or opportunities for employment
advance. Furthermore, it builds upon and in turn helps to sustain an
unequal division of labour within the home which perpetuates women's
unequal position in the labour market in a self-reinforcing and cumulative process (Bruegel and Perrons, 1995). These working arrangements
prioritize employer needs which, besides reinforcing female dependency, in the longer term depress male wages as new employment
opportunities are constructed along similar lines (Bruegel and Perrons,
1995). Yet wage competition in a global economy can lead to competitive
undercutting until wages decline to the least well-paid worker anywhere in the world (Freeman and Soete, 1994). Low wages also undermine the incentives for employers to invest, leading to the development
of a low-skilled labour force, and in the contemporary high-technology
world an uncompetitive economy. Clearly this is only one form of
flexibility – defensive rather than adaptive flexibility – which relates to
the employees' ability to develop new skills in accordance with economic change (Bosch, 1995). However, it is this numerical or defensive
flexibility which seems to be the most dominant form in the EU at
present.

The European Commission recognizes that job flexibility can be
regarded as a risk to the extent that it generates job insecurity, low
incomes and gaps in social protection. Thus, if flexibility is to be the road
to equal opportunities, then flexible employment needs to be regulated
which is surely a contradiction, at least in terms of the UK Conservative
government of the 1980s and 1990s reading of flexibility. However, other
states such as Austria support some degree of control over flexible
working given that it leads to workers being unable to finance their
reproduction (Commission of the European Communities, 1997b). In
Spain, however, labour legislation has been changed to permit greater
flexibility and reduced hours of work owing to the high levels of
unemployment and the sense that any job is better than no job at all
(Commission of the European Communities, 1998). Thus, within the EU
there are differences of perspective about how to move forward. These
differences need to be drawn upon and more progressive strategies
identified and supported, rather than to allow a single monolithic
strategy which may conform to the Maastricht criteria, but not to equal
opportunities, to dominate the debate.

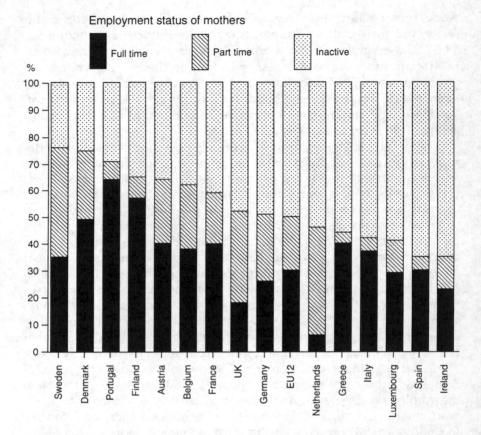

FIGURE 8.3 *Working patterns of mothers with dependent children, 1995*
(Commission of the European Communities, 1997b)

Flexible working and reconciling work and family life

Besides promoting competitiveness, flexible working is also thought to facilitate the reconciliation of work and family life. Whether or not flexible working is necessary for this reconciliation also depends on the availability of care. There is no direct relationship between the amount of care available in any society and the employment patterns of carers (Meulders et al., 1993) and there are considerable differences across the EU in the working patterns of mothers with dependent children (see Figure 8.3), although in almost all countries and regions this figure is increasing.

In Sweden, Denmark and Finland the presence of children seems to make little difference to women participating in paid work. One of the reasons for this is the comparatively generous level of parental leave as well as subsidized childcare (Jonung and Persson, 1993). In both Belgium and France, provision of childcare is also comparatively high and the participation rate of mothers declines only when there are three or more dependent children (Commission of the European Communities, 1997b).

In Germany, Ireland, the Netherlands, Greece, Spain and the UK, the participation rate of mothers with one child is significantly lower than women without children and participation declines as the number of children increases. In Italy and Portugal, there is little difference between the participation rates of childless women and women with one child but the presence of two children leads to a lower participation rate. Qualifications also lead to an increase in labour market participation (Glover and Arber, 1995). However, it is important to examine the extent of participation. The UK has low childcare provision and comparatively high participation rates and this is linked to the high degree of part-time work. Additionally, a high percentage of mothers work unsocial hours and at the weekend when their partner or other relative can take care of the child (Ferri and Smith, 1996). The participation rate also increases when the youngest child reaches school age (Glover and Arber, 1995). For low-paid workers a similar situation of unsocial hours and family care is also found in countries with relatively good childcare provision, in particular, in Sweden and France. The provision of childcare simply does not match the new flexible working patterns and, furthermore, even when subsidized and income related, the costs of childcare can still exceed the budget of low-income parents (Commission of the European Communities, 1998).

Even within the limited liberal agenda of equal opportunities, in mainstreaming, the Commission has recognized the need to address more fundamental questions, such as the division of domestic labour between women and men. However, there is a wide gulf between policy intentions and legislative change. The EU has specified minimum conditions through its Parental Leave Directive of July 1996, following the 1992 recommendations on enhancing childcare provision and family leave. The minimum period is three months, which can be taken up until the child is eight years old, and further leave should be made available for urgent family circumstances. However, there is no requirement that this leave be paid and it is not specified how it should be treated for social security reasons. Similarly, the idea that the leave be non-transferable, so that some leave has to be taken by fathers as well as mothers, is at the discretion of the member states. There has been no directive in relation to childcare.

Given the minimal nature of these recommendations, the scope for wide variations between states will remain. For example, three months' parental leave is permitted in Greece compared to 36 months in Germany and France. In the UK, there has only recently been introduced parental leave but at a minimal level. Furthermore, there are variations in the way leave can be taken. In the Netherlands, parental leave takes the form of a reduced day in order to encourage both parents to remain in the labour market and combine care with paid work. Similarly, in Sweden and France, parents have the right to work a reduced day but this right is, almost universally, exercised only by women, whereas in Belgium, Germany, Austria and Greece the employer's consent has to be obtained (Commission of the European Communities, 1997b). And this is one of the few practical measures designed to 'adapt the organisation of society to a fairer distribution of men's and women's roles' (Commission of the European Communities, 1997b: 15–16). This weakness reflects the different views of member states in relation to questions of equality and difference between women and men which, in turn, have been theorized in a number of ways: see Sainsbury (1994) for a review of the gendered welfare regimes perspective; Pfau-Effinger (1995) for the gender arrangements perspective; Walby (1994, 1997) for a discussion of differentiated patriarchy; and for a general overview see Duncan (1996) and Perrons and Gonas (1998).

Equal opportunities and cohesion

Equal opportunities initiatives have been incorporated within the structural funds, especially the European Social Fund. Specific initiatives include NOW (New Opportunities for Women), launched under the Third Action Programme for equal opportunities in 1991 and continued under the Fourth Programme. Other community initiatives such as URBAN, LEADER and INTERREG also include equal opportunities measures; for example, women's cooperatives in agriculture, traditional products and tourism have been set up under LEADER.[15] These projects are small scale and, while individuals undoubtedly benefit, their contribution to long-term changes in structural inequalities or equal opportunities has not yet been monitored or researched.[16] Under Agenda 2000 the objectives of the structural funds and the areas to which they apply have been consolidated. Equal Opportunities is included in the reformulated Objective 3, concerned with the modernization of labour markets, education and training and combating social exclusion. Similarly, the Community's own initiatives have been consolidated and reduced from 13 to three and the third area of interest, human resources, is intended to pay special attention to equal opportunities (Commission of the European Communities, 1997c).

At the policy level, therefore, there is now some 'cohesion' between equal opportunities policies and other cohesion policies. But so far I have assumed that equal opportunities and social cohesion are complementary and I have not problematized the concept of equal opportunities. However, the limitations of this liberal position have to be recognized. Equal opportunities policies are essentially concerned with attaining parity between women and men within a hierarchically structured society. Neither the hierarchical structure, nor many of the structures that generate inequalities between individuals, are challenged.

As a consequence, the policies have largely benefited a minority of well-qualified women who have secured a stronger position in the labour market, while other women have remained confined to lower-paid and casualized jobs (see Bruegel and Perrons, 1998). The importance of the different social positioning of individuals is graphically illustrated by Tessa Jowell, the UK Labour government's minister for public health:

> When I had my baby, there were five babies lined up in their cots like runners in a race – but the most important thing had already happened to them, the circumstances they were born in. One was going straight in to care, because his brother had been sexually abused at home; one was going back to a bed and breakfast; the father of one had just lost his job with the closure of a steel works; and two were going cosily back to well provided homes. (Jowell, cited by Whitehorn, 1997)

Clearly, if these babies are treated by the state in the same way then they are scarcely being provided with equal opportunities in terms of their life chances (Langan and Ostner, 1991).

In addition to individual adversities and gender differences, there are other social cleavages between women, in particular concerning race. At present, EU equal opportunities policies do not address the situation of many of the women in the worst situations, migrant women especially from the transitional states in eastern Europe and from less-developed countries, some of whom have only indirect rights of citizenship via marriage and some of whom have been attracted into Europe by promises of jobs in tourism and catering, which often turn out to be jobs in the sex trade (Morokvasic, 1991; Humbeck, 1996). These women face triple exploitation as workers, as women and as members of an ethnic minority (Van Boeschoten, 1997). Recently measures have been considered to grant these women primary rather than secondary citizenship, so that they have a right to remain within the EU after divorce, but so far no legislation has been passed. Thus within the policy agenda there are clear limitations as to the extent to which equal opportunities policies will bring about social cohesion.

Alternative economic strategies and social well-being

Alternative economic frameworks emphasize the importance of monitoring and measuring real rather than nominal economic performance; that is, paying attention to what is happening to industries and people, rather than focusing only on nominal measures such as inflation and interest rates. Furthermore, while recognizing that the market may provide the most efficient means of allocating scarce resources and setting prices, they emphasize the importance of the state for preventing major macro-economic instabilities, securing full employment and promoting innovation (Boyer, 1996). These analyses are put forward by the Cambridge (UK) School (Kaldor, 1970; see Hall, 1991 for a review); the French Regulation School (Aglietta, 1979; Lipietz, 1985; Jessop, 1990; Boyer and Drache, 1996) and long wave theorists (Freeman and Soete, 1994).

While these theories differ in many respects, two common features relevant to this case are, first, that the state needs to remain a key actor in shaping the political–economic trajectory and, secondly, that real productivity increases and their more widespread distribution are central to economic growth.[17] In the Fordist era, productivity increases were obtained through technical and organizational changes in working practices organized by firms but the gains were shared, at least to some degree, with the employees through collective bargaining agreements and by the provision of collective goods by the state. In this way the skills of the workforce, and the necessary demand to sustain investment and technological advance, linked to expanding output could be maintained.

Clearly, there are important differences between the 1990s and the period when many of these theories were first developed. In particular, globalization has opened economies. Thus products have to be internationally competitive and the internal market no longer constitutes the prime source of demand; that is, the organic and self-sustaining growth mechanism between internal demand, productivity increases, output and growth has been broken. In these conditions it would be very difficult for any single country to pursue an expansionist Keynesian policy without generating a huge influx of exports from countries following austerity policies, as the French Socialist government found in the early 1980s (Boyer, 1987). Even so, it is even more important to retain control over the specificity of national economies, especially in maintaining levels of innovation, knowledge and training, protecting the environment, providing the necessary framework within which markets can operate (Boyer, 1996) and, more generally, by setting political priorities, including questions of distribution[18] and greater equality between women and men. Otherwise the logic of the present policies is for wage levels to fall to the lowest level prevailing in any part of the globe (Freeman and Soete, 1994; Scharf, 1996) and for welfare measures to be residualized leading to further increases in regional disparities and fewer opportunities for female workers.

The present deregulated labour markets have led to a polarized labour force rather than full employment, and the high interest rates associated with monetary stability have led to low investment. Furthermore, as argued above, the expansion of female employment has been associated with the development of jobs paying only partial wages, as women are assumed to be dependent on a male breadwinner. However, given the deregulation of the labour market, these low wages also lead to reductions in male wages (and male inactivity) which promotes the low-investment, low-skills cycle characteristic of the British economy (Bruegel and Perrons, 1995, 1998). One way out of this impasse is to provide circumstances in which women and men can share caring and paid work responsibilities. However, as market failure pervades both the promotion of equal opportunities and childcare facilities, both regulation and state finance will be required. The private costs of family-friendly policies can often be lower than those associated with a high staff turnover, depending on the state of the labour market (Business in the Community, 1993; Holtermann, 1995). However, these policies entail risk; for example, firms without such facilities may poach trained and retained female employees and are less likely to be cost-effective for smaller firms.

Looking at the comparative performances of economies within the EU in the post-Fordist era, it is clear that those countries that least adhered to market principles, such as Germany and Sweden, have actually performed better (Commission of the European Communities, 1996f). Furthermore, high levels of productivity and high levels of employment co-exist in the more prosperous regions.[19] These regional economies are often those with a high level of female participation in paid work.[20] However, the European Commission fails to consider the mechanisms through which this combination is realized and it is far from clear that this desirable outcome has been attained by the pursuit of the economic policies currently recommended. In many EU countries, the distribution of personal income has become more uneven, and in virtually all member states the share of wages and salaries (pre-tax) in total income has fallen, while the share accruing to capital has increased (Commission of the European Communities, 1996f). While the European Commission comments that 'these developments underline the contribution that European labour has been making to the restructuring of the economy to meet the challenges posed' (Commission of the European Communities, 1996f: 45), it is not clear that this sacrifice is actually conducive to long-term growth and certainly not towards cohesion. Within the alternative economic framework, this shift in the distribution of income would provide some explanation for the failure to move on to a higher growth trajectory as well as for the failure to meet the cohesion objective.

Despite the mainstreaming of both cohesion and equal opportunities, the orthodox economic paradigm has retained supremacy. The question is why. This dominance is very difficult to explain. The individualistic values of Thatcherism and Reaganism seem to have been very powerful and the radical critiques, referred to earlier, have been marginalized. Even in the formerly radical urban and regional literature, attention is increasingly being paid to developing regional competitiveness and marketing places successfully without apparent recognition that in the context of overall slow growth the success of one region, city or locality doubtless has negative implications for others. Moreover, in effect these strategies often depend on transfers of funds from the general population to a minority of shareholders in a zero sum competitive struggle. Certainly, market societies have in the past been comparatively success- ful in providing the majority of consumers in the advanced countries with rising standards of living (Boyer, 1996), although, as argued above, the more successful of these countries have not adhered to the con- temporary economic orthodoxy. However, tensions are emerging within and between the original EU members, especially between Germany and France, about the desirability of these policies, given their inability to sustain existing levels of welfare, and these tensions are likely to intensify if the EU continues to widen its boundaries.

Conclusion

Overall, while 'any notion of European cohesion is inevitably inter- twined with that of citizenship, democracy and solidarity' (Commission of the European Communities, 1996f: 47), macro-economic stability policies and increased employment flexibility lead to widening regional disparities and to polarization in the labour market, in which dis- advantaged groups such as women and ethnic minorities occupy lower positions. Consequently, there is a discontinuity between the objectives of empowering people as citizens and the reality of disempowering them as workers (Marquand, 1997). Both cohesion and equal oppor- tunities policies have been mainstreamed and, at the policy level, the cohesion objectives are quite explicit, but conflicts between these objec- tives and recommended economic policies have not been adequately recognized or addressed.

Much of the new flexible employment is part time. Part-time work provides only a partial income, so the recipient must be able to draw upon other sources for maintenance – implicitly a male main bread- winner. Thus, in effect if not in intention, the strategy rests on gender inequality or at best the presumption that the genders play different roles, and this is clearly at odds with the idea of social cohesion. While 'cohesion should not be confused with harmonisation or uniformity' (Commission of the European Communities, 1996f: 45), economic dependency can hardly be thought of as being compatible with 'widen- ing opportunities'. Furthermore, given the changing social patterns and to some extent the individualization of European society, it is very important that everyone be given the right either to work or to a level of income sufficient to maintain themselves and their children. Indirect access to welfare through a partner is not desirable because 'there can never be a right to a husband for life' (Langan and Ostner, 1991: 138).

As long as societies remain predominantly capitalist – that is, based on a wage relation and the pursuit of profit – there will be inequality between people and between regions. Furthermore, as long as patriar- chal ideas and attitudes remain in place, there will be unequal relations between women and men. In addition, as long as societies continue to be racist – attitudes which seem to be strengthened by a Fortress Europe mentality which highlights the differences between insiders and outsiders or ourselves and others – then inequalities will also have a racial dimension (Castles, 1995). However, the nature of all of these 'structures' can be modified by social policies which remove their extreme forms. There can be varieties of capitalism: for example, the different pathways out of the crisis of Fordism (Leborgne and Lipietz, 1991). There can be differences in the nature of patriarchy (Walby, 1997) and in gendered welfare state regimes (Duncan, 1996; Sainsbury, 1996). It is important to identify and advocate those policies that are more conducive to social inclusion in order at least to challenge, if not overcome, these structural inequalities and so move towards a less divided Europe.

NOTES

1 In fact, the UK's Labour prime minister argued that the priority objective on employment was deregulation to create an adaptable labour force. Paul Webster comments that 'While the French were spared from agreeing to the word "flexibility", they signed up to a euphemism, inspired by Britain, to encourage policies that would allow the jobs market to "react to economic changes"' (Webster, 1997: 16).
2 At the IGC in Amsterdam the French did propose an alternative economic strategy in which greater attention was given to compatibility between macro-economic stability and growth and to increase the employment intensity of growth, as well as to increase investment from the European Investment Bank and a growth fund to finance high tech projects but this was strongly opposed by the new Labour government 'with strong approval from Helmut Kohl', the German chancellor (Webster, 1997: 16).
3 Many of the efficiency gains for example have been estimated from rather dated technical relationships (Thompson, 1993).
4 The successful East Asian countries have for example based their economic strategies more on the works of Schumpeter (1934) and List (1909) where emphasis is placed on the creation of new resources through innovation rather than on the neo classical model and its concern with the allocation of scarce resources (Maleki, 1997).
5 In relation to equal opportunities policies 'mainstreaming' is defined as 'the systematic consideration of the differences between the conditions, situations and needs of women and men in all Community policies, at the point of planning, implementation and evaluation as applied to Europe, the industrialised and the developing countries' (Commission of the European Communities, 1996g: 8). The strategy was made explicit in a Commission Communication in February 1996 (COM (96) 67 final of 21.2.1996).
6 GDP measures the output produced in a particular region/country; GNP measures the output retained in the region after allowance is made for imports and exports. The difference between these measures is important. The output figures, i.e. GDP can be inflated by companies operating transfer pricing and no account is taken in the GDP figures for the profits that are repatriated.

7 For example during the period 1989–94 the number of children's nurses in day-care centres decreased by 20 per cent of which 97 percent were women. Somewhat contradictorily in January 1995 in Sweden, a new law stipulated that the local authorities have to provide care for all children aged between 1 and 12 years if the parents are studying or working. Thus, either the cost to the parents will rise or the quality of services is likely to fall (Gonas, 1998). In late 1997, one year before the next election the Social Democratic Party began to expand public expenditure, especially on welfare and schooling, and full employment was given precedence over maintaining a budget surplus (McIvor, 1997).

8 Eurostat provided harmonized but very limited data on earnings. In relation to hourly earnings, on average women earn 80 per cent of the male figure. There are significant variations between sectors and the gap in the formal sector tends to be lowest where employment regulations are most effective: for example, in Greece and Portugal (Perrons, 1995).

9 In the period between 1990 and 1994 the total number of jobs fell by 3 per cent in the EU. Part-time jobs increased by 9 per cent and full time jobs declined by 6 per cent. By 1994, 30 per cent of all female jobs in the EU 15 were part-time compared to 28 per cent in 1987. Part-time work accounts for 5 per cent of male employment in 1994 (Commission of the European Communities, 1997b).

10 Temporary jobs accounted for about 14 per cent of all jobs in the EU in 1995 (about half the number of part-time jobs). In 1995 about 15 per cent of women and 13 per cent of men were employed in temporary jobs. There are substantial variations between countries. Temporary work tends to be more prevalent in countries where employment is highly regulated, for example, Spain where 33 per cent of men and 38 per cent of women were in temporary jobs, compared to 6 per cent of men or less and 10 per cent of women or less in Belgium, Luxembourg, Italy, Austria and the UK. The most precarious jobs are part-time temporary and about 39 per cent of women temporary workers were part time (Commission of the European Communities, 1997b; see also Meulders et al., 1994).

11 This report is based on a comparative qualitative study across six EU member states of the experiences of flexible working in the retail sector among people with caring responsibilities. This study was financed by the EU under the Fourth Action Programme on Equal Opportunities (Commission of the European Communities, 1998).

12 Employers would have to pay £11.20 National Insurance a week for employing one full-timer working 40 hours at a rate of £4 per hour compared to £4.80 for two part-timers (Meadows, 1997).

13 The actual combination of paid work; state support in cash or in the form of subsidized childcare varies between EU countries. These issues are being investigated further under the Fourth Action Programme.

14 In the UK these low wages are supplemented by the state in the form of Family Credit: in this case the competitiveness of the firm is being sustained by the other taxpayers.

15 One specific example was the village of Velvendos in Greece. This is a women's cooperative that produces traditional cakes. It was established in 1985 and now produces a range of traditional produce such as jams which are marketed throughout the country; 20 women in the village are now employed. The cooperative has been involved in activities such as marketing, designing logo and packaging and is supported by LEADER (Commission of the European Communities, 1997b).

16 For example Sweden has included quantifiable equal opportunities components in all its objectives. For example Objective 6, which covered sparsely

populated regions is designed to maintain the presence of women in the region.

17 Real productivity increases are obtained when more outputs are obtained for the same or a smaller amount of real inputs. These are to be distinguished from the 'efficiency' gains arising from many of the privatization and contracting out of public sector activities in the UK which have arisen predominantly from paying workers lower wages for the same or higher levels of output.

18 For example more egalitarian taxation policies could finance wage increases in the non-traded sectors without raising public expenditure.

19 The recent decline in the German economy can be attributed to particular problems encountered through reunification and perhaps also by the extreme difficulties Germany has faced in meeting the Maastricht criteria in these new circumstances.

20 It has to be recognized, however, that employment segregation and income inequality are higher in the more developed regions where high-level jobs are found.

9

Ethnicity, Racialization and Citizenship as Divisive Elements in Europe

Paul White

The notion of Europe as a continent is a social construction: in truth, Europe is but the western end of the great Eurasian land mass that stretches from Cape St Vincent to the Bering Strait. Similarly, the concept of 'Europeans' or of a 'European civilization' is in reality imprecise. Archaeological evidence tells us that most 'European peoples' arrived from elsewhere, particularly from the East, with waves of arrivals occurring over several millennia up to the spread of Ottoman influence into south-eastern Europe around 500 years ago.

Yet during the 200 years since the Enlightenment, coinciding with the colonial and empire-building period that spread the influence of 'European' people and ideas across the whole of the globe, the concept of 'European-ness' has hardened into a familiar set of identity markers based on notions of 'us' and 'them', and on hierarchies of power and legitimacy operating between the groups so defined. Europeans have seen themselves, in relation to many of those from other parts of the world, as intellectually and morally superior, economically more powerful, and politically more advanced. Leaving on one side the complex histories of European views of China and Japan, the only notable exceptions to these identifications of European superiority (and then only partly so) have been the views of areas settled by Europeans themselves. Although colonialism was a complex phenomenon operating in different ways by different European powers, there was almost always an element of what the French defined as their *mission civilisatrice* (civilizing role) present.

In the words of a commonly used phrase, more recently we have seen 'the empire strike back' (Centre for Contemporary Cultural Studies, 1982). Over the course of the past century, population migration has transformed the demographic and cultural scene throughout Europe. 'Other' people from 'other' places have become 'other' people 'here'.

To a general European culture that has emphasized the 'difference' of the European tradition, this has brought a number of tensions and the potential for the transformation of many aspects of both official and everyday views, debates and discourses.

However, Europe itself is not as homogeneous as the discussion so far suggests. In particular, the progress of state-building and the evolution of national consciousness in Europe (Gellner, 1983) was predicated (as with distinctions between Europe and elsewhere) on a series of identifications of 'difference' and the elevation of these as markers of separate identity. The ways in which these processes operated, and the detailed outcomes for the actual form of national consciousness, were highly variable and continue to play an important role to the present day (van Amersfoort, 1991).

The arrival in many European countries of significant numbers of migrants, both from 'other' European countries and even more so from the rest of the world (particularly those labelled as 'non-European'), has not therefore had the same outcomes everywhere, since the circumstances of the European destinations were themselves varied. The objective of this chapter is to explore the varied positions of migrants and ethnic minority communities of migrant origin within today's European states. Attention is focused on the extent to which the political cultures of European countries permit or prevent the possible inclusion of sectors of society defined by race or ethnicity. The emphasis throughout is on groups that have a history of migration within the past century, thereby complementing the discussion in Chapter 4 of social integration, exclusion and multi-culturalism as they affect more long-standing 'national minorities' and regional identities within Europe. While the prime focus of this chapter is on issues within western Europe, to which the principal migration flows of the twentieth century have occurred, attention is paid in a later section to the significant new migration strands in the complex post-communist scenarios of central and eastern Europe.

The significance of race and ethnicity

The history of migration to and within Europe over the past century has been extremely complex, with many motivations for large-scale mobility. In the post-war period, however, one of the dominant flows has concerned labour migration set in motion by the needs of certain economies for relatively low-skilled workers in Fordist production units, or to act as replacement labour in unskilled service tasks. More recently, new international migration into southern Europe (see Chapter 6) has once again related to economic opportunities, although within a more post-industrial context. Whereas migrants arriving in France or Germany in the 1960s took employment in large-scale manufacturing such as in Renault or Siemens, recent migrant arrivals in Spain or Italy have found (or created) work in petty services and trade within new consumption markets (Misiti et al., 1995; Knights, 1996).

It might appear that the particular history of the evolution of international migration systems throughout western Europe would have created a particular geography to contemporary patterns of racial exclusion. The early post-war period saw two basic kinds of movement. The first occurred into the colonial or ex-colonial metropoles (such as Britain and France, and to some extent the Netherlands) where significant proportions of migrants came from overseas possessions, often in circumstances of relatively unhindered passage (at least in legal terms). The second type of movement involved countries with no 'obvious' colonial sources of labour supply, who turned instead to nearby labour-rich countries, initially elsewhere in Europe, with very careful legal controls on all aspects of the systems: key exemplars would be Germany or Switzerland, operating the *Gastarbeiter* system. Early post-war labour migration (in the period up to the mid-1970s) had a very specific geographical imprint, with the core countries of northern and western Europe (France, the United Kingdom, Belgium, the Netherlands, Luxembourg, Switzerland, Austria, Germany and Sweden) witnessing migration from southern Europe (Portugal, Spain, Italy, Yugoslavia and Greece) and from further afield (King, 1976). It should be noted, however, that many more 'local' patterns of labour recruitment supplemented the general patterns just outlined: for example the United Kingdom and Sweden both drew significantly on traditional labour sources in Ireland and Finland respectively, with little formal control on these flows. France attracted migrants from non-colonial origins (such as Spain and Italy) as well as from its overseas possessions, while Belgium and the Netherlands actually recruited more labour from third-party countries than from their ex-colonies.

The history of large-scale migrant arrival has been much longer in northern and western than in southern Europe. Arrivals in southern Europe did not take off until the end of the 1970s, by which time primary flows into other parts of Europe had been much reduced in scale. Nevertheless, despite these evolutionary scenarios affecting different parts of the continent at different times, with migrants from different sources arriving under varying legal frameworks, geographical factors at the international scale do not play the primary role in determining the outcomes in terms of. ethnic or racial exclusion. Instead, social constructions of 'otherness', often supported by official definitions and policies, need careful examination.

While some early commentators (for example, Böhning, 1972) saw the mass labour flows almost entirely in demographic, political or economic terms, others noted that much popular discourse regarded the migration (erroneously) as entirely involving 'blacks' and set out to show that the circumstances of immigrant workers were determined by their class position rather than through their 'race'. For example, Castles and Kosack

(1973: 482) argued that the labour movement needed to recognize the coincidence of interests for all workers, immigrant and indigenous, and that the 'false consciousness' that creates prejudice and discrimination 'can only disappear when it is supplanted not merely by a correct understanding of the position of immigrant workers, but by a class consciousness which reflects the true position of all workers in society'. The succeeding years, however, have led to a modification of these positions for a number of reasons.

The early labour flows have diversified in several ways, through greater family movement, through increased flows of entrepreneurial migrants, through more international exchanges of skilled rather than unskilled workers, and in recent years by increased flows of refugees (King, 1993b; Council of Europe, 1996). The composition of the ethnic minority populations of most European countries, certainly those with the longest migrant histories, has become much more complex.

Research on the living circumstances of immigrant and ethnic minority groups, from a variety of original migration streams, has considered the relative balance of class and racial explanations of differences in life chances and living circumstances and has suggested that, while the two are often present together, the balance is often such that race or ethnicity is the more important factor (for example, van Amersfoort, 1988; Pirani et al., 1992; Seifert, 1996). The evidence so produced clearly leads in the direction of identifying certain negative impacts of racial or ethnic labels, and suggests that processes of racialization act more strongly for certain groups than for others. This is even reflected in academic research where specific groups have been intensively considered, with the neglect of others. Such activities have in many ways reflected wider public and political interests in particular sections of the ethnic minority populations of individual countries, but at the same time they add to the processes problematizing those groups above others. Perhaps the most extreme case occurs in France, where the focus is much more strongly on Algerians, with relatively little recognition of the fact that there are more Portuguese resident in the country (650,000 against 614,000 Algerians at the time of the 1990 population census). Such a tendency reflects the issue raised in the introduction to this chapter: that those who have been seen as most significantly different are those from outside the social construction of Europe.

It is clear today that in many European countries there are strongly differential processes at work that lead to distancing between the fortunes of different groups. Ideas derived from American ecological traditions of migration and social analysis would expect such differentiation to occur on the basis of dates of arrival, through the operation of assimilation processes among the earliest arrivals. Such ideas are of little contemporary relevance in Europe, where early migrant groups and long-standing communities have often been shown to be more disadvantaged than recent arrivals: for example, this explains the greater focus in France on the Algerians, who arrived early, than on the later-arriving Portuguese. This immediately raises questions concerning societal processes of racialization and discrimination, as well as individual and community responses among those affected.

Concepts of ethnicity, race and racialization have so far been used without comment, but one of the complexities of analysing issues concerning minorities in Europe is that such terms are not only somewhat flexible within the English language, but that there are great difficulties in translation between languages. In contemporary English discourse (at least at the academic level), racism is normally defined in terms of the expectation of behaviour, generally with a negative attachment, related to attributes that are supposedly biological (Jackson, 1987; Miles, 1989). Ethnicity, in contrast, is seen to relate to common characteristics and feelings of identity originating in language, religion, culture or history. The distinction between 'race' and 'ethnicity' is less easy to make in many other European languages. The French *Robert* Dictionary shows the word *'ethnie'* but the word is in practice not used in relation to immigrant communities or their descendants, and refers to a group rather than to a concept that can be felt by individuals. German and Italian have no original equivalents to 'ethnicity': in both cases the English word has recently been adapted as *'Ethnizität'* and *'etnicità'* respectively, but again the currency of these words is extremely low.

In many ways English discourse is exceptional in Europe for reasons (to be discussed below) relating to citizenship. In most European countries, high proportions of 'others' of immigrant origin remain 'foreign' or non-citizens, and can be described simply in those terms. In the United Kingdom, this possibility does not exist since most of the major communities resulting from immigration are composed of British citizens. Interestingly, within mainland European countries there have been shifts through time in the extent to which particular migrant groups have been seen as 'different'. As Castles (1993) has commented, in the early days of mass migration similar attitudes to immigrant labour appeared to exist throughout the destination countries, whatever the migrant origins. However, '[b]y the late 1980s, there appeared to be a much higher degree of social acceptance of intra-European migrants, which contrasted with strongly exclusionary attitudes towards immigrants from the south and minorities who were phenotypically different' (Castles, 1993: 26).

New linguistic ways of identifying different immigrant groups are springing up: for example, Italian now has the word *'extracomunitari'* to identify those originating from outside the European Union (formerly the European Community), as opposed to the older words for foreigners, *'stranieri'* and *'forestieri'*. In Germany there has been considerable discussion (at least in academic and political circles) about terms for the foreign population. The usual term *'Ausländer'* etymologically means 'outsider' (although it is usually translated as 'foreigner' and is used in that sense within Germany), and this is problematic when it is used to describe those of the second or subsequent generations who have been brought up in Germany but who are still thus labelled as excluded (Piper, 1997).

While 'racism' is generally seen in official discourse as a negative feature of modern societies, it is an almost inevitable outgrowth of the processes of 'othering' described earlier, and which were bound up in the emergence of modern nationalisms within European nation-states (Brubaker, 1992b). A consequence of the establishment of the ideological superiority of a nation was the parallel establishment of a view of the superiority of its people, of the construction of a view of who those people were, and the relegation to inferiority of those who were 'outsiders'. Among such outsiders, however, further labelling processes have operated to attach racial labels to some (and hence make them subject to racism) while failing to label others. These patterns of racialization involve the social attribution of significance to particular categories (Miles, 1989), resulting in much clearer social constructions of certain groups. Researchers have shown the existence of selective racialization in many European societies (for example, Girard, 1977; Hagendorn and Hraba, 1989); for example, through hierarchies in the acceptability of particular foreigners as neighbours or marriage partners for offspring.

However, it is common throughout Europe to meet with official denials of the existence of racism, either as an ideology or in terms of racist outcomes of social and economic processes (van Dijk, 1993). One of the most extreme cases of this occurs in German discourse, where despite the many publications using 'Rassismus' in their titles (for example, Memmi, 1987; Butterwegge and Jäger, 1992), the official view is that racism does not now exist since this word is only officially used to describe the anti-Jewish ideologies of the Nazi period. The 'milder' word that replaces it is 'Ausländerfeindlichkeit' (or 'hostility towards foreigners'), an expression that in some ways accentuates and perpetuates the condition of otherness through its use of the epithet 'Ausländer'.

Since processes of racialization have been historically determined through colonial views, patterns of more recent labour flows and elements of cultural absorption, it is unsurprising that the outcomes, in terms of social constructions of particular groups, differ from country to country within Europe. The only common element tends to be the particular identification of Islamic populations wherever they may be, largely ignoring the divisions within Islam (Gerholm and Lithman, 1988). The growing recognition of the distinctiveness of the circumstances of different groups, referred to earlier, has both a positive and a negative aspect: it reduces the over-definition of all migrants with a single label, but at the same time it aids the process of hierarchization in the minds of the non-minority population. Since the more subtle recognition of ethnicity (with groups self-defined by feelings of cultural affiliation) is not yet common in Europe, it is only in the United

Kingdom that there is a question on ethnicity in the census and in government surveys. Elsewhere the crucial issue is normally whether or not one holds national or foreign citizenship. Those with national citizenship are (in theory at least) part of the national polity on a par with others: those without such citizenship (whatever their place of birth, process of socialization, educational background or cultural affiliations and interests) remain as 'others' or outsiders to the national polity. It is because of this pivotal effect of citizenship that the notion of skin colour as a universal European indicator of 'otherness' has to be partly discounted, since metropolitan France is home to significant black residents from the overseas *départements* of Guadeloupe and Martinique who are full citizens and, at least in theory, not susceptible to 'othering'.

The borders between 'nations', 'citizens', 'races' and 'ethnicities' in contemporary Europe are generally becoming more complex. The ideologies driving legislative and social processes in certain countries (such as Germany) may attempt to maintain clear-cut relationships between the definitions of these different sets, but such attempts are everywhere doomed to fail in the longer term. The French nation is no longer coterminous with a white French race. The deconstruction of any vestigial notion of Belgian-ness, given the rivalry between Flemings and Walloons, today must add consideration of the Moroccan and Turkish ethnicities that are now embedded within the national territory. Nevertheless, as will be seen later in this chapter, concepts of inclusive citizenship, and its inevitable exclusionary implications, are still very much present in western Europe and in some cases have been reinforced in recent years. In contemporary central and eastern Europe issues of citizenship are particularly salient, with profound threats to certain excluded groups.

The social exclusion of racial and ethnic groups

As a result of processes of racialization and the operation of racist ideas, often at the everyday rather than ideological level, ethnic minority groups of immigrant origin are particularly susceptible to the operation of social exclusion mechanisms that reduce the life opportunities for significant numbers of people in contemporary European societies. Processes leading to social exclusion can operate in a number of ways (White, 1998), and their applicability to ethnic minority groups can vary at a number of geographic scales.

First, minorities can be excluded from participation in various spheres of life through legal restraint (for example, through definitions of eligibility). Secondly, they can be excluded through ideologies of 'othering' which result in a failure to accord legitimation to their interests or even to their very existence. Thirdly, they can be excluded through a failure to provide group-specific services for them while other sub-groups of society have their particular needs met. And, fourthly, they can be excluded from full participation in consumption activities through poverty and economic exclusion.

From place to place different elements within these mechanisms of exclusion can take on greater or lesser importance. Exclusion through legal restraint, for example, is a specific and important issue in those states where minorities retain the status of 'foreigners' lacking citizenship, and where this exclusion from citizenship can then be used to perpetuate distancing between 'us' and 'them' along legally enforceable and therefore highly formal lines. This may then affect activities in a number of spheres, such as the inaccessibility of certain jobs where national citizenship is required (for example, the civil service), full rights of personal mobility (for example, at the European Union level) or rights to political activity (it has only been since 1981 that foreigners in France have had the right to create political associations; Wihtol de Wenden, 1995). Elsewhere, for example in the United Kingdom, minority groups may in fact have full legal rights including citizenship such that these legal exclusion mechanisms are not operative.

The second mechanism of exclusion, operating through de-legitimization, is most egregiously associated with racist political parties and movements. However, levels of everyday attitudes perpetuated in ordinary discourses among those who would disavow any identification with the racist politics of the *Front National* in France, the *Republikaner* in Germany or the *Vlaams Blok* in Belgium are also part of a wider culture that questions the rights of many minorities to live how and where they do. The existence of racist sentiments in Europe is undeniable, but there are unanswered questions concerning, for example, the 'background' level of such feeling (or the way it is embedded into socialization processes), the importance of empirical catalysts, and the mechanisms for the translation of racist sentiment into racist politics. To take the last issue, it would clearly be unwise to suggest that French or Austrian social discourse is more racist than that in Britain or Sweden simply on the evidence of the stronger electoral support of parties with partly racist platforms in the former two countries. National maps of extreme right voting in Europe during the period 1971–91 (Vandermotten and Vanlaer, 1993) demonstrate a general tendency for the extreme right to fare best in cities and regions with high proportions of immigrants or minority communities: for example, Bavaria in Germany, or the Provençal coastal areas of France. Working-class urban areas throughout Europe have a higher propensity to support the extreme right. However, it must also be observed that the attraction of the extreme right is not simply over its views on race or immigration but on a variety of other issues tied up with nationalist conservatism, such that certain regions in which race and ethnicity are of low importance (for example, rural Flanders in Belgium) also produce strong extreme right support. The regular *Eurobarometer* surveys, for example, produce a more nuanced picture based on wider measures of attitudes (Vandermotten and Vanlaer, 1993).

There are certainly great complexities in the relationships between racist sentiment and action. Liberal democratic notions such as freedom, equality and human rights are commonly used to support both racist and anti-racist positions (Verkuyten et al., 1994). There are also possible links over time with earlier racist feeling. Schor (1996) has commented that the operation of racism in today's France is very similar to that of the 1930s, although there is less open expression of racist ideas. In the 1930s the targets were the Jews; today they are North Africans, and predominantly those identified as Algerians. Consideration of the historical continuity of racism/*Ausländerfeindlichkeit* (see above) in Germany is a highly sensitive and rarely discussed subject, although Kromschröder (1983) has demonstrated the vernacular existence of a number of racist anti-Turkish jokes that either make direct comparisons to Jews or connect to the Holocaust.

In a number of countries there are indications that overtly racist sentiment is increasing. For example, in Italy, where large-scale immigration has been a relatively recent phenomenon, anti-immigrant rhetoric has played a part in the electoral success in recent years of both the right-wing MSI (*Movimento Sociale Italiano*) and the northern separatist movement, the *Lega Nord* (Zincone, 1993). Enoch (1994) has suggested that in Denmark intolerant attitudes have probably been latent for years and have only surfaced now that 'target' groups are present. The number of racist incidents has been seen to be increasing in many countries (for example, Solé, 1995; Witte, 1995), although, as R. Miles (1994) has pointed out, this may reflect a rise in willingness to report incidents rather than an underlying rise in racism.

In Germany, the apparent increase in racism has complex relationships with re-unification and the re-definition of 'German-ness', particularly for those living in the former East (Treibel, 1994). Here discussion of what the new Germany might be like has provided the Republican Party with an opportunity to rehearse arguments about *Volksgemeinschaft*, the community of German interests in which Germans are themselves seen as belonging to a closed ethnic community such that Germany has nothing to offer immigrants (M. Evans, 1996). As will be discussed later, such rhetoric is not discouraged by the nature of the German constitution, which applies a particular concept of inclusionary citizenship: as always, inclusion for some implies exclusion for others.

The third social exclusion mechanism (the failure to provide for group-specific needs) is in many countries closely related to governmental attitudes. In this respect there are considerable differences between different countries and a number of typologies have been suggested to describe the observed variations (for example, Entzinger, 1994; Castles, 1995). Most of the needs that are commonly identified (such as specific educational measures, access to public service broad-casting or the delivery of information services in minority languages) concern the state (local, regional or national) in some way. There is often a relationship to the nature of the state's welfare regime which, as Esping-Andersen (1990) has outlined, generally involves, in western Europe, a mixture of social democratic, corporatist and (neo) liberal elements (see Chapter 1). The Swedish welfare regime could serve as an example of the social democratic model, Germany as a corporatist example, and the United Kingdom of the Thatcher years as moving towards the incorporation of more (neo) liberal elements. It is only where social democratic approaches to welfare provision are predomi-nant, emphasizing the objectives of reducing inequalities and creating mechanisms of social inclusion, that policies have commonly been fully developed for the specific provision of social goods to ethnic minority groups (as also for other potential 'out-groups' in society).

There are, however, potential dangers in moving too strongly towards differentiated systems of social goods provision. Experience in the United States has shown that if the delivery of particular programmes (such as the implementation of welfare schemes) is shifted too far towards the community level and away from universalism, then policies cease to be class or income based and become class/ethnicity based. In these situations racialization is very difficult to fight against, and the affected ethnic groups become further labelled as 'distinct' and 'other' (Faist, 1995).

The fourth mechanism of exclusion is a more general one and relates to economic opportunities in which the fact of being drawn from a 'different' ethnic group may reinforce patterns of disadvantage that are experienced by wider groups affected by, for example, economic restructuring, social polarization or poverty relating to demographic circumstances. Inevitably, for ethnic minorities, this fourth exclusion mechanism may in part result from the operation of the other mech-anisms already described which circumscribe their opportunities further.

As part of the 'underclass' debate in a number of European countries, researchers have sought to identify whether there are any signs of an emergent ethnic underclass. In France, Wacquant (1993) has specifically related this issue to that of the marginalization of the great French social housing estates (the *grands ensembles*), but has suggested that the devel-opment of outcast status among the residents of such areas accompanies an addressism that applies to all residents of whatever origin and, as other commentators have emphasized, such estates are far from being accurately described as ethnic ghettos (Vieillard-Baron, 1994). Research in the Netherlands (Roelandt and Veenman, 1992) suggests that there is as yet no ethnic underclass, but that Turks and Moroccans (the two biggest 'foreigner' groups) are in danger of being trapped in future in a vicious circle of marginalization from which little upward mobility would be possible. Nevertheless, as a number of studies have shown, the material circumstances of foreigners actually tend to be quite good in the Netherlands, particularly because of the highly developed nature of state housing policy within (and beyond) a social housing sector that is among the biggest in Europe (Cortie and Kesteloot, 1998). In Germany, by contrast, in a situation with lower minority group access to a welfare state that has been historically based more on corporatist principles, the prospects for the creation of an ethnic underclass appear stronger, particularly in a situation of increased deregulation (Faist, 1993).

It is possible to debate the extent to which public policies of one sort or another, often related to the various mechanisms of social exclusion, contribute to the perpetuation of the 'othering' of ethnic groups, or to the potential generation of an ethnic underclass. In one very important respect state policies play a pivotal role in these issues, and that is through their definition of citizenship criteria, which may lie behind many other aspects of exclusion. Much political action among ethnic minority groups in a number of countries has been geared towards campaigning for the reform of citizenship laws to increase the rights of individuals whose citizenship status relates to their, or their ancestors', migration histories. The general trend of legislative revision in recent years, however, has been in the reverse direction (with some notable exceptions) with the acquisition of citizenship becoming harder rather than easier. This issue will be returned to later.

However, the emphasis on the role of legal citizenship as a vital element in reducing the distancing between ethnic minorities and wider societies must be questioned. The granting of citizenship as a precursor to full participation in civil society may be a necessary condition (although even that is not certain) but it is far from being a sufficient condition. Throughout Europe there are many who hold the citizenship of their country of residence and yet who are still disadvantaged through racist ideologies, discriminatory practices and circumstances prejudicial to their insertion in many aspects of economic and social life. The position of many ethnically defined sections of UK society testifies to this point: the vast majority of 'Indians', 'black Caribbeans' or 'black British' (among other groups) hold (and in many cases have always held) British citizenship, yet their continued degree of 'otherness' and exclusion is maintained in other ways.

The continued 'otherness' of even those ethnic groups holding citizenship should not, nevertheless, be related solely to conditions imposed by the 'host' society. Forces at work within the minority group may also be of some importance, leading to the retention of 'distancing' and to low levels of participation even where it is legally fully possible. Thus in Sweden, with a liberal naturalization regime by European standards, attitude surveys have shown that among Turks taking Swedish citizenship many do so for pragmatic reasons rather than as a result of a commitment to Sweden. A major factor is the ease that adding

Swedish to Turkish citizenship brings to the possibility of continued circulatory movements between the two countries (Icduygu, 1996). Similarly, despite the fact that foreigners with three years' residence in Sweden have had the vote in local elections since 1975, actual electoral participation rates have remained very low (Soininen and Bäck, 1993). In the Netherlands, research with those who had naturalized to Dutch citizenship has produced similar findings, and indicates that one of the important reasons for not fully feeling 'Dutch' is the fact that the respondents believe that, despite their new citizenship, they are still regarded by others as foreign and as immigrants (van den Bedem, 1994). Legal citizenship is not enough to guarantee 'inclusion'.

Issues of citizenship today potentially concern the international arena, at least within the European Union, where they inevitably become entangled with discussion over the freeing up of European internal borders. At the international scale there are also wider concerns over the possibility of the operation of anti-racist legislation and the creation of genuine monitored equal opportunities policies. Such legislation exists in a number of countries, for example the United Kingdom, France (Costa-Lascoux, 1994) and the Netherlands (Rodrigues, 1994), but not in others. The original treaty setting up the EEC prohibited discrimination on the basis of nationality, but this referred only to those holding the nationality of one of the member states (Vermeylen, 1993). Wider issues of racial or ethnic equality have not been tackled at the EU level, and they are not explicitly written into any of the social chapter provisions of post-Maastricht EU policies (Costa-Lascoux, 1995). Nor is this likely to change in the continuing situation where certain major countries (most notably Germany) continue to regard 'foreigners' as temporary residents excluded from the rights only applicable to 'citizens'. It is because foreigners remain juridically apart that there has been no general introduction of racial equality measures in the EU.

The European Union, via initiatives such as the Trevi Group, the Dublin Convention on asylum seekers and the Schengen agreement, has been relatively successful in moves to harmonize the ways in which immigration and asylum issues are dealt with. Indeed, the third pillar of the Maastricht Treaty is concerned with harmonizing asylum procedures and co-ordinating police activity to ensure the internal security of the Union. Such moves, however, highlight the 'problems' associated with the arrival of 'others' and reinforce elements of exclusion. Immigration legislation and laws concerning immigrants are not the same thing, but discussions of immigration policies inevitably foster the impression that immigration is generally not desirable, and that by retrospective implication the population movements that brought the immigrants of the past (and that created today's ethnic minority communities) were unfortunate. Problematizing immigration also leads to the problematization of immigrants, such that politicians and others often claim that a major cause of racist incidents lies in perceived high levels of immigration – which is tantamount to blaming the victims.

Recent European Union policies have arguably not just ignored the social exclusion of ethnic minorities (by failing to promulgate policies of racial equality) but have actually exacerbated other aspects of exclusion. In the drive towards the establishment of the single currency, many countries have adopted deflationary measures in an effort to ensure that budgetary deficits are reduced to below the threshold criteria agreed for the introduction of the euro. Unemployment levels have risen in many cases (most notably in Germany), disproportionately affecting ethnic minorities who remain concentrated in economic sectors that are vulnerable to recession. At the same time, reductions in welfare entitlements and deregulation in the labour market (for example, in France and Italy) have enhanced the income squeeze on the long-term unemployed (see Chapter 2), among whom are found many individuals of ethnic minority background (Faist, 1995). Finally, it is possible that these economic difficulties may be reflected in greater electoral support for extreme right parties (Baimbridge et al., 1994).

Excursus on central and eastern Europe

Issues of ethnic minority communities of recent migrant origin were of little significance in state socialist central and eastern Europe because international migration was unimportant. However, the period since the political upheavals of 1989–92 has brought about an increase in ethnic issues relating to migration in these eastern European states, resulting from a number of different factors (see also Chapter 4). However, by far the greatest problems of ethnic exclusion today concern long-standing settled minorities (such as Turks in Bulgaria, Hungarians in Romania and in the Vojvodina province of Yugoslavia, and gypsies throughout central and eastern Europe) rather than recent migrant groups. Nevertheless, certain elements of the western European scene are becoming established even here.

One important event has been the break-up of the former Soviet Union, which has resulted in the potential re-conceptualization of certain past 'internal' migration streams as 'international'. The 25 million ethnic Russians living outside Russia in the early 1990s, for example, have seen their citizenship status threatened in a number of new countries, particularly the Baltic states (Brubaker, 1992a). Inter-ethnic attacks have also been reported, indicating the likely existence, or development, of patterns of racist attitudes analogous to those observed in western Europe, except that those subjected to such attacks were not originally seen as migrants. Particularly in Estonia and Latvia, there have been repeated suggestions from local politicians that citizenship of the new republics would only be automatically granted to those who could prove descent from individuals who lived in the inter-war independent states, something that would be impossible for the vast majority of Russian residents who arrived during the post-war period (Aasland, 1996). Migration flows are currently operating in such a way as to reduce the impact of any exclusionary measures or racist developments that might occur in the future: between January 1993 and June 1996 there was a net flow of 127,000 people from the Baltic states to the Russian Federation, over 90 per cent made up of Russians (UNECE, 1996). It is notable that while some of the smaller states emerging from the break-up of the former Soviet Union have flirted with exclusionary definitions of citizenship, the new constitution of the Russian Federation itself, approved by referendum in December 1993, refers to 'We, the multi-national people of the Russian Federation' (Dixon, 1996).

The break-up of Yugoslavia also created the conditions in which significant ethnically defined populations were left in the 'wrong' country and thus became susceptible to harassment and exclusion. The 'ethnic cleansing' that has occurred in the Bosnian civil war, and in parts of Croatia, has been a particularly intense manifestation of historic and long-standing processes of racialization giving rise to precise exclusion mechanisms. A significant outcome has been mass population movement, in the form of refugee flows, to leave behind 'purer' ethnic territories.

Elsewhere changes in international migration regimes following the political liberalization of central and eastern Europe have brought large numbers of new migrants to a number of countries, and especially to their capital cities. For example, between 1989 and 1995 the proportion of foreigners in the total populations of Prague and Budapest rose from close to zero to around 10 per cent (Prague) and 5 per cent (Budapest). A proportion of these are made up of western professionals, but large numbers consist of central and eastern Europeans and Asians, and of illegal transit migrants who have failed to find a route into western Europe (UNECE, 1996). There are already signs of labour and housing market segmentation occurring, with clear elements of illegality in various activities. The pre-conditions for the development of exclusionary processes appear to be in place. However, the actual trajectories of exclusion could well be more drastic than those applying in western European countries over the past 50 years. In particular, the arrival of migrants in Hungary, the Czech Republic, Poland and elsewhere does not relate to buoyant labour demand there (unlike the position in western European immigration countries in the 1960s and early 1970s) but rather to even poorer labour market conditions elsewhere (see Chapter 7), coupled with restrictions on movement further west. In addition, the period since the fall of communism in these states has been marked by a resurgence of national identity, with consequences for the likelihood of future exclusionary tendencies for those who do not 'fit'.

Responses to the presence of ethnic minorities

It is clear that a number of processes are involved in the 'othering' of ethnic minority groups of migrant origin. 'Othering' need not necessarily involve social exclusion, yet in many cases this appears to be an outcome. Certainly, a number of originally migrant groups, or individuals, have taken routes of upward social, occupational or economic mobility, but for many the greater likelihood has been marginalization and, in the circumstances of economic restructuring and the crisis of welfare states, the growth or perpetuation of disadvantage in a number of spheres of life.

It could be argued that the forces of social exclusion and the contextual issues in contemporary European economic and political life discussed earlier in this chapter are present and operative throughout the continent. However, there are important variations from state to state (just as there are in the philosophies of welfare regimes) which produce a series of varied responses and state objectives in the face of significant ethnic minority communities. In looking to the future of ethnic or racial divisions within Europe it is necessary to consider these different responses in further detail. Although most articulated through state policies (and therefore part of the first and third mechanisms of social exclusion discussed above), these responses are part of wider attitudes to 'others' that lie rooted in historical factors and political and cultural evolution at the 'national' scale and which relate to ideas of nationhood. Castles (1995) and Entzinger (1994) have put forward three models of response. To take Castles' labelling of these, they might be termed 'differential exclusion', 'assimilation' and 'pluralism'. Entzinger (1994) refers to the first as a 'guestworker' model, and the third as an 'ethnic minority' model.

Certain European societies have regarded the presence of immigrants and later ethnic minority communities in an extremely instrumentalist fashion, seeing them as fundamental to labour supply and also wishing to maintain them in that role rather than allowing their wider participation in society (which might result in their role changing). The countries involved have been those that have operated guestworker systems during the post-war period (and which are sometimes re-creating them today, as Germany is for Polish workers on short-term work permits). Castles (1995: 295) describes differential exclusion as 'a situation in which immigrants are incorporated into certain areas of society (above all the labour market) but denied access to others (such as welfare systems, citizenship and political participation)'. Exclusion then takes place both through legal mechanisms and through the delegitimization of the interests of the minorities, resulting in their racialization and discrimination and racism against them.

Germany is perhaps the classic case of this response to the presence of immigrants, and the attitudes involved can be traced back to the history of German nation-building in the nineteenth century. The formation of the German state from a fragmented set of smaller political units was crucially based on rhetoric stressing the unity of the German people (*Volk*) and an official view of national homogeneity (Kurthen, 1995). The roots of German citizenship, as indicated earlier, lie in a view of German-ness, regarded as a shared ancestral and cultural identity, such that the concepts of the German nation, and German citizenship, have ethno-cultural groundings (Halfmann, 1997). The paradox of such views has been well illustrated in two developments of recent years: the return to Germany of the *Aussiedler*, and German re-unification in 1990.

The return of the *Aussiedler* from various parts of central and eastern Europe (chiefly Poland, the Soviet Union and Romania) showed that the German constitution, under Article 116(1) of the Basic Law (*Grundgesetzt*), and the German state were more prepared to give favourable treatment to people of German 'origin' (often dating back several centuries, and with individuals unable even to speak German) than to young people born in Germany, brought up in Germany, speaking German, having gone through the German education system, but whose parents or grandparents happened to have migrated from Turkey or elsewhere before their birth. The fact that vernacular German attitudes to the arriving *Aussiedler* were not always favourable raised obvious questions about the nature of German views of the historic basis of citizenship (Herdegen, 1989).

Re-unification in 1990 added a further strain on the concept of German homogeneity, when people brought up through nearly 50 years of completely separate political, economic and cultural development were brought together under an umbrella definition of unified citizenship but with, in reality, a number of schisms and different views of nationhood (Kurthen, 1995).

Partly in response to these obvious paradoxes, German naturalization law was relaxed between 1990 and 1995 to permit the acquisition of citizenship on the basis of residence (effectively *ius domicilii* in place of the traditional *ius sanguinis* through bloodlines), and with easier naturalization for foreigners educated in Germany to continue beyond 1995 (Halfmann, 1997). The government, while disapproving of the mass of racial incidents that marked the period 1991–3 (Willems, 1995), could also cite the feelings behind these as 'proof' that German society was not yet ready for too drastic a reform in the laws on citizenship: in fact, opinion polls did not suggest any underlying increase in hostility at this time. However, breaking down the old and exclusive definitions of German-ness will not be easy (Vertovec, 1996a), and the continuation of the model of 'differential exclusion' appears assured in Germany, possibly more so than in other guestworker countries such as Switzerland where naturalization (although based on a very extended period for residence criteria) is actually easier.

The second major response to the presence of immigrants and ethnic minority communities is labelled 'assimilation' by both Castles (1995) and Entzinger (1994). This involves policies seeking to incorporate minorities into the host society through encouraging them to adapt. The host society itself is not expected to change. France is the major example of a European state (and society) officially taking this view (Commissariat Général du Plan, 1988; Mitchell and Russell, 1996). The basis is an idea of French-ness relating to a wider concept of citizenship than exists in Germany, but with a parallel feeling of exclusion at the edge. As Brubaker (1992b: x) has pointed out, '*Vis-à-vis* immigrants, the French citizenry is defined expansively, as a territorial community, the German citizenry . . . restrictively, as a community of descent.' The crucial difference between the French and German cases lies in the willingness of the French state to incorporate 'new' elements, as long as they conform to a view of 'French-ness', and to be born and brought up in France is a crucial route towards the acquisition of that characteristic (Hargreaves, 1995). In theory, French-ness is defined without reference to race or religion: thus the black French born in Guadeloupe, Martinique or Réunion (France's overseas *départements*) are as French as the scions of peasant families in Normandy or Limousin. One of the major figures in Le Pen's *National Front*, Huguette Fatna, is indeed a black French West Indian, who can campaign against the settlement of other black people in France (or for their repatriation) on the basis that they are not French by birth and have not assimilated to French cultural norms. French citizenship has historically depended on being born to French parents within the French state (wherever in the world that may be). The acquisition of citizenship, if one's parents were not French, has been straightforward for those living from birth to the age of majority in France (through the concept of *ius soli*), through an extended period of residence as an adult, or through marriage. A crucial clause in the regulations, however, explicitly states that one of the necessary conditions for naturalization is that the candidate must 'demonstrate his or her assimilation to the French community' (*justifier . . . son assimilation à la communauté française*). The latest legislation on this issue, dating from 1993, provides for a faster track to naturalization for those from Francophone countries who are therefore seen as more easy to assimilate (Costa-Lascoux, 1993).

A certain concept of France – *la France, une et indivisible* (France, united and indivisible) – lies at the root of the French rejection of the third type of response to the presence of 'other' groups: that of according them equal rights as ethnic minorities, or allowing 'pluralist' structures. The question on self-defined ethnicity in the 1991 British census is unthinkable in France, where the crucial issue remains whether a person is legally French or not. Elsewhere in Europe, however, official policies have sometimes sought to promulgate the acceptance of immigrant or ethnic minority communities as 'separate' entities. Such a policy of pluralism 'implies that immigrants should be granted equal rights in all spheres of society, without being expected to give up their diversity, although usually with an expectation of conformity to certain key values' (Castles, 1995: 295). This is a statement that has been accepted in multi-cultural policies in various parts of the world (such as Australia) but which has not been fully developed in any European country. Elements of the concept have, however, been appropriated and used in a number of situations, most commonly in the Netherlands and Sweden, but also to some extent in Belgium and the United Kingdom among other cases. It is also notable that in certain countries (the UK is an example) multi-cultural policies have been introduced more fully at local government than at national levels, with inevitable geographical variations.

The Netherlands has a long tradition of religious and cultural pluralism (relating to the historically 'split' nature of the country's society), and in recent years has developed extensive state policies promoting equal rights for minority communities irrespective of citizenship status, for example in housing and welfare rights (Entzinger, 1994). While calls for the recognition of Islamic schools have fallen on deaf governmental ears in the UK until recently, they are fully recognized in the Netherlands (Dwyer and Meyer, 1995). Multi-culturalism would appear, in many ways, as the most effective of the three models for combating or preventing social exclusion, when operated within a social democratic welfare regime, while also presenting minority groups with the possibility of the maintenance of elements of separateness and self-determination that are frowned upon by assimilationist outlooks. It is also important that multi-culturalism is flexible enough to recognize the emergence of 'new' ethnicities among established groups, for example as seen in the development of 'beur' (second generation North African) identities in France, or the 'black British' identification in the UK.

However, multi-culturalism is not without its problems. These are in part philosophical in nature. Just as assimilationist views appear to accept as ideal the situation where all minority groups and individuals lose their distinctiveness and merge into a 'national' unit, so pluralist or multi-cultural visions seem to legitimate the retention of distinctiveness and of division. The problem with both scenarios lies in the fact that vernacular attitudes to ethnic minorities are generally well established and are often negative. Thus, in the assimilationist model the movement from outsider to insider is qualitatively regarded as a change from an inferior to a superior status. Those who have not followed this course are subjected to exclusion. In the pluralist model, different groups may theoretically be accorded equal legitimacy, but this is not the way in which hegemonic power relations (which dominate in western societies) operate. Hegemony necessitates social control, and that implies the existence of norms and rules that are applied by certain elements in society and are to be accepted by other elements, among whom ethnic minority groups have an important place. Multi-culturalist programmes designed to foster equality have often excluded minorities from real incorporation into the wider public domain by segregating their interests and working against any 'cross-over' of activity between different elements in society (Vertovec, 1996b). Pluralism or multi-culturalism are strategies which, for their success in incorporating minority groups into wider society, require a readiness to see a transformation in the whole basis of that society. Such a willingness tends, however, to be selective and unequal in operation. Certain groups are accorded privileged status within such policies, while others are not.

Multi-culturalism may encourage fragmentation among minority groups, where each pursues its own agenda rather than acting together in pursuit of common causes, for example over citizenship issues. This suggests that a real pluralist policy can only be successfully implemented once there is legal equality over rights, accompanied by equal opportunities policies that are adequately monitored and acted upon. Where such general frameworks do not exist, ethnic mobilization throughout Europe has been very much group-specific and with little co-ordination between groups (Fijalkowski, 1994; Rex, 1994). A concern for local issues has been very common, and attempts to create bridges between such local groups via national organizations (such as *SOS Racisme* in France) have met with limited success. As Wihtol de Wenden (1995) has commented, localism has hindered attempts at wider integration.

Although the three models of state attitudes identified as differential exclusion, assimilation and pluralism (or multi-culturalism) are relatively clear cut, there have been inconsistencies in their application in individual countries, and there have also been shifts in recent years. In the United Kingdom, for example, various legislative measures since 1971 have introduced notions of 'patriality' (or lineage) into questions of citizenship and have therefore moved some way towards the German model of the 'nation'. In France, new regulations in 1993 tightened up the conditions for naturalization, particularly among young 'foreigners' born in France.

Yet, while tightening up on citizenship criteria in a manner that may enhance aspects of exclusion, at the same time the United Kingdom has fostered ideas of multi-culturalism and the legitimacy of ethnic difference through such everyday practices as ethnic monitoring in all aspects of public service provision (such as education and health) and in job applications, and in the generally more favourable (at least in recent years) media coverage of major ethnic celebrations such as the Notting Hill Carnival or Chinese New Year processions. An index of the difference between French and British official attitudes on multi-culturalism and assimilation can be seen in the comparison of the high degree of officially sanctioned (and financed) 'visibility' of Chinatowns in London and Manchester (with their oriental-design street furniture, road signs in Chinese and so on), and the absence of any such indicators in the much larger Chinatown of south-east Paris.

In central and eastern Europe few countries have as yet determined what might in the future be seen as a stable model of citizenship and the inclusion of ethnic minorities. The early post-communist period has brought strengthened discussion of local nationalisms, and in these circumstances the greater likelihood is of enhanced processes of othering and of the operation of exclusion mechanisms, particularly given the widening of income disparities and the increase in social polarization attendant upon the dissolution of long-established welfare regimes in the widest sense. Methods of ensuring inclusionary citizenship for migrant minorities are not high on the agenda, nor are they likely to rise in salience while migrant communities are small and new. However, the basis for problems of racialization and exclusion in the future is already being set. All forms of pluralism, whether political or ethnic, were off the agenda throughout the communist period: rapid transition to such systems is difficult.

Future issues: beyond formal citizenship

Political demands for citizenship rights for minorities may mislead in terms of what such citizenship might be able to deliver. The attitudes to minority citizenship of the three models discussed above are varied. Under differential exclusion the response is 'we don't want you to have it'; under assimilation it is 'you can have it on our terms'; while under pluralism the response is 'citizenship doesn't matter: we shall legitimate your presence (if we choose) by other means'. In each case it is the state that decides on the degrees of freedom available to the minorities.

In practice, however, for most members of ethnic minority groups of migrant origin in Europe, it is not citizenship status that is the prime determinant of everyday living conditions but their actual material circumstances as influenced by the existence of everyday racism (Jäger, 1992) and the discrimination that results in the continued restriction of opportunities for advancement and for self-determination. And boundary drawing between groups is supplemented by limited spheres of action dictated from within minority communities themselves. Nowhere are these more present than for young females of ethnic minority background. A number of studies have shown how the attitudes and aspirations of such girls are little different from those of their 'white' school friends, but also how such aspirations are likely to remain unfulfilled through the two sets of pressures just described: from outside the group and from in-group influences (Roux, 1991).

In seeking to reduce the ways in which ethnicity divides societies in contemporary and future Europe, there is certainly a need to incorporate state policies that do not themselves perpetuate or accentuate ideas of distinction and difference and which thus contribute to racialization and exclusion. Even at this level, however, there is no European consensus about what the ultimate objectives of state policies for ethnic minorities should be, as the three competing models discussed above illustrate. Although the word 'integration' is commonly used as a desirable outcome, what is meant is highly variable between different cases (Miles, 1993: 175–85).

Beyond consideration of formal citizenship and rights there is also a continued need to consider the extension of full substantive citizenship to all European residents (Hammar, 1990). Such a concept concerns the ability of individuals to participate fully in all aspects of society, and to enjoy the same overall standards of welfare as other citizens, taking a broad definition of welfare to cover health, education, work, income and quality of life (Roche, 1992). In order to achieve such participation in European societies, the prime need is for a reduction in all those processes that create or perpetuate 'othering'. The task is for the elimination of racism as a set of ideological prejudices towards individuals or groups on the basis of their supposedly identifiable characteristics. Equal legitimation is needed instead.

However, this is a particularly difficult task to envisage in Europe. European national history has been built up on processes of identification of in-groups and others, and the strong currents of resistance throughout the continent to the real deepening of any identification of a pan-European identity demonstrate the continued vibrancy of the historical legacy. The continued sidelining of Turkey's application for membership of the European Union partly reflects the perception of Turkish society as consisting of un-European 'other' influences. The emergence of a pan-European identity would anyway carry a further danger for Europe's ethnic minorities, many of whom would not 'fit' any image of Europe as white, based on Christian values, and with particular shared elite cultural traditions in literature, art and music.

The issues surrounding ethnic divisions within European societies are therefore not to be regarded on their own or as working autonomously within each separate country. Although certain aspects of racialization in the recent past have been similar throughout Europe, the historical and evolutionary circumstances of individual countries, along with state policies, have also been crucially important. The future inclusion or exclusion of Europe's ethnic minority communities derived from international migration will be conditioned not just by these past factors: such issues are today also intimately bound up with questions of the future of the whole of Europe as a (series of) political, cultural and social unit(s).

10

Divided Responses to an Ageing Population: Apocalyptic Demography, Ideology and Rational Social Administration

Anthony M. Warnes

This chapter examines the response of European governments to the problems of financing social welfare programmes during the past quarter century. It focuses particularly on their response to the rising cost of support for pensions and other old age income benefits. The escalation has led to widespread policy debates and media discourse of varying levels of sophistication. In simpler treatments, an important attributed cause is 'demographic ageing'. The chapter examines the appropriateness of this explanation and some of its negative consequences on the regard for older people within Europe. Recent reforms of social security are also evaluated.

Several divisions within Europe are revealed in the analysis, the starkest being between the 'problems of maturity' in the social welfare programmes of north-west European countries, and the problems of replacing the collapsed social welfare arrangements of eastern European countries after 1989. Another dimension is the stage of development of the social security systems of northern and southern European nations, and the difficulties of achieving convergence (or harmonization) during periods of poor economic growth. Finally, there are contrasts among European countries in their receptivity to the rampant wave of neo-liberal economic thought.

The chapter has three main sections. The first sets out the context, presenting some details of the ageing of European populations and the nature and size of their social welfare problems. The second examines two aspects of the response: the policy discourse and specific policy and administrative measures. The final section discusses the appropriateness of these analyses and responses, and seeks to clarify the principal policy requirements for promoting the welfare of older people.

European population ageing: causes and timing

A profound age structure change has been taking place for a century in Europe and a further surge in the growth of the absolute and relative size of the older population will undoubtedly occur during the first quarter of the next century. The transition has been described and explained many times. The first wave of academic accounts and official inquiries appeared in Europe and America during the 1930s (see Clark and Spengler, 1980: 3–5). The first international comparative study, commissioned by the United Nations Organization (1956), appeared over 40 years ago. Such studies are now commonplace (for example, Council of Europe, 1985; UNO, 1985, 1993; OECD, 1988b; Commission of the European Communities, 1996d).

The century-long ageing of European populations has resulted from the succession of the long period of above-replacement fertility during the nineteenth century by, from the 1920s, generally much lower birth rates. At times these have fallen below replacement levels. Until recently, each ten-year birth cohort reaching 60 years of age has been larger than its predecessor, and as birth rates have fallen, inevitably older people have formed a larger share of the total population. This underlying foundation for the growth of the older population will evaporate after 2040, as the low birth cohort of the 1970s attains its seventh decade.

The decline in fertility has been irregular, with variations by nation and region (Chesnais, 1986; Coale and Watkins, 1986). Two stages have been most widely shared: a fall from the last quarter of the nineteenth century to around replacement levels by around 1930, and a sharp fall in the early 1970s after two decades of relatively high birth rates. By the 1980s, in most countries of Europe, birth rates were lower than ever previously recorded. There are, however, considerable differences in the degree to which European country populations have already aged, and in the likely rate of increase of the older population over the next three decades (Figure 10.1) (Warnes, 1993a). In the countries of north-west Europe, the share of the population aged 60 years and more is projected to reach 25–30 per cent by 2025, and many southern European countries, presently undergoing more rapid growth, will reach similar levels. In most central and eastern European countries, the 60+ years share is expected to reach only around 20 per cent (Velkoff and Kinsella, 1993: ch. 3).

A factor of growing influence during the past 30 years has been accelerating falls in mortality in the older age groups (a large share of earlier mortality improvements was accounted for by reduced infant death rates) (Caselli, 1994; Caselli and Lopez, 1996). Many commentators believe the trend will continue, and so improving survival in old age will become more important as a cause of age-structure change, but recent trends in British adult mortality indicate irregular progress.

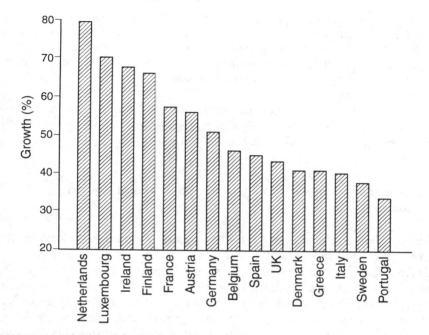

FIGURE 10.1 *Anticipated growth in the 60+ years population in the EU, 1995–2025 (after Commission of the European Communities, 1996d: fig. A)*

According to OPCS (1993: 8), mortality rates are tending to rise at present for men between the ages of 24 and 46, largely due to deaths arising from HIV infection, but there are also increasing numbers of suicides and accidental deaths. The cohort change is sufficiently established to have been adopted in the assumptions of the official UK population projections.

Despite the wealth of data and the detail of the reports, popular and policy references to 'demographic ageing' not surprisingly simplify the process and represent it, if only by default, as a new but relentless phenomenon. An alarmist attitude to ageing is also widespread, by which the growing number of old people and lengthening survival are 'problematized', as in the frequent references to the 'grey tide'. The reasons are to do with macro-economic management, fiscal policy and government expenditure, and disputes about the role of the state in providing social protection, raising social welfare and 'nation building', but both the financial strains and the policy disputes arise from profound social, economic and political changes. Either to blame or to regret the lengthening of life is perverse, as Richard Titmuss (1955) and others commented long ago:

It is difficult to understand why the gradual emergence . . . of a more balanced age structure should be regarded as a problem. What we have to our credit . . .

is a great reduction in premature death . . . Paradoxically, however, we are alarmed by our success . . . we speak about the 'crippling' burden of old age, forgetting that the (former) youthful structure entailed a phenomenal growth in numbers and was accompanied by great losses from morbidity and mortality among children and young people.

Given the consensus that to reduce world population growth is desirable, and that a long life and the absence of disease are worthy aims, to lament that 'we are all living longer' is obtuse.

Representations of societal ageing in popular and policy discourse

The greatest financial impact of a more aged population on government expenditure is through state-managed old age income support and payments for health services. Age-structure change has altered the ratio of social security contributors to beneficiaries and corrective changes are required, but the change is not catastrophic and is often exaggerated. For example, in the EU, the age-dependency ratio will increase by 50 per cent between now and 2020; with existing rules and benefits, this will absorb an estimated 5–7 per cent of GDP (Commission of the European Communities, 1993a: 24). Population ageing does not automatically imply crisis:

> In part, the cost of ageing depends on long-run productivity growth [and] is subject to political management. Many countries today are reversing a decades-long policy of lowering retirement age . . . it is decisive whether social policy encourages low female employment and early retirement (as in the EU nations), or maximum participation (as in Scandinavia) . . . It is also decisive whether, as in southern Europe, the incidence of informal, black market employment is high and growing. The spread of irregular work in countries like Italy [forms] a negative spiral: the heavy social burdens incurred by over-loaded income maintenance programmes stimulate informal employment which, in turn, further erodes the tax base. (Esping-Andersen, 1996b, 7)

The problematic rendering of the issues has been more common in the United States than in Europe, partly because its rates of anticipated elderly population growth are high, partly because the federal entitle-ment programmes have been generous, and partly because the con-stituency for a minimal government role in social welfare is powerful. American attitudes have been infectious, particularly among and through international agencies, several of which are US-based and even, as in the case of the World Bank, US-controlled (Walker, 1990).[1] Demographic analyses were among the earliest contributions to fan the alarmist American flames, largely through the translation of age-group projections as economic dependency ratios (Preston, 1984). A recent apocalyptic statement from Lester Thurow (1996a) was of sufficient public interest to be reprinted in the *New York Times Magazine*.

A new class of people is being created. For the first time in human history, western societies will have a large group of affluent, economically inactive, elderly voters who require expensive social services like health care and who depend upon government for much of their income. It is a revolutionary class, one that is bringing down the social welfare state, destroying government finances, altering the distribution of purchasing power and threatening the investments that all societies need to make to have a successful future . . . As the numbers of elderly rise, pensions costs will soar and health care spending will explode, particularly as we employ expensive new technologies to help the elderly live even longer.

Thurow's longer statement adds 'we have met the financial "enemy" and he is the elderly "us" in both the public and the private sector.' This ranting is supported mainly by references to similar doom-laden articles in the news magazines *Business Week*, *Fortune* and *USA Today* and the Washington newspaper *International Herald Tribune* (Thurow, 1996b). Similar 'dire projections' have been made by Daniel Callahan (1987, 1994) in a sustained argument that our societies cannot afford to provide open-ended high-technology medicine to people in advanced old age.

Some international agencies have, however, recently adopted a less problematic stance on 'ageing', and with reference to older people now specify particular policies for attention. An OECD (1996) conference issues publication, *Beyond 2000: the New Social Policy Agenda*, begins, for example, with the assertion that 'policies have been influenced by a feeling that the strains on the social protection system are the result of economic disequilibria. The evidence does not support this view' (1996: 2). It continues:

Old age no longer implies low incomes. Current pension arrangements and asset accumulation have significantly reduced old-age poverty in most countries, but the burden on the working population of supporting the elderly is becoming painfully high. A majority of the elderly will continue to depend on public pensions, but reform is required in countries were even those with high earnings are provided with pensions in excess of their contributions . . . An immediate issue is the extent to which the financing of the current pensions systems (including disability and early retirement benefits – which finance 'premature ageing') are 'crowding out' arrangements which enable parents to combine child raising and careers. (OECD, 1996: 2, 19)

The economics and politics of social welfare programmes

Delivered social policies are a compromise between the ideal, the possible and the partisan. They often include objectives such as 'eliminating poverty', 'creating equal opportunity' and 'eliminating discrimination', but their implementation is always high cost. A common component is to raise the income of an economically inactive or disadvantaged group, and 'making a difference' is achieved only through substantial expenditure. There are always calls to do more, and as the level of living of one group is improved, the claims of others are raised. Partisanship, or patronage and coercion, also come into play because, in any set of expenditure choices, the values of the group in power are revealed. A disinterested, 'technical' consensus on the priorities of social administration is impossible. Rather, social welfare has become a leading battleground of ideological contest, and the struggle focuses on social security (Klein and O'Higgins, 1985; Esping-Andersen, 1996a).

This is because both class-based interests and macro-economic principles and practice are engaged (see also Chapter 1). First, large social security funds are a major factor in taxation and redistribute income among socio-economic and age groups, between the sexes, among regional populations or across the life course of individuals. Often the redistribution is intended but there are many unintended consequences and, whichever way, some groups benefit and others pay. Social security schemes are financed unequally by individuals (employees) and organizations (employers), and by different income, age and gender groups. They inevitably excite self-interest and encourage struggles between the major interest groups in the resolution of fiscal and welfare policy. Explicit contests between advocates for children and 'the elderly', or between generations, became prominent during the 1980s in the United States, Britain and New Zealand (Quinn, 1996). Competing lobby groups formed, as with Americans for Generational Equity and the National Committee to Preserve Social Security and Medicare (Pampel and Williamson, 1989; Thomson, 1996).

Secondly, many schemes are so large that they are important in macro-economic management and require levels of employer costs that restrain the supply of jobs. Some ideologues see the promotion of economic growth and modernization as the leading responsibility of government, and regard the diversion of substantial sums to state-managed funds as damaging because it either reduces the sum available for investment in production or increases the cost and rigidity of labour. Others object because large social security budgets imply high personal, corporate and sales taxation, which they believe is a disincentive to effort, enterprise and economic growth. Proponents of the privatization of pension funds point to the growing disparity between the allegedly low yield from state-managed, pay-as-you-go schemes and dividends from equities. It is not, however, straightforward to calculate the former because of the long periods involved, and the rate varies for different cohorts. As the turbulence on world stock markets during the last quarter of 1997 highlighted, the greater risk of investment in stocks and shares, and the possibility of substantial falls in capital value, must be weighed alongside the possibility of higher returns.

Expanded welfare programmes are only likely during favourable periods when government revenue grows and expenditure decreases, as during periods of high economic growth. More generous programmes sow the seeds of later budget problems, as when an economic recession follows. Sophisticated social security schemes therefore build in the capacity to vary contributions and benefits, but the responsiveness of even flexible schemes is poor because electoral considerations restrain the extent of cuts. The more that an administration has to respond to special-interest constituencies (and the media pressures they incite), and to short-term electoral advantage, the more difficult it becomes to sustain a long-term, collectivist and utilitarian social policy.

It is the great disadvantage of spending on old age income support (and on health services for older people) to be in the small category of spending programmes for which forecasts beyond two decades have substantial plausibility. By contrast, capitation-led changes in secondary education spending can be anticipated no more than 11–18 years ahead, while even the most sophisticated macro-economic models claim little ability to forecast the level of unemployment beyond five years. A quantitative basis for long-term forecasts of contributions and expenditure and the accumulated deficit or reserve relies on simplistic conversions of the numbers in the working and the retirement ages into 'old age dependency ratios'. The economically active rates of the older working age groups are, however, unstable and have recently been subject to two clear effects. First, the rates rise and fall with greater and lesser economic growth, the variations being greatest among the youngest and oldest workers (Table 10.1). Secondly, the recent trend has been for male economic activity rates to decrease during the last third of the average life. Until the 1960s, the falls were mainly in the recognized (or statutory) retirement ages, but more recently there have been sharp decreases in workforce participation among men in their fifties (see Chapter 2). The trends for women have been more complex, but their economic activity rates in late working age have recently been decreasing (see Chapter 8). The decreases were slowed but not reversed during the 'boom' of the late 1980s.

The ability to make predictions of pension spending so far ahead makes it a common target for those who wish to undermine support for social security. A shrill chorus of alarm demeans the worth of older people to the extent that many self-identify as burdens on the state (Warnes, 1993b). The intensity of the debate in the United States was highlighted in January 1997 by the inability of the cross-party 1994–6 US Advisory Council on Social Security to agree a strategy for maintaining the fund in surplus after 2029 (Kleyman, 1997; Thomas, 1997). Five of its members recommended only modest changes, including a gradual increase in the 6.2 per cent level of employee contributions and some tightening of eligibility for benefits. A second group backed an increase in contributions to 7.8 per cent of earnings, with all the additional funds

TABLE 10.1 *Economic activity rates of older men and women in the UK, 1951–94*

	1951	1961	1971	1975	1981	1985	1990	1994
Females								
55–59	95.0	97.1	95.3	93.0	89.4	82.6	81.5	76.1
60–64	87.7	91.0	86.6	82.3	69.3	55.4	54.4	51.2
65+	31.1	25.0	23.5	19.2	10.3	8.5	8.7	7.5
Males								
55–59	29.1	39.2	50.9	52.4	53.4	52.2	55.0	55.7
60–64	14.1	19.7	28.8	28.6	23.3	18.9	22.7	25.6
65+	4.1	4.6	6.3	4.9	3.7	3.0	3.4	3.2

Source: United Kingdom censuses and UK Labour Force Survey; 1951–90, compiled by Walker and Maltby (1997: table 1)

going into trust funds. A third group proposed radical reductions in benefits and much more means testing, raising employee payments to 6.96 per cent, and putting all the workers' contributions into a personal account. As Thomas (1997: 7) puts it, 'with more young Americans believing in the existence of UFOs than in the chances of receiving social security on retirement, the pressure to act is mounting.'

The strident attack on state-managed social security has produced a spirited reaction. Complaints about the simplistic reasoning of many analyses of demographic ageing and pension funding have been prominent in 'critical gerontology' for at least a decade, and use of the term 'apocalyptic demography' now signals the author's disparagement of the usurpation of social policy for economic ends (Walker, 1990; Robertson, 1997). Strong defences of the welfare efficacy of social security schemes are also evident (Hills, 1993), as well as contributions from the perspectives of older people. For example, the European Union has encouraged several organizations that represent older people's interests and established an Observatory on Ageing. One result has been the following welcome but unusual alignment of ideas: the comments 'ageing must not be confused with a decrepit society [and] in no way assimilated to the concept of regression occasionally associated with biological ageing' are followed by complaint about the very low birth rate when 'Europeans want more children than they actually have' (Commission of the European Communities, 1996d: 5). One may expect to hear more pro-natalist expressions which will at least divert the blame for adverse dependency ratios towards young adults and their attitudes to procreation.

Old age income support in Europe

France, Germany and the Nordic and Benelux countries established the earliest modern welfare states, and after the Second World War most western European countries considerably developed their social security systems (Gordon, 1988). Their long evolution and variations in the economies, religious affiliations and social structures of European nations have produced considerable diversity not only in entitlements and financing but also in broad principles and goals. Convincing generalizations are elusive but five models can be described. Esping-Andersen's (1990, 1996b: 10) typology of three social welfare designs in western Europe has been widely adopted (as in modified form in Chapter 1).

The Nordic or Scandinavian model stresses social inclusion and income support for all households especially those with children, producing exceptional benefits for female-headed households and young adults. The model depends upon high labour-force participation among women and low private-sector supplementation of pensions and health care. Esping-Andersen's 'continental European model' organizes state support largely through the full-time, long-term employed male householder, and has been generous neither to the unemployed nor young independent or maternal females. Pensions are managed through employment-based social insurance schemes, with private-sector supplementation, and therefore produce greater income inequalities among older people than the Nordic schemes. Many analysts subdivide the group between the 'Bismarckian' or Germanic states, where a wide range and high values of social welfare benefits have been developed but fewer social services than in the Nordic group, and the poorer 'southern European' countries, where both services and benefits are relatively few but family and church-organized provision is comparatively high. The difference may be one of stage of development rather than principle or goals, and related to these countries' earlier stage of industrialization and lower national wealth. Italy is a special case, closer to the German model than that of its neighbours. The fourth group is the 'Anglo-Saxon' model of Britain (and Canada, the United States and New Zealand). It is distinguished by fewer (and reducing) entitlements during sickness, pregnancy, early retirement and disablement. The trend is towards an increasing role for private-sector insurance particularly for pensions and health-care costs, with the state role progressively targeting benefits and moving towards the maintenance of a 'safety net' rather than social security or state-managed insurance schemes to raise material standards.

There are two kinds of state old age income support in the EU: basic pension schemes and supplementary benefits. 'The dominant pension model couples an earnings-related basic pension with a voluntary occupational one' (Walker and Maltby, 1997: 45). The main exceptions are Denmark, Finland and Sweden, which have universal flat-rate pensions that are provided as a right of citizenship (defined in terms of residence) and without taking into account employment records, while they also have relatively weak private occupational pensions. Hansen (1997) has summarized further contrasts in the provisions among northern European countries (Table 10.2). This brings out the relatively selective

TABLE 10.2 Public pension schemes in six northern European countries, 1995

	Denmark	Sweden	Finland	Germany	UK	Netherlands
Basic public pension scheme						
Statutory pension age (yr)	67	65	65	65[1]	60/65[2]	65
Flexible retirement	No	Yes	Yes	Yes	Only delay	Yes
Eligible groups	All	All	All	Employees, some self-employed	All[3]	All
Pension dependent on work history/income	No	No	No	Yes	No, but level yes	No
Pension dependent on marital status	Yes	Yes	Yes	No	Yes[4]	Yes
Means testing	Partly	Partly	Partly	No	No	No
Supplementary public pensions						
Exists	Yes	Yes[5]	Yes	No	Yes	No
Eligible groups	Employees	Employees, self-employed	Employees, self-employed	–	Employees	–
Pension dependent on work history/income	Only work history	Yes	Yes	–	Yes	–
Net replacement rates[6]						
Former single worker	56	67	67	76	53	47
Former working couple	56	69	69	72	58	43
Former non-worker	50	41	37	0	16	47
National cost of public pensions[7]						
As % of GDP	9.9	11.6	10.3	10.8	9.5	9.8
As % of govt spending	24.8	28.1	34.3	34.4	24.1	17.9

[1] See section on Germany later in chapter for further details.
[2] 65 years for men; 60 for women, to be increased to 65 over ten years starting in 2010.
[3] The basic pension for those who have been working differs from the income support for those who have not.
[4] Only the non-contributory element (for people who have not previously been working) is dependent on marital status.
[5] A new pension scheme will replace the existing arrangements, with gradual implementation from 2001.
[6] Replacement relative to the annual disposable income of the 'average production worker' (or non-worker or combination).
[7] Refers to individual years between 1985 and 1992.

Sources: World Bank (1994: table A.5); Hansen (1997: tables 2.8, 2.9)

coverage of the German scheme, and its greater emphasis upon an income-related pension as the foundation (or basic) scheme. The UK scheme has a similar emphasis upon income or work history benefit levels but has a more developed safety net for non-contributors (although the real value of the basic and widow's pensions has considerably reduced since 1979).

The fifth model is of the central and eastern European countries, of chaotic and unstable social security arrangements. They are experiencing 'an era of socio-economic experimentation in which all forms of "welfare system" are being considered and to some extent tried, whether by design or [through] *ad hoc* responses to specific crises and conflicting objectives' (Standing, 1996: 225). The background to this has been the debates concerning how best to improve social protection for the increasing numbers in need, while reducing total social expenditure in response to real or perceived resource constraints.

European responses to social security 'crises'

Rising unemployment during the 1970s sowed the seeds of actual and perceived crises in western Europe's state social welfare funding, although many other factors have played a part, including the pressures of increasingly global manufacturing and capital markets to increase labour productivity, and therefore to reduce either unit labour costs or employment (see Chapter 2). A crisis indicates an excess of promises over capacity: 'the dizzying levels of post-war economic growth are long gone' (Esping-Andersen, 1996b: 1) but, even so, real GNP has increased by 45 per cent in the wealthier OECD countries since the mid-1970s.

The resolution of social security tensions has been complicated by several special factors, some widely understood but some disguised. They include the collapse from 1989 of the legal and administrative social welfare frameworks of eastern European countries. Secondly, pressure has been growing among EU member states to harmonize not only national economic and government finance indicators but also social and health care benefits: the Maastricht Treaty raised the prospect of the southern states emulating the range of social benefits in the north. Thirdly, the wave of neo-liberal advocacy from the United States and Thatcherite Britain has questioned many precepts of the post-war European social welfare consensus and has particularly targeted the transition countries. While to date radical change has been resisted, more and more political interests in north-western Europe have come to accept, and even to promulgate, some of the underlying assumptions of the 'new right', that personal taxation is too high and must be reduced, that the range of welfare benefits has grown too large for the good of the economy and for the recipients, and that citizens should take more responsibility for making financial provisions for their own old age. The

TABLE 10.3 *Public preferences for alternative approaches to financing old age pensions*

	Mean rank[1]		
Policy option	Czechoslovakia	Netherlands	Switzerland
Abolish early retirement programmes	–	2.5	2.8
Raise monthly taxes or social premiums on income	2.6	3.0	3.6
Raise the retirement age from 65 to 67 years	2.6	3.0	3.6
Force children to support elderly parents financially	3.5	4.6	3.9
Make old age benefits dependent on number of children	2.9	4.8	4.1
Lower benefits payable to older people	3.8	4.0	4.3
Sample size (persons aged 20–64 years)	1,257	1,307	2,765

[1] Respondents were asked to rank the six options (five in Czechoslovakia) on a scale from 1 (for most preferred) to 6 (5) (for least preferred).

Source: Moors and Palomba (1995: table A.13)

complexities of each nation's schemes are well detailed in the official and academic literature (OECD, 1988a, 1994).[2] 'The Anglo-Saxon nations have favoured deregulation, but with varying degrees of commitment to equality. [Continental] Europe is bifurcating into a vaguely distinguishable renovation of the Nordic social democratic welfare state amidst crisis, and essentially "frozen" continental European welfare states' (Esping-Andersen, 1996a: x). Germany, France and Italy have induced labour supply reduction while maintaining existing social security standards. The 1990s have seen widespread measures to reduce early retirement benefits, the measure most favoured by public opinion (Table 10.3). To present more of the divided responses to the conjunction of ageing and economic recession, with the various complicating circumstances, attention is directed specifically to Sweden, Germany, Spain and Poland.

Sweden

The Swedish basic public pension (*basbeloppet*) is open for all, requiring only attainment of 65 years of age and a minimum residence period in the country. The supplementary earnings-related scheme for employees and the self-employed requires a minimum period of work (or contributions) and, in comparison to other nations, forms a large share of total pension income (Hansen, 1997: 10–11). Sweden's pension schemes are both generous by international standards and have exceptional 'reach', that is, they benefit virtually all older people in the country (Table 10.2). For this reason they have been intensively studied, while their 'crises' during both the 1970s and 1990s, when the social security account came under considerable strain, attracted much international interest (Rix and Fisher, 1982; Gordon, 1988; Thorslund, 1991; Daatland, 1992; Stephens, 1996).

Apart from precipitate increases in unemployment at the beginning of the two decades, one source of the problem was the adoption a generation ago of supplementary earnings-related pensions which turned out to be inadequately funded. These had been demanded by trade unions from the 1930s, but when it was proposed to introduce them through a state-managed fund, employers strongly resisted (Gordon, 1988: 49). After considerable controversy, including an inconclusive referendum and the fall of one government, new legislation was passed in May 1959. It provided for the basic and supplementary pensions to provide most workers with about two-thirds of their previous earnings (with a ceiling). The pensions could be claimed as early as 63 years, and could be deferred beyond 67 years. The normal retirement age was lowered to 65 years in 1976. Financing the supplementary scheme was entirely through an employer payroll tax on earnings (between a base and the ceiling), initially at 3 per cent but from 1981 at 12.5 per cent. In 1982 the calculation was changed to total earnings, reducing the rate to 9.6 cent.

Stephens (1996: 43–51) gives many details of the intricate measures that were taken by the coalition governments of the 1970s to maintain the pensions system and other aspects of social security, but budget deficits mounted. In 1982, the Social Democrats returned to power and made swingeing macro-economic changes to preserve Sweden's welfare state. There were currency devaluations of 10 and 16 per cent in successive years and a strong wage restraint policy was introduced. With the international economic upturn of the mid-1980s, a boom in profits and capital formation occurred and the government's deficit reduced, but around 1989 the boom ended. Many economic indicators quickly worsened: unemployment increased from 1.6 per cent in 1990 to 7.7 per cent in 1993.

Even before the downturn it had been clear that a funding shortfall faced the earnings-related tier of the pensions system, and both annual downward adjustments in payments and a cross-party approach to reform had begun. In 1993 the real value of pensions in Sweden was lowered by 2 per cent, and trimming occurred in the following two years. In 1995 and 1996, the inflation uprating was only 0.6 of an index of prices (Hansen, 1997: 39, 49). On reform, a consensus was reached that in the new system, enacted in 1994, 'each generation would have to pay for its own pensions' (Stephens, 1996: 46). A full pension will now require 40 years of contributions with no special consideration for 'best earning years', and the guiding principle determining payments has been changed from 'defined benefit' to 'defined contribution'. The reformed pensions will be funded equally by employers and employees at 9.5 per cent of the payroll (rather than exclusively by employers). As a consequence, the new system benefits full-time, full-career workers, that is predominantly men, and there are correcting 'pension points' for those engaged in childcare, higher education and conscript military service.

The Swedish approach to correcting its projected deficits in the state-managed social security scheme, or 'public pillar' to support pensions, has neatly inverted the alarmist response to the costs of an ageing population. It has both applied a wide range of adjustments, to contributions and payments, and put in place more substantial rule changes which raise the long-term revenue to expenditure ratio. In this way, it has preserved all the essentials of the pensions scheme, and maintained the prospect of rising real incomes in retirement. Vulnerabilities necessarily remain, and the long-term financial health of the arrangements depends upon a continuing relatively high level of lifetime employment.

Germany

In Germany, 'as in most European countries, the principles of the current pensions scheme lie in post-war reconstruction' (Casey, 1996: 36). German federal expenditure supports social insurance schemes and direct payments which fund pensions, health-care expenses, unemployment benefits and the long-term care of older people. The Pensions Act of 1957 established 'social insurance schemes' on Bismarckian principles (Table 10.2). Benefits are related to contributions, so pensions are redistributive across the individual life course but not among social groups. They cover 80 per cent of the adult population, including the self-employed but excluding civil servants (who are separately covered). Being earnings-related, German pensions have in the past perpetuated income differentials into old age. The highest income quintile among pensioners receives almost 30 per cent of total public pensions transfers. Pension ceilings have now been introduced, which will increase their redistributive effect but also encourage high earners to increase their holding of private pensions or investments.

Most commentators agree that the state's social security programmes are generous, especially in the encouragement to retire early. One scheme of 1984 encouraged retirement at 58 years but scarcely was it introduced than the labour minister discovered the 'demographic time-bomb' and 'was obliged to call upon [workers] to consider working later' (Casey, 1996: 38). The underlying problem was (West) Germany's very low fertility, which implied a reduction in the population from 56 to 42 million between 1990 and 2030, while the population aged 60 years or more would increase by some 4 million, increasing the ratio to the working aged 20–59 years from 0.36 to 0.81. The cost of Germany's social security is very high. Employers' social insurance contributions make up 42 per cent of wage costs, and pensions contributions alone will comprise 21 per cent in 1998 (Munchau, 1997b). Pensions payments are financed directly through contributions, supplemented by government subsidy which in 1997 reached DM68 billion. Expenditure on pensions accounts for one-third of all federal social expenditure.

Having recognized the need to act, a political process began to build a consensus for reform (Hinrichs, 1991). Casey (1996) directs attention to five of its components. First, pension increases were to be indexed to net rather than gross earnings, so that the previously rising (earned income) replacement rate of 70 per cent at the end of the 1980s would be held. Secondly, opportunities for early retirement were curtailed: the flexible retirement age of 63 years will be abolished by 2006. Thirdly, reduced pensions for early retirement are being introduced (and the increase of pensions for 'late starters' has been reduced). Fourthly, a partial pension has been introduced for those aged 62–64 years who continue to earn (and to accumulate pension rights). Finally, the state's contribution to the system will be linked to the growth of contributions in addition to gross earnings. The Pension Reform Act was passed by parliament on 9 November 1989. That very evening the East German authorities opened the country's frontiers and the Berlin Wall was breached, eclipsing interest in pensions for some years.

By the mid-1990s, with the worsening of Germany's unemployment (12 per cent in 1997), the reform of social security regained high political priority. Legislation of 1997 (to be implemented in 1999) accelerated reform but effected 'technical adjustments' rather than radical change. The 'net replacement' rate of the maximum pensions will be gradually cut from 70 to 64 per cent of net income. One aim is to reduce employers' contributions to less than one-fifth of wage costs. Consideration is still being given to raising the pensionable age.

Germany has its neo-liberal adherents to the World Bank (1994) recommendations that a country should not develop its old age income support arrangements entirely through the 'public mandatory pillar' of state-managed social security or social insurance schemes, and should encourage both a 'private mandatory pillar' of occupational and personal pensions with a true investment portfolio (on the model of Chile's reforms) and 'residual savings'. The Independent Institute for Economy and Society (IWG) in Bonn proposes replacing the current social insurance schemes with a (minimal) tax-funded basic pension and private pensions (Munchau, 1997a). But there is little support from politicians, employers or the social insurance institutes. To move to an equity investment based system would generate an impossible government bill for the pensions of older people who are presently and shortly to be retired (and who have few private investments). There are also legal obstacles to reform, for radical change would likely increase the living standards of one generation at the expense of another (today's retirees) and would risk being ruled out by the Federal Constitutional Court (Munchau, 1997a).

Casey's (1996) view is that the German reform has been one of the clearest attempts to adapt a state pension system to the requirements of population ageing, while Munchau (1997a: 5) believes 'that the present system is likely to survive, in reformed guise'. Germany seeks to restrict early retirement and marginally to reduce benefits rather than to privat- ize. 'The German pension reform was a coherent, but reactive, response to the perceived financial problems caused by population ageing. There is scope for governments to be active rather than reactive, to adopt positive labour market policies that prevent a worsening of the depen- dency ratio instead of policies that limit or spread the costs of an increasing pensioner population' (Casey, 1996: 42).

Spain

Until very recently, the normal form of material and instrumental support for frail or widowed older people in Spain has been co-residence with younger relatives. As recently as 1988, only 6 per cent of men aged 60 years or more lived alone, the lowest in western Europe (the com- parable percentages for Italy, Germany, Hungary and Denmark being 13, 15, 11 and 20) (Dooghe, 1991: 98, table 4.3). Among older women, 19 per cent lived alone, while in the comparison countries the shares were 32, 51, 30 and 47 per cent. Several reasons for the perpetuation of multi-generational households can be advanced, including the persis- tence until recently of low incomes, Roman Catholic family values and a large farm population (see also Chapter 6). Until the demise of the Franco dictatorship, other important structural factors were the paucity of community health and social services and the weakness of pensions. 'Care facilities for the young, the old and the disabled or disadvantaged, where they existed, consisted almost entirely of large, dehumanised institutions . . . referred to as asylums, a name which reflected their degree of isolation [of the clients] from the rest of society' (Ross, 1997: 205).

The changes in both social forms and social policy since the restora- tion of democracy have been substantial. By 1997, Ross observed that it had become far less common than it had been in the past for younger relatives to be living in close proximity. Even when they did so, in common with young people elsewhere in Europe, they were increas- ingly reluctant to take on caring roles within the family home. Ross's conclusions should be qualified, however, for changes in living arrange- ments are unreliable indicators of the willingness to care and may reflect a combination of increased affluence (allowing more housing space and independent living – the universal preference of all age groups) and the decline of farm or rural households. None the less, the increasing number of very old people and rising affluence will themselves lead to increasing demands for residential and nursing home care, both in the private sector for the higher-income groups, and supported by the state for low-income older people who are without family carers or suffering from physical or mental incapacities.

There have also been ambitious plans to develop social services. The post-Franco local authorities quickly established domiciliary provision and, to support and co-ordinate these efforts, the central government in 1978 set up the National Social Services Agency. Impressive programmes to develop residential homes, day centres and even supported vacations (comparable to subsidized spa treatments in Germany and France) have been developed. Shortages of funds have retarded their growth and some programmes, such as 'vacations for third agers', are provided on a very small scale (Moragas Moragas, 1991; Bazo, 1997).

Christopher Ross (1997: 198-9) has recently published an informative synopsis of the development of state old age income support in Spain. While the country's earliest pensions were introduced at the beginning of the century, only under Francoism did a genuine social security system emerge. However, the unusual circumstances of the development left a legacy of major problems. In 1970:

> flushed with economic success . . . reforms improved a number of existing benefits and extended coverage to new areas. In the medium term, their impact was disastrous. One reason was that they further accentuated the complexity and overlap which . . . characterised the Franco system. It included numerous special arrangements for particular groups of workers, as well as other anomalies. Consequently, while significant sections of the population had no social security cover, some individuals were receiving multiple pensions. (Ross, 1997: 199)

Another major problem of the 1970s reforms was their high cost, not least because they had failed to predict economic events. From 1973, the recession brought about a sharp fall in the system's income, which was derived solely from employers' and employees' contributions. Unemployment also increased rapidly, while the reforms had increased the benefits and coverage for this group.

By 1979, the system's outgoings exceeded income, and in 1986 the annual deficit was 24 per cent of the budget. The deficit gave rise to much talk of a 'crisis' in the social security system. The incoming socialist government of 1982 was left with 'little option but to cut back on some existing benefits' (Ross, 1997: 199). Given that pensions represented the largest component of state expenditure on social security, they were the obvious starting point for the reforms. In 1985 some pension payments were cut back, while the number of qualifying years (of contributions) was increased. In 1990 a Social Security Act was introduced, both to rationalize duplicative and generous benefits, and to increase the coverage of the scheme to disadvantaged groups. The

legislation introduced non-contributory benefits, including basic pensions for older people and the permanently disabled. Ross (1997: 200–1)
concludes that the:

> rationalisation has ironed out some of the system's structural problems.
> However, it has not resolved those caused by the high level of employers'
> contributions. Moreover, by increasing welfare coverage it has aggravated the
> underlying financing problem which is exacerbated by mass unemployment.

The social welfare system that Spain is building most closely follows
the 'continental European' model in that it is 'service lean' but 'transfer
heavy' and benefits are most generous for those with full-time employment records. There is comparatively little support for single-parents or
children, which Esping-Andersen (1996c: 78) argues stems from the
continuing strong influence of Roman Catholic teaching on family
responsibility and is implicated in Spain's exceptionally low fertility (the
world's lowest after Italy): 'The real demographic problem in continental
Europe is not ageing but low fertility and low activity rates . . . women
face a rather powerful trade off between careers and family formation.'

Spain's economy and social formations have been changing rapidly:
any improvement in the rates of employment and economic growth are
likely to accelerate the modernizing trends (Miguel, 1996). The interaction of more older people, *enbourgeoisement* and rising female employment will progressively raise the demand for institutionalized care for
nursery children and frail older people. Ross (1997) argues that the most
pressing demands which Spain's welfare services face result from the
rapid growth of the older population. The Spanish National Health
Service has rapidly developed a first-rate acute hospital network, and
primary care is improving fast, but there is relatively little provision of
domiciliary or residential community health and social care services –
the sector most oriented towards older people.

The pace of adjustment in Spain to new social and economic forms
over the past 30 years has been impressive, particularly given its persistent economic difficulties. As their pensions have been largely
unaffected by cutbacks since 1985, older people have been the major
beneficiaries of the expansion of social provision. The 'grey vote' is, as in
other western countries, an increasingly important factor. In the so-
called Toledo Pact of 1995, all the major parties agreed not to make
pensions an issue in the prospective general election. Once elected prime
minister in 1996, José María Aznar explicitly excluded pensioners'
benefits from the proposals to reduce public spending (Ross, 1997: 207).
Despite its economic weaknesses, a leading social policy achievement of
democratic Spain has been to raise the material standard of life of its
older population. The next challenge will be to raise the quality of
life and of extra-hospital care, particularly for older people of low
income and without family carers.

Poland

Not all countries in central and eastern Europe have similar economic indicators or approaches to the construction of a new social welfare system. The countries range:

> from the affluent enclave of Slovenia to the impoverished nuclear military states of Ukraine, Belarus and the Russian Federation. In the early and mid-1990s, the countries of eastern Europe, and in particular those of the Commonwealth of Independent States of the former Soviet Union, were in much worse structural crises than the main countries of central Europe, led by the Czech Republic, which by 1995 had minimal open unemployment and considerable economic dynamism, and Hungary and Poland, which had experienced very high levels of open unemployment and poverty, while having a dynamic private economy. (Standing, 1996: 225–6)

Poland is distinctive in other respects: more than any other country of the region its population affiliates to the Roman Catholic church and the population is ethnically and linguistically homogeneous.

Before 1989, Poland's social welfare system provided old age pensions, health care, family allowances and disability, work injury, maternity and sickness benefits (Les, 1992; Maret and Schwartz, 1994). Many social benefits were administered through the employing organization and, indeed, eligibility depended upon being an employee. The system, as in the other Comecon nations, was 'service heavy' and 'transfer light', pensions being the main exception to this rule. Their provision was through a virtually exclusive public system (private pensions did not exist) of three social insurance institutions; for farmers, for the military and the police, and, by far the largest, for the general population. By the 1980s a system had evolved which encouraged early retirement but also paid retirement pensions at a relatively young age and to those who continued working. There were several other systemic problems:

* liberal eligibility for early retirement and disability pensions and generous stipulations on the right to receive pensions while employed;
* insufficient financing;
* weak administration;
* high average replacement rate (of former earnings);
* poor cost of living or inflation adjustments;
* persistent structural abnormalities, as with occupationally specific privileges, for example, early retirement for people in specific industries (Maret and Schwartz, 1994: 58).

Whatever the merits of the system during the early post-war decades, during the 1980s the real value of social security benefits, and particularly pensions, declined substantially. Rapid inflation occurred but pensions were under-indexed. After 1989, with the collapse of many state enterprises, the basic administrative framework for the distribution of benefits no longer existed, and measures were rapidly introduced to provide basic income support. Similar crises soon affected the other eastern European countries, attracting the strong prescriptions of the World Bank (1994). In Poland, as in the rest of central and eastern Europe, pension reform cannot be divorced from the 'relentless external pressure to cut back the state, and it is in that context that variants of the "Chilean model" have been advocated' (Standing, 1996: 241). The World Bank advocates a 'two pillar' approach, in which a low state-managed, tax-funded, basic pension is supplemented by mandatory earnings-related private or occupational pensions (with the contributions from the second tier invested in equities and therefore not available to pay current benefits). It argues against the 'continental European' social insurance approach.

'The revolution that has been taking place in central and eastern Europe is the first in history in which social policy has been shaped and influenced by international financial agencies' (Standing, 1996: 230). Standing argues that this is not necessarily a criticism, but instead recognizes the pressures on governments in the region. However, others are more critical, for the World Bank recommendations, while strong on sound macro-economic management, neither appreciate the demonstrated achievements of social security funds in western Europe in reducing poverty and inequality, nor have much time for the Bismarckian or Nordic goals of promoting social inclusion and 'nation building' (Lloyd-Sherlock and Johnson, 1996).

During the 1990s, subsidies and in-kind benefits were sharply cut back, while cash benefit payments increased rapidly, especially for pensions and unemployment; as a result, the increased financial burden on the social insurance system made it 'basically unsustainable' (Maret and Schwartz, 1994: 53). Pensions were indexed to average wages in 1990, and radical changes to the eligibility rules and the calculation of benefits were legislated in 1991. The number of pensioners has increased and the costs have spiralled, by 1993 reaching 15 per cent of GDP and approaching one-half of all government expenditure (Table 10.4). In an attempt to contain the mounting deficits, in 1994 the pension age was raised to 65 years for men and 60 years for women.

The partial privatization of pensions implies that the social policy goal is to create only a 'residual social safety net', which targets resources on the most needy, and leaves individuals themselves and capital markets and financial institutions to raise the level of living of most old people out of subsistence. As Standing (1996) emphasizes, there is a strong likelihood that pensions systems will become an important source of socio-economic differentiation in the next few years. 'The threat is that there will be persistently high levels of poverty, accompanied by widening inequality . . . [however] the opportunity is still there to create

TABLE 10.4 *Pensions expenditure in Poland as percentage of GDP, 1988–93*

	1988	1990	1991	1992	1993
Expenditure on pensions	7.1	8.1	12.2	14.8	15.0
Transfers from state budget (all outputs)	1.4	2.8	4.4	6.3	6.6
As percentage of total expenditure	14.6	34.0	35.5	42.7	43.8
No. of pensioners and contributors[1]			1989	1991	1993
Millions of pensioners			2.2	2.8	3.0
Millions of contributors			14.7	13.8	13.0
Ratio of contributors to pensioners			6.7	4.9	4.3

[1] To the general social insurance scheme (not the farmers' or military schemes).

Source: Maret and Schwartz (1994: table 1 and figure 1)

a new system that builds on the sense of social solidarity and desire for redistributive justice that are still prevalent' (Standing, 1996: 251). But the newly rich have tended to opt out of the old and the new social security systems, and there is a danger that many will cease to see themselves as stakeholders in the welfare system. In practice, the debate is finely balanced. While there are proponents of the western European models of social insurance, others reject this route because of its reliance on what is likely to be a declining industrial employment model, yielding dwindling resources, and a host of economic weaknesses – in labour productivity and the strength of the informal (and fraudulent) sectors.

Discussion and conclusion

From this short analysis of the imperatives for social policy in Europe, it can be argued that the demands arising from demographic ageing are no more than the zest in a cocktail of socio-economic trends and radical, even revolutionary, structural political and economic changes that are challenging governments and institutions. There are both severe funding strains and crises of social protection and inclusion. These are mostly the consequence of the revolution in eastern Europe. In the West:

> many of the difficulties that welfare states today face are caused by market failure: that is, badly functioning labour markets produce an overload on existing social programmes . . . there is possibly also welfare state failure: that is, the edifice of social protection in many countries is 'frozen' in a past socio-economic order that no longer obtains, rendering it incapable of responding adequately to new risks and needs. (Esping-Andersen, 1996b: 2)

The latter remarks apply most obviously to Germany which, despite its immense economic power, has created a panoply of social welfare entitlements that have over-reached the nation's capacity and which

appear to be a structural impediment to raising the international competitiveness of labour. But the case of Germany, and its recent reforms,
remind us that social welfare systems are flexible and can be fine-tuned.
A short period of funding strain does not mean that the entire system is
misconceived or incapable of reform. As the World Bank (1994: 102)
observed, 'the problems facing public pension plans stem partly from
poor design features such as early retirement and over generous wage
replacement rates that favour high-income groups.'

There have been other important sources of change. Most widespread
has been the collapse of low productivity mining and manufacturing
employment and the other profound changes in the availability of paid
work, which, particularly through higher unemployment and earlier
retirement, both increases demands on a pay-as-you-go social security
fund and reduces contributions to it. In southern Europe, and particularly the countries that have emerged from dictatorial regimes,
political ambitions and rapidly changing economies have prompted
governments to a fast pace of social welfare development. They are
pressing their tax-raising capabilities to the limit of their administrative
capacity and the tolerance of their electorates.

An additional element in the equation are European-level policies,
which during the 1990s have increasingly impinged on the social welfare
domain (Walker and Maltby, 1997: 43). The importance of financial
security as an element in European citizenship for older people has been
recognized by the EU in the Charter of the Fundamental Rights of
Workers. This was adopted at the Strasbourg European Council in
December 1989 by all member states except the UK, was reaffirmed in
the July 1992 Council recommendation on the convergence of social
protection policies, and was incorporated in the social chapter of the
Maastricht Treaty. Two paragraphs are especially pertinent:

> 24. Every worker of the European community must, at the time of retirement,
> be able to enjoy resources affording him or her a decent standard of living.
> 25. Every person who has reached retirement age but who is not entitled to a
> pension or who does not have other means of subsistence must be entitled to
> sufficient resources and to medical and social assistance specifically suited
> to his needs.

Most commentaries on pensions and social security systems feature
their current and projected funding scenarios. Much less is said about
their achievements in reducing poverty or promoting equality and social
solidarity. Twentieth-century social welfare in Europe has not removed
all social ills or produced utopias, but its contribution to raising the level
of living has been immense. Among the remaining problems which
social policy should address are continuing concentrations of poverty
among very old people, particularly widowed women. Poverty among
older people divides EU countries roughly into three groups. Denmark,
Germany, Ireland and Luxembourg have relatively low (less than 10 per

cent) poverty rates. The middle group of Belgium, France, Italy and the Netherlands have poverty rates of 10–29 per cent, while higher poverty rates are found in Greece, Portugal, Spain and the UK. A worrying trend is the rise of inequalities in retirement income (Walker and Maltby, 1997: 53).

The most widespread response to the projections of further age-structure change has been a change of direction in old age income support policies during the past 15 years, notably the reversal of tendencies to encourage retirement at younger ages. Pension reforms have recently been introduced, or are planned, in Germany, Greece, France, Italy, Austria and the UK (for women only), which effectively increase the retirement age, or extend the working life in order to qualify for a full pension (Eurolink Age, 1997: 19). There have also been 'awareness campaigns' to promote the interests of older workers and to end age discrimination in employment (in the UK both the Bank of England and the Confederation of British Industry have been active).

In western Europe, the most radical reforms have been undertaken, sometimes too quietly, in the United Kingdom. From the earliest months of the Conservative administrations that began their long run in 1979, measures have been taken to reduce the real value of the basic state retirement pension to a niggardly (and eventually inconsequential) level. The 1986 Social Security Act dismantled many of the ambitions of the legislation, established by cross-party consensus eleven years before, which was focused on the social policy goals of eliminating poverty in old age. The 1975 Act established a state earnings-related pension scheme (SERPS), vastly improved the entitlements of widows and other dependants, and reduced the ability of individuals to withdraw their contributions to occupational pension schemes or neglect the financial claims of spouses and dependants to their accumulated private pension rights. Macro-economic and ideological considerations were the main reasons for that reversal. In both the Nordic and the 'continental European' countries, a greater consensus for social policy continuity has prevailed, and technical adjustments have been introduced rather than a sea-change. But the ideological and macro-economic pressure to privatize is growing and will be fuelled by the world-wide hunger for investment capital, the international agencies that see the world through macro-economic lenses, growing affluence, the self-interest of the rich, and the financial services industry. It remains to be seen whether European social policy can continue to be driven by social objectives or, as in North America, is increasingly steered by fiscal and financial interests.

NOTES

1   Walker (1990) refers to reports on the public expenditure implications of demographic ageing from the International Monetary Fund, International

Labour Office, Council of Europe, International Social Security Association and the Organization for Economic Co-operation and Development. The list may now be supplemented by the World Bank.
2 Two journals are particularly useful sources: the *Journal of European Social Policy* and *International Social Security Review*.

11

Boundaries of Social Exclusion in Europe

Lila Leontidou and Alex Afouxenidis

The theme of social exclusion incorporates a kind of 'nostalgia without memory' (Gates, 1995: 214), for this has a substantial tradition which is set aside in most of the current debate. In this long tradition, the borders between the excluded and included/incorporated/integrated/assimilated populations have shifted, and they are still shifting endlessly. As we will argue here, these moving boundaries have to be taken into account and incorporated into theoretical language, which is today rather static and mono-dimensional.

The list of 'exclusions' is endless: besides the most significant ones such as poverty, unemployment, homelessness and hunger, there is exclusion from knowledge and information, new technology, access to welfare, free trade and the city, and exclusion by language, age, sex and gender. The theme of drawing boundaries and excluding is integral to the history of human development: 'and the life of man, solitary, poor, nasty, brutish and short' (Thomas Hobbes, *Leviathan*). From the Greek city-state (*polis*), where wealth and power were distributed among those who were pre-defined members of society, to the Middle Ages and beyond when Europe expanded cruelly, every expansion generated refugees or outsiders, established the idea of what it means to be poor and 'dispossessed', and gave birth to all contemporary boundary and border metaphors.

The conceptual confusion between 'social exclusion' and 'poverty' limits many of the current approaches to the latter. The present chapter proposes to cross existing boundaries in order to move towards an understanding of social exclusion. The remarkable research tradition on poverty has been condensed today into a combination of productivist and welfarist explanatory models and the perceived 'objectivity' of globalization. However, territoriality, the Muslim resurgence, xenophobia, religious conflicts and terrorist attacks point to inter-subjective tensions and to forms of social exclusion which are inadequately addressed, except in the field of migration studies (Castles, 1993; King, 1993b, c; Pugliese, 1995; Weiner, 1995; Rees et al., 1996).

The neglected spatial and geo-political dimensions of social exclusion are now taking their revenge by rising to dominance in the European Union. Spatial location is inextricably linked and interacts closely with perceptions of exclusion, peripherality, territoriality and its construction, as well as movement and fixity. The spatial dimension imposes different forms of imperatives from the productivist one, especially in territorial skirmishes, conflicts and wars. In this chapter, we will combine an analysis of imaginary *boundaries* and real *borders*. Both concepts will be used to further the debate conceptually and theoretically (boundaries) and spatially (borders) in Europe. This is an exercise to investigate the *socio-cultural construction* of boundaries and borders defining *limits for social exclusion*, as well as territorial limits.

The '*longue durée*' of social exclusion in Europe

'Hark! Hark! the dogs do bark, The beggars are coming to town' (old nursery rhyme, in Kamen, 1971: 427). Braudel's *longue durée* is an appropriate framework for the study of social exclusion. From the fifteenth century, 'in western Europe, it is probable that at least one-fifth of a town's population would consist of the wholly poor' (Kamen, 1971: 429). Today it is one-eighth: over 53 million are impoverished in Europe, of whom 20 million are unemployed, comprising about 15 per cent of the population of 360 million (*Le Monde Diplomatique*, February 1995). The following extract sounds like modern London or Paris; or could it also be a description of contemporary urban living for the homeless in the UK?

> From the late fifteenth century onwards the laws against vagabonds and begging multiplied in volume . . . the poor were a major problem. Chronic vagrancy was almost exclusively an urban phenomenon . . . The chief cities of Europe were inundated with beggars, many of whom were not native to the region. The situation in Rome was so serious that it led Sixtus V . . . to complain bitterly of 'these vagrants wandering through all the streets and squares of the city in search of bread'. (Kamen, 1971: 428)

In ancient times, exclusion was associated with citizenship, but European citizenship is hard to define. Europe has never been a continent with definite borders (Leontidou, 1997). It has been a construct, starting from the ancient belt around the Mediterranean and constantly moving, until it shifted from 'being a mere geographical entity to being a synonym for Christendom' (Szucs, 1988: 292). Great states created, formed and reformed borders and with them ideas on laws, individual freedom, sovereignty and power: ideas which were discussed long before Hobbes, Locke or Rousseau (long before the French Revolution). In the Middle Ages (thirteenth century) 'such ideas were . . . at the centre of political theory' (Szucs, 1988: 295). So the 'West' early on separated the concept of 'society' from that of the 'state'.

In Britain, in the eighteenth century, 'the "common herd", the "black masses", were generally rated as not much higher in the social scale than the cows and pigs which shared their hovels' (Langer, 1996: 125). After the Enclosure Movement incorporated open land, the poor people lost the: 'right to glean the fields, to gather faggots in the forests, to run their cow or pig on the common . . . By the mid-nineteenth century there were in England and Wales some 700,000 families of landless agricultural laborers, representing about 1/5 of the population.' In Europe (except France), 'the same process was under way . . . It left a fifth of the population of Sweden and Norway and two-fifths of the Finnish population landless . . . similarly in Sicily . . . Hunger and cold led to a great increase in the consumption of hard liquor by the lower classes of both country and town.' An American observer, commenting on the condition of the peasantry in the nineteenth century, said 'they are not slaves, but they are not free' (Langer, 1996: 126–7).

From the Industrial Revolution onwards, economic exploitation occupied centre stage. Capitalism grew and thrived in the midst of misery, alienation, marginalization and exclusion of a large section of the population. 'Every day that I live I thank Heaven that I am not a poor man with a family in England' wrote an American in 1845 (Hobsbawm, 1984: 95). Images of the early twentieth century are similar. Pember Reeves described how 'not the poorest people of the district' (Lambeth) wondered how families of six could survive on no more than 20s (£1) a week in 1910; but 'behind the barriers the communal spirit was enhanced, as the inhabitants sought to evade the moral and physical sanctions of polite society'. Aid is not something that is provided *only* by the 'welfare state', since 'the total amount of aid the poor gave the poor . . . far exceeded the efforts of all outside agencies' (Vincent, 1991: 14–15).

Poverty and exclusion were gradually posed as a purely *political question* related to the role of the state (and also as an issue related to ethics and morals). 'Welfare statism' (which probably originated during the regime of Queen Elizabeth I which issued the first poor laws in 1601), categorizes the population, accepting directly the pregiven fact that some people will always be poor (Kaufman, 1996: 294). In the UK, the 1601 regulations were not put into practice, but in 1834 a new law was enacted which created a comprehensive system of asylums and similar institutions controlled by the government (Fraser, 1982). The idea, however, was that people must 'return' to work. The moral issue in question was related to the idea that the poor 'naturally' disliked work.

In France, the betterment of the poor was never really an issue for the government (Kaufman, 1996: 300), and so the debate focused on issues of job insecurity with specific attention to industrial workers and their families. Industrialization led to poverty because of the radical break with the former feudal system of economic relations. The problem was how to find a means of wealth redistribution which would also resolve problems of over-production and low consumption. From the nineteenth century, the French founded a 'Keynesian' model of development. In the mid-nineteenth century, French businessmen introduced the 'family wage' and prototypical forms of insurance contributions. Therefore the problem of industrial poverty 'was linked to family insecurity and not, as in the case of Germany, to labour issues' (Kaufman, 1996: 301).

Early German welfare began in Prussia in 1794. After 1854 (and Bismark) a wide network of social support was created, based on contributions from employers and workers. An important difference between the British and the German approach is that the latter was not solely concentrated on managing poverty but on generating security networks for labour (a corporatist model, with a stronger role for the state). Additionally, the German system was (is) based 'on the consolidation of earners into ordered status groups with different regimes of social insurance, for example, the white collar employees as opposed to manual workers' (Lash and Urry, 1994: 181; see also Chapter 1).

In Scandinavia, the welfare system grew out of concerns for poverty in old age. In 1889, and then again in 1912, laws were passed which made *everyone* eligible for insurance (Kaufman, 1996: 302). This particular model did not, in theory, 'exclude' any member of society from its obligation towards the rest.

Essentially, western welfare systems generated a universal vocabulary of marginalization. Recent post-modern interpretations (Cross and Keith, 1993) related to the nature of identity and/or boundary formation, are crucial in demonstrating the cultural significance of the nation-state as a 'moulder' of social inequality. In this sense, historical evidence does not corroborate any causal link between the construction of the 'welfare state' and 'Fordist' development. Contrary to a variety of writings (Harvey 1989a; Lash and Urry, 1994), which view changes in the organizing principles of the labour process as central, borrowing from Gramsci, an understanding of 'civil society' and 'stateless society' is needed. There are major gaps in comprehending the organizing principles (economic as well as non-economic) which promote divisions across different modern European societies. The state (and through it nationalism) has always created dichotomies and divisions, continuously shifting the ground with respect to generating new boundaries and borders which justify its existence (Anderson, 1997). In this multi-faceted context, a renegotiation of theory with respect to the character and the constitution of the state would be particularly useful.

From industrial Manchester and Victorian London, through inter-war Greek refugee settlements to the post-war black and Hispanic American urban underclasses and German *Gastarbeiter* communities, social exclusion has always accompanied economic growth and migration waves. However, the boundaries between inclusion and exclusion shift in time and space: spatial borders, economic restructuring, social dynamics, cultural divisions have moved centre stage in different times and localities. In the period of European Union integration and socialist disintegration, there is a tendency to make fetish of globalization as the main cause of social exclusion (Brown and Crompton, 1994; Musterd, 1994; Knox and Taylor, 1995), while some references to gender and cultural dimensions have also appeared (Stichter and Parpart, 1990; Sibley, 1995). Meanwhile, 'categories and measurements' of poverty and social exclusion, such as the following, abound, and the picture of Europe that they provide is not very encouraging:

- Europe, with 20 per cent of world population, has over 53 million people impoverished (a 40 per cent increase since 1975, with the standard definition of under half the average per capita income), of whom 20 million are unemployed. (*Le Monde Diplomatique*, February 1995: 9).
- In 1992, only 60 per cent of the working age population of the EU had found employment (compared to 75 per cent in Japan, 70 per cent in the USA) and in, for example, north-east England only 50 per cent (*Le Monde Diplomatique*, February 1995: 6).
- More than 20 per cent of EU population live in areas or peripheries where GDP is below 75 per cent of the EU average (Cole and Cole, 1993: 234).
- Southern countries record poorer scores on such indicators. Italy has over 15 per cent of its population impoverished, around 9 million citizens. In Spain, more than 24 per cent of the elderly are regarded as poor. In Portugal, this figure rises to 45 per cent, and in the slum suburbs around Lisbon over 50 per cent of the inhabitants are illiterate (Michel, 1995: 10–11). In Greece it is around 29 per cent (EKKE, 1996: II, 88).
- However, there are seemingly different but ultimately similar problems in the north: in the UK it is estimated that over 2 million people are homeless (which means that they are excluded from the most basic of civic activities, that of voting). In London, people employed at the 2,000 small workshops in the East End experience appalling working conditions (Pearson, 1985: 54).
- In the domain of work, moving beyond the simple unemployment figures, an estimate from France (1989) shows that about 39 per cent of available job positions do not require any special training or qualifications. But people are still not regarded as 'employable' because 'management considers reasons like low school grades, where you live, skin colour, foreign surnames, as important for taking on people as unskilled or low skilled workers' (Clerc, 1995: 15).

Much has been written about the elusiveness and even deceptiveness of indicators when cross-cultural comparisons are attempted. They are as loosely defined as the concepts they try to 'measure': 'masses', 'underclasses', 'dangerous classes', 'marginals', in conditions of poverty, 'polarization', 'new' urban poverty (Brown and Crompton, 1994; Morris, 1994; Hadjimichalis and Sadler, 1995; Bailly et al., 1996). The extent of poverty and the degree of social exclusion are confused in indicators and 'measurements' (Godfrey, 1986; Symes, 1995; Hausserman and Mingione, 1996). Recently, the informal sector and migration have been added as indicators, despite the fact that informality may generate substantial incomes, and that spatial fixity may be more conducive to exclusion than mobility (Mingione, 1991, 1993, 1995; Leontidou, 1993).

Current concepts and approaches

On the margins of the 'globalization' narrative, other previously neglected dimensions of social exclusion, such as gender, have taken centre stage in the current debate (Stichter and Parpart, 1990; see Chapter 8), but studies of *simultaneous spatial and social exclusion* have yet to influence European research and policy. References to the topological isolation of marginal groups in the context of network theory (Bailly et al., 1996) do not offer any solid spatial dimension to the debate. Discussions of space usually revolve around national differences and over-represent global cities. There are interesting contributions on national variations of poverty (if not social exclusion) between European countries today. The social class/work dimension is stressed in the UK because of its tradition of wage labour combined with a welfare state, which is now in crisis (Morris, 1994; Afouxenidis, 1998). In southern Europe, family and informality form the main dimensions in countries with a multitude of family enterprises and sizeable self-employment and home working (Mingione, 1991, 1993; Leontidou, 1993; Hadjimichalis and Sadler, 1995). There is also a migration dimension in these countries (for example, King, 1993b, c; see also Pugliese, 1995, Bailly et al., 1996).

We also encounter in the existing literature a lopsided treatment mainly directed towards two types of localities. On the one hand, *polarization studies* focus on hi-tech environments such as global cities, which disadvantage certain populations. Manifestations of poverty in core European cities have given rise to concerns about their 'Americanization' or the possible return of the 'dangerous classes' (Morris, 1994; Musterd, 1994). On the other hand, *locality studies* opt for de-industrialized regions hit by redundancies and hence unemployment and poverty, or depopulation through massive emigration. References to a third category, namely poverty in the rural underdeveloped hinterland, become progressively scant, with the spatial switch in southern Europe from predominantly rural poverty and unemployment, push factors for urbanization and emigration in the past, to urban poverty (Leontidou, 1993).

Less-favoured regions and localities outside the core of Europe are usually treated as 'losers' in EU competition or place marketing, with the usual underlying hypotheses on peripherality: the friction of distance, impaired accessibility, economic disintegration, including both stagnant production structures and de-industrialization or failure in productive restructuring (see Chapter 2). These are all *productivist* postulates, reflected in concepts such as 'glocalization', 'spatial switching', spatial 'mismatch' and so on (Musterd, 1994; Bailly et al., 1996). 'Disintegrated' localities are in fact disembedded within an overall context of post-Fordist footloose industries, new divisions of labour between professional and service workers, redundancies in the factory and mining sectors, and neo-liberal privatization and the demise of the welfare state. In addition, developments in central and eastern Europe since the beginning of the 1990s, have shifted entrepreneurial (and research) interest to the 'new south', post-socialist Europe: it is here that earlier disadvantages which used to characterize traditional low-wage countries of southern Europe are now sought.

The research tradition which prioritizes *productive economic* restructuring is one of two traditions which dominate studies of poverty and social exclusion. This school associates poverty with unemployment after de-industrialization and new divisions of labour in post-Fordist Europe. In the discourse of the second influential school, another aspect of globalization comes to the forefront: *policy-relevant* changes, especially the demise of the welfare state. This *welfarist* school discusses individuals and social groups within localities, and targets problems of ageing, gender, ethnicity and unemployment (Brown and Crompton, 1994). It considers planning and policy gaps and changes as causes of social exclusion and 'local economic disintegration', and analyses privatization and neo-liberal policies as the processes whereby a variety of forms of social exclusion are created around Europe (Mingione, 1993; Brown and Crompton, 1994).

The tendency for national typologies follows Esping-Andersen's (1990) intervention. His work is increasingly cited to highlight the difference between the central European (Germany, France) and the Anglo-American welfare model (see Chapter 1). The different policy backgrounds of countries and the different types of kinship systems in Europe are identified, but such typologies diminish them, to the extent of including southern Europe in the German model. The least one can say about this is that it is insensitive to southern experiences such as the *Gastarbeiter* phenomenon: the fact is that, if anything, the Mediterranean has been for several decades a labour reservoir for the German 'economic miracle' and exclusive welfare state. Now that southern Europe has changed from a region of emigration to one of immigration (King, 1993b, c; see Chapter 6) new realities and distances are created, equally different from the German model. Moreover, the welfarist–corporatist model (Germany, France) only functions with respect to certain parts of the population. Flexible working practices include illicit work. Walraff (1988: 26) described how over '200,000 Turks, Pakistanis, Greeks are illegally employed in the building industry alone' in Germany. The sub-contractors working for the government are basically never prosecuted for 'if the authorities had shown any desire to be informed of what was going on, this would have been sufficient to bring the business to an end' (Walraff, 1988: 26–7). This is a replica of the landowners illegally sub-dividing and selling land on the urban fringe, who were never prose-cuted in the unauthorized settlement networks, because these networks 'solved' the housing problem during years of massive rural-urban migration in Mediterranean Europe (Leontidou, 1990).

The creation of a single European market does not incorporate any substantial institutional changes to guarantee conditions for workers. 'Exclusion' in this respect works in two ways: first, racism or marginal-ization is expressed semi-officially through laws that are only applied to a certain category of paid employment but not to all; and, secondly, employment in itself may be regarded as a segregatory rather than integratory type of activity. In many respects, the system is based on people's silent participation while consistently failing in the provision and guarantee of any reasonable sense of security and human dignity (Afouxenidis, 1990).

It is interesting that the binary distinction between productivism and welfarism outlined above, reflects the clash between monetarism and Keynesianism, and mirrors familiar dualisms of private/public and market/planning. This *dualism* in the proliferating literature on poverty and exclusion in contemporary Europe is receding as theory becomes more sophisticated: the two schools interact and influence each other, but keep prioritizing globalization. However, they have been recently intercepted by the insertion of a *third* category, namely *reciprocity*. Alternative models proposed by researchers in southern Europe start from the premiss that in countries where the welfare state has traditionally been weak, dual schemes of private/public in the creation of social divisions must be substituted by triplets such as private/*reciprocal*/public, market/*reciprocity*/association (Mingione, 1991; Leon-tidou, 1993, 1995) and further triplets, such as structure/socialization/agency. The triplets introduce an alternative logic, incorporating a bottom-up element in studies monopolized by the top-down logic of globalization and welfarism.

The border as a field for the construction of a *third* theoretical model

Problems of poverty in the core of Europe have thus generated the productivist and welfarist explanations, forming the *political-economy* perspective when they interact, recently embedded within *regulation* theory, and the globalization trend (Amin and Thrift, 1994). However, even if all localities are embedded within the same globalization process, this does not render de-industrialization, privatization and polarization equally important for all of them.

Reciprocity adds yet another model to evade the impasses of the political-economy perspective, but does not deconstruct it. In deconstructing the claims of the two schools, their combination and their variations, we must *not* abandon them, as modernism and the erection of grand narratives would dictate; but we must *relativize* them, showing their limited relevance in the context of a *third* theoretical model (Moulaert and Leontidou, 1995; Leontidou, 1996). In fact, social exclusion cannot be understood effectively within the logic set out in standard poverty theory, nor does it exhaust itself in forms encountered in polarized global cities, for that matter. Social exclusion dynamics in Europe are varied, based on diverse spatial divisions of labour, local histories and cultures, memories and interpretations (Carter et al., 1993; Sibley, 1995). *Inter-subjectivity* must move centre stage in interpreting the shifting realities encountered in Europe today.

The political-economy interpretations are tailored after forms and processes of social exclusion in most regions of the *core* of Europe. If we look slightly further, to the east, south, and north-west of the European development corridors, the political-economy perspective is automatically relativized within a comparative inter-cultural framework, which is *not* one of its strengths, and different categories and concepts spring to the foreground. For example, it is still to be thoroughly researched and established that there are poor populations in European localities today who have *never* seen Fordism, and have *never* experienced a proper welfare state. Several localities of southern Europe are not particularly affected by types of restructuring, de-industrialization, welfare state demise and the various types of 'decline', which the political-economy perspective has to offer, and yet are profoundly affected by severe instances of social exclusion.

Within an alternative model dealing with societies where homelessness and unemployment are relieved by reciprocity within the family and the community, the resilience and the prospects of *informal provisioning* rather than standard indicators of disadvantage become important. Urban analysis in Greece and Italy (Leontidou, 1990, 1993; Kazepov, 1995) has consistently revealed suburban rather than inner-city poverty and a low incidence of homelessness which, however, is now rising in the inner-city streets and metro stations. Poor populations of peripheral localities suffer double exclusion, as they cannot possibly compete in European neo-liberal place marketing, for lack of organizational resources and managerial skills (Leontidou, 1993, 1995). It is the most affluent among competing localities which attract further funding. And, among populations of communities hit by poverty, it is the most enterprising who migrate, leaving the rest on the margin and further draining these regions. These and other findings in peripheral Europe highlight the *importance of space, spatiality, the community and the family* and of the *local (grassroots) level* for individuals and cultures besides top-down 'decisions' of enterprises and welfare agencies. Last but not least, they underline that, besides economic restructuring and policy changes, there are also other objective forces which marginalize populations and localities: peripherality, for example, and culture.

In order to break away from the over-exposed and often misplaced north/south division as the sole example of the relevance of peripherality, we introduce here the external European border as a counter-example to prevalent social exclusion models. Spatial peripherality in its acutest form involves many places on the margin (Shields, 1991; Sibley, 1995). Among them, along the shifting EU border, changes are rapid, with migrations, tensions and wars. Events since the fall of the Berlin Wall in November 1989, when post-socialism reshuffled borders, not always peacefully, have raised complicated questions. 'Post-wall' Europe:

> is producing the same effect as that of a house of cards: a chain reaction of destruction . . . today, as the cries of those who proclaimed the definitive triumph of liberal society are slowly drowned out, the final ripples of the wall's collapse are provoking the break-up of the old post-war regimes for which the Berlin Wall was a buttress. (Ramoneda, 1994: 54)

The process of European integration has created two (internal and external) types of borders and sets of contrasting processes therein. A *third* type of border has now appeared as borders among EU nation-states have loosened after the Schengen agreement in the core of Europe (Acherman, 1995). At the same time, cross-border co-operation along external borders, between the EU and non-member states, becomes harder as external borders tighten in the 1990s. The debate on 'Fortress Europe' has intensified (King, 1993b; Blacksell and Williams, 1994; Hadjimichalis and Sadler, 1995; Hall and Danta, 1996). Destabilized enclaves of conflict, war and civil strife approach the EU border, which has been constantly re-defined and diversified by the admission of new members. When problems on the external border overwhelmed Europe, once more, in 1995 (Ireland and Greece) and 1996 (Cyprus), many unanswered questions re-emerged. Social exclusion in this area has strong cultural, geo-political and spatial components.

Different EU policies reveal their restricted theoretical and spatial scope when the border is in question. EU policy has a strong productivist emphasis, from the structural funds (Objectives 1, 2, 5b) to Community initiatives, as well as a welfarist emphasis as in Objectives 3 and 4, and has turned increasingly to monetarism after Maastricht (Commission of the European Communities, 1993c, 1994d). Other dimensions are rare and occasional. The INTERREG Community Initiative is exceptional, but hardly interacts at all with social exclusion policy. Moreover, it includes many dissimilar places on both the internal and external border (though it excludes Sicily) but is not equally effective for them all. Cross-border co-operation has already raised stimulating issues and policies *within* Europe, and it has become evident that INTERREG is most beneficial for *internal* EU borders. Around them (for example, Dutch–Belgian, Belgian–French and so on), there may be discontinuity in social, infra-structural and economic indicators, but there is increasingly peace, prosperity and co-operation, especially after the Schengen agreement was signed, and in the context of the high-speed railway system and EUREGIO policies giving rise to new co-operation, solidarities and alliances. Having benefited abundantly from European integration and Community policy, localities around internal borders are already experi-encing cross-border co-operation and interaction: commuter flows, flows of migrants and consumers, economic transactions and networks. Some of them are already growing to be the new dynamic regions of the EU.

By contrast, policy on the issues concerning localities on the external borders is long overdue, as is research on them. It is thought that border problems are to be targeted by political agents or types of institutions other than research institutions (such as defence, migration control and so on). And yet, these multi-cultural regions challenge conventional wisdom on many levels, including the socio-cultural one and bring new dimen-sions into the discussion of social exclusion. Problems of ethnic tension, illegal migration, asylum seeking, hostilities, escalation of security prob-lems, instability repelling enterprises, depopulation and brain drain, economic decline and cultural identities create an environment which sustains social exclusion, objectively and *inter-subjectively*. They often culminate in war and terrorist attacks. Moral questions linked to immigrant and refugee populations receive increased intensity: 'do all countries, irrespective of their size, density, economies, polities, social structure, and political ideologies, have the same moral obligations with respect to whom they ought to admit?' (Weiner, 1995: 169). The various policies followed raise issues of fairness with respect to citizenship and issues linked to the overall question of 'open' or 'closed' borders.

Socially excluded groups and poverty take an acute form on the diverse external EU borders, but also present particular configurations, due to spatial peripherality and multi-cultural localities (Wilson and Donnan, 1998). Besides populations who arrive, move, pass through or trespass, there are vagrant minorities, such as gypsies. The external border thus becomes an appropriate research milieu for the investiga-tion of unexpected dimensions and 'hidden' causes of social exclusion, which are manifested through the damage they create in the process of European integration, peace and co-operation. Excluded groups, forms and causes of exclusion in border areas can *not* be studied within the same discourse as the ones hitherto targeted in core/global cities and underdeveloped rural areas. Unfortunately, it is only with hypotheses that we can deal with such *cultural* causes of social exclusion at the present stage of research.

How is the border represented, and what is its meaning in the perception of populations living near one? How do they experience their citizenship and sense of belonging and the negative construction of identity of 'others'? Do they contrast themselves with the 'otherness' of populations beyond the border, or with their social exclusion from host societies? What are the relevant inter-subjective perceptions in major cities closer to the core, which finally 'host', or exclude, the cross-border movers? Besides racism, which has attracted attention (Cross and Keith, 1993; Sowell, 1994), there is a multitude of possible interactions between host and migrant populations (Goldberg, 1995; see Chapter 9). After all, the 'idea' of a border does not only have to do with geographical location, as the case of so many undefended cities amply demonstrates. In this context, what is the meaning invested in borders *within* cities and how does the population experience their demolition? What do the Berlin Wall and the Belfast peace line signify (Buckler, 1991; Wilson and Donnan, 1998)? The concepts of *territoriality* and *spatiality* have yet to be used effectively in combining border cultures with social exclusion studies.

It is not known whether there is any axis of separation between the aspiration for 'permanence' and vagrancy. The former is reflected in residence applications by asylum seekers, or in the quest of poor people for 'assimilation' to host societies. Vagrancy, however, points to another type of marginality: people cultivating cultures of dissent, such as gypsies, may be dismissive of dominant cultures. How far do they value the skill to contravene regulations and develop informal activities, crime, drug-trafficking? Are any groups creating alternative cultures of self-imposed marginality or constant movement? This point can be extended to the distinction between *voluntary* and *imposed* socio-cultural exclusion. An example from the contrast between Nordic and Mediterranean societies is indicative. Scandinavian individualism, manifested in solitary walks in the countryside and Euro-scepticism, points to cultures which, in a way, value exclusion from the dominant norm. These form a stark contrast with popular Mediterranean neighbourhoods run by families who fight for inclusion in the city via infrastructure development, or others who seek inclusion in the European Union, for that matter.

The above question also has a bearing on the dynamics of spatial fixity and mobility among socially excluded populations. Are population movements in border localities similar to migrations in the rest of Europe, or are there differences in turnover, temporality and composition? Albanians change their names into Greek ones in order to be given visas and settle indefinitely in Greece. Others are but transient populations passing through marginal localities as intermediate stations toward the better fate sought in city lights. Are these geographies of exclusion, or geographies of mobility and ephemerality?

Last but not least, there is a whole chapter on tension and war in post-socialist Europe, which cries out for a connection with social exclusion research. Security is not a matter of military defence, but of a sense of integration within a multi-cultural social world by the populations in border societies (Gow, 1991). Social cohesion is not a matter of economic integration alone, or of an efficient welfare state: it is a matter of socio-cultural and political processes of *legitimation* and *trust*. When we talk of 'society' do we mean everyone – all the people or all the people's communities in co-existence? The latter requires some intermediate form of legitimacy: communities of immigrants or of ethnic minorities who are not recognized officially may either not exist at all or may not form part of 'society'. Welfare is one form of legitimacy 'from above' for certain groups of people who are regarded as excluded from 'society'. Hence the state legitimizes 'society' and by so doing it legitimizes itself. But welfare (and hence the 'state') cannot be regarded as the sole contributor of a sense of co-operativeness in society. This is opposed to Titmuss's (1979) view of welfare as the major stabilizing element/ instrument of modern society.

At the grassroots level, the crystallization of collective identity and cross-cultural co-operation, which is more difficult in societies divided by borders, presupposes but also creates a sense of security related to social cohesion. Collective identity can break down in periods of crisis of the process of legitimation (Gow, 1991). Confused identities and the breakdown of trust create insecurity, especially among the more vulnerable populations, whether unemployed, poor, elderly or ethnic-language minorities. Insecurity is an outcome of the breakdown, or the weakening, of the processes of legitimation and trust, not just the outcome of 'foreign' presence or terrorist attacks. The theoretical 'repossession' of these dimensions in social exclusion studies is especially urgent now, in post-socialist Europe.

Conclusion: boundaries and borders in flux

In understanding social exclusion across Europe, there is no such thing as *one* representative reality, except perhaps in the minds of those who attempt to construe (and construct) the 'real'. The rigidity of the political-economy perspectives and the neglect of questions of ephemerality and inter-subjectivity are manifested in a new type of 'borderland epistemology' (after 'borderland phenomenology' in Aronowitz et al., 1996: 27), which is seemingly establishing itself as yet another dominant grand narrative of the 1990s explaining the 'new realities'. Perhaps by establishing ourselves 'on the border' rather than in what we think of as 'the world reality', we shall eventually be situated where it matters: along and across many different dimensions which 'constitute both the unified self and the multiple others' (Aronowitz et al., 1996: 27).

Are border/boundary/barrier synonymous metaphors (for they exist only as metaphors) of exclusion?

> A border crossing . . . is a violation, not an act of trade, or love, or harmonious association. The border is not a skin which can be caressed, it is a barrier. In relation to the inviolate body, all 'others' are simply potential 'wetbacks', 'illegal immigrants', terrorists travelling on forged documents. (Wilson, 1996: 222)

European integration, which increasingly polarizes the problems in internal/external borders on top of the core/periphery polarization, cries out for the revision of notions of social exclusion adopted by the political-economy model. Crucial aspects of social exclusion outside the core are yet to be investigated: space, peripherality, culture, ephemerality, vagrancy, war and so on.

The complex realities in the various corners of Europe, the objective world, are *mediated through culture and inter-subjective realities* in order to 'construct' boundaries of social exclusion; because the latter is a human creation and a grassroots reality distinct from the often-studied strategies of enterprises or state agencies (Dworkin and Roman, 1993; Silver, 1994; Sibley, 1995). Moreover, it is an inter-subjective reality in flux:

> places and localities are increasingly ephemeral in our rapidly-moving epoch. Places are produced historically and intersubjectively, but these identities are in flux. 'Culture is contested, temporal and emergent' (Clifford et al., 1986: 18–19). Rationality and certainties are shaken as never before in Europe today, especially after the demolition of the Berlin wall, which seems to have taken down with it many certainties and many dominant cultures which had long ceased to be hegemonic. At the start of the 1990s, the world realised once again that history is not linear, but moves as in Benjamin's last phrase in *Illuminations* . . . 'For every second of time was the strait gate through which the Messiah might enter' (Benjamin, 1973: 266). (Leontidou, 1996: 191, 193)

A set of methods for the study of ephemerality and shifting identities has not been crystallized yet, and has to be considered as a matter of the highest priority in the European research agenda. Europe at the turn of the millennium is constantly surprised by its periphery. A multi-faceted and locally sensitive approach to social exclusion is necessary after the Cold War, which collapsed along with many grand narratives. It is important to understand the rapidly changing cultures of people who may remain marginal to society at present, but *can* bring about change, if appropriately empowered. The political relevance of such a starting point is obvious.

CONCLUSION

12

A Divided European Future?

David Sadler

This volume has been concerned with the causes and consequences of the present, highly divided state of Europe. It began with the apparent paradox that Europe is becoming an increasingly homogenized political–economic space, and yet is also characterized by deepening social divisions. Subsequent chapters both problematized the notion and extent of that political–economic convergence and chronicled key dimensions of social inequality, stressing the ways in which space is a fundamental constitutive element. Such is the extent of these inequalities that many commentators, including some contributors to this volume, have begun to question whether the system which produced them is sustainable over the long term. This concluding chapter seeks first to draw together the common threads of the volume as a whole, and then to consider the implications of the evidence and analyses presented within it for the future of Europe.

Divided Europe

In their introduction, Ray Hudson and Allan Williams stressed the significance of national states and the limits to European regulation. They also posed the question 'what will be the future outlines of a divided Europe?', identifying three possibilities. In these different scenarios, (a) market forces and national states could remain decisive; (b) the European Union could become more powerful, perhaps with a more socialized conception of the market; or (c) a third way could emerge based on devolution, decentralization and principles of association. In each of these, however, it is likely that there will continue to be a role for interpretations which stress competing national models of capitalist development. They concluded that within the constraints of a market economy, there are structural limits to state policies, and in consequence the future map of Europe is likely to continue to be marked by inequality. That is to say, even presuming the will to tackle a divided Europe on the part of national states and the EU, there are fundamental limitations in terms of capacity within the boundaries of what are presently deemed to be legitimate policy interventions. This chapter will return to these contrasting scenarios in its closing remarks.

Key structural features of the new European economy considered by Ray Hudson (Chapter 2) included the rise of the service sector, jobless growth and high unemployment, and flows of foreign direct investment. Rather than a linear sequence of different forms of production organization succeeding one another over time, it was argued that the European economy is better characterized instead in terms of the simultaneous persistence of different forms of the organization of production. None the less, an important contemporary trend is towards labour process intensification through new forms of high-volume production and the reorganization of work in services. This has created new threats *and* opportunities for organized labour, although, as Hudson stressed, the former are particularly powerful – not least because present-day models of work organization are critically predicated on there being *no* return to conditions of full employment.

One selected aspect of economic process was explored further by Adam Tickell (Chapter 3), the role of financial services. The 1990s has been characterized as a 'momentous decade' for the European financial sector, with a change from separate national systems to one dominated by the prospect of a single European currency and a European Central Bank. It was argued that harmonization of regulation in this sector as part of the '1992' single internal market had a significantly smaller impact than the EU itself initially anticipated. Instead, the key driving force of change has been the globalization of financial markets. On the other hand, Tickell argued that the single currency project *will* dramatically affect the European financial sector, and that these effects will prove to be associated with even more sweeping geographical variation than occurred in the early 1990s. In particular, the single currency will encourage still greater spatial concentration within the sector. The main question thus becomes *where* growth will be focused rather than to what extent, with the most likely contenders for the role of leading European financial centre being London and Frankfurt.

In Chapters 4 and 5, the emphasis switched to political processes. Colin Williams identified 1989 as a watershed year in European history (comparable to 1789, 1848, 1918 and 1945), marking a break from one epoch to another. His central concern was nationalism, arguing that this was remarkably successful in the 1990s partly because of its flexibility, in the sense that nationalist movements made common cause with other social and political forces, through a range of shifting and often temporary alliances. In cautioning against essentializing nationalism (giving it undue explanatory weight as a causal process in its own right), however, he also pointed to another and potentially contradictory trend: the move from a collectivist to an individualist conception of social order.

This theme was picked up by Joe Painter (Chapter 5), who questioned both the extent to which east European societies attained a truly democratic status in the immediate aftermath of the post-1989 reforms, and the very meaning of democracy in Europe more generally. Debates about democracy have conventionally been conducted in an aspatial framework, partly because of their roots in a universalist project. Painter proposed instead that democracy should be viewed as a set of spatially situated social practices, the geography of which undermines attempts to apply a universal model. Thus Europe is presently witnessing the emergence of a polycentric system of governance, characterized by some as a 'new medievalism' (Anderson, 1996), in which power and influence will increasingly be distributed widely among overlapping, discontinuous and multi-layered forms of political organization.

Territory formed a common theme for Chapters 6 and 7, which were concerned with the Mediterranean and central Europe respectively. Russell King and Marco Donati suggested that the Mediterranean represents the most problematic flank of Europe, simultaneously a divide (between north and south) and a source of unity. They charted the changing relationship between Europe and the Mediterranean over time, paying particular attention to past and recent patterns of migration. The chapter concluded with an examination of a number of problems with this relationship. These include a refusal within Europe to recognize that there might not be a universally applicable development path, neatly echoing concerns expressed in the preceding chapter. Allan Williams and Vladimir Balaz (Chapter 7) focused on the interaction between processes of re-internationalization and privatization, and the reconstitution of social and spatial divisions, in central Europe. They concluded by pointing to the likely significance of the EU's Structural Funds in this region in future, particularly under *Agenda 2000* (Commission of the European Communities, 1997d; see also Fothergill, 1998; Hall, 1998), once again highlighting the increasing re-engagement between east and west. In different fashions, then, these two chapters emphasized the role of Europe in a wider world, and the ways in which this role has changed over time.

Key dimensions of social inequality in Europe were explored in the remainder of the book: gender, ethnicity and the life course, followed by a review of the history of the concept of social exclusion. Perrons (Chapter 8) challenged the idea that there is an inherent opposition between efficiency and equity, pointing out that regional disparities have declined during periods of strong economic growth, and that gender equality in terms of employment is higher in the more developed regions. She argued also, however, that economic growth is not a *sufficient* remedy to inequality in itself, and that addressing problems of gendered inequality would therefore involve considering the distribution of paid and unpaid labour. This in turn would involve re-evaluating macro-economic priorities, particularly the emphases on low inflation and limits on public sector expenditure associated with the Maastricht convergence criteria. Perrons developed concerns (which are also expressed in Chapter 2 by Hudson) over the nature of contemporary labour market restructuring, which she described as being driven by a search for 'defensive' rather than 'adaptive' flexibility. This is based on an acceptance of gender inequality in that the stress on part-time employment (a disproportionate share of which is taken by women), offering wages which are below reproduction levels, is predicated on an assumption that men and women play different and distinctive roles with respect to productive and unproductive labour. Perrons also criticized the gulf between objective and outcome with respect to the incorporation of equal opportunities principles into the work of the European Commission. Finally, she questioned why genuinely radical critiques of the EU's current development path, and the exposition of alternatives, have both been marginalized. This is an issue which is explored further in the final section of this chapter.

Paul White (Chapter 9) began with the observation that European colonialism set in motion a set of identity markers and hierarchies of power and legitimacy which were further reinforced through processes of immigration to Europe. He explored the varied positions of those migrants and of ethnic minority communities of migrant origin in Europe. Focusing on the process of 'othering' of ethnic minority groups of migrant origin, he considered three different models of state response: differential exclusion (an instrumentalist view under which migrants are seen as temporary or guest-workers), assimilation (which seeks to incorporate migrants into the host society by encouraging them to adapt, while the host society itself is not expected to change) and pluralism (which accords migrant groups equal rights within the host society as ethnic minorities without any expectation that they be forced to adapt). Favouring the latter, the chapter argued that the according of equal rights to 'other' groups requires the elimination of racism as a set of prejudices.

Anthony Warnes (Chapter 10) examined the responses of governments to the problems of financing social welfare programmes during the last quarter of the twentieth century, focusing especially on the rising cost of support for the elderly. He reviewed four contrasting situations: Sweden, Germany, Spain and Poland, highlighting the differences that exist between states in this regard. The problem of state support for the elderly forms just one facet of the broader question of inequality on the grounds of age, which takes on spatially variable forms within Europe. Finally, Lila Leontidou and Alex Afouxenidis (Chapter 11) reviewed the long history of social exclusion in Europe, which they suggest has been set aside in much of the current debate. In this long tradition, they suggest, the borders between excluded and included populations have shifted, and they are still shifting endlessly. These moving boundaries and identities should therefore be incorporated into contemporary theoretical debates, which are criticized for being static and uni-dimensional.

Divided futures

The review and summary of the contents of this book identified significant areas of common ground between contributions, and also some points of disagreement and debate. This final section seeks to amplify these issues of consensus and contention, as a precursor to some reflections on the likely dimensions of a divided European future.

The most significant point of contention concerns whether Europe is presently on a course of political–economic convergence or differentiation. As indicated at the outset of this chapter, the editors introduced the volume by pointing to an *apparent* paradox between ongoing processes of political–economic homogenization (as signified for instance in the Treaty of Maastricht) and social fragmentation. Perhaps, however (as hinted at by the editors), that paradox is itself false because Europe is instead becoming increasingly dominated by conflicting national models of political and economic organization? Several contributors stressed the significance of competing types of welfare capitalism. These include the distinction between social-democratic Scandinavian (with a strong role for the state in the provision of welfare services), liberal Anglo-Saxon (heavily market-based) and conservative central European (with a key part played by the family and voluntary agencies) forms of welfare provision (Esping-Andersen, 1990). Additionally, the 'southern European' variant (Mingione, 1995) is characterized by relatively dynamic family enterprises and self-employment, non-wage contributions to household livelihood strategies, and the limited formation of a working class engaged in manufacturing industry. Thus there is substantial evidence for the persistence and, indeed, deepening of historical differences between forms of capitalist development within Europe. Put another way, it could be argued that the social fragmentation which Europe is presently witnessing is not in itself paradoxical alongside political–economic convergence (the evidence for which is at best tenuous), but is instead intimately related to the emergence and reinforcement of a *divergent* array of political–economic processes, particularly but not exclusively in terms of the differences between national states.

Even putting aside differences between states for one moment, there are indications that the political map of Europe is becoming more rather than less fragmented. This is an argument made by Painter (Chapter 5), who characterized Europe in terms of a polycentric system of governance, and by Colin Williams (Chapter 4), who indicated that a drift is taking place from collectivist to individualist conceptions of social order. In a related fashion, King and Donati (Chapter 6) also cautioned against the adoption of universal models of development in the context of their analysis of Europe's changing relationship with the Mediterranean.

That social differentiation is *integral* to current political–economic system(s) within Europe is not in dispute among the contributors to this volume, however. On this point there is substantial agreement. Perrons, White and Warnes (Chapters 8–10) showed how dimensions of inequality such as gender, ethnicity and age are inherently bound up with the nature of current (low-)growth paths. This was most clearly demonstrated by Perrons in her analysis of labour market flexibility, which examined how this aspect of economic change is dependent upon a particular conception of social order, in this case an unequal distribution of responsibilities concerning paid and unpaid labour. A similar point was made by Hudson (Chapter 2) when he argued that current labour market trends are based on the assumption that there will *not* be a return to full employment. Thus insecurity in the workplace and gendered inequality in employment are outcomes of a particular model of political–economic organization and its many different variants – not *necessary* outcomes, but ones which are only made *possible* through certain political choices.

Additionally, the contributors to this volume have demonstrated in different ways how space is central to processes of exclusion and inequality. This is most explicitly evident in the chapters by King and Donati (Chapter 6) and by Allan Williams and Balaz (Chapter 7). These recorded both how rates of growth have been sharply differentiated *within* southern and eastern Europe respectively, and how in their different ways these European peripheries have themselves seen highly distinctive development paths when set against the rest of Europe. Leontidou and Afouxenidis (Chapter 11) made the point that there is a long history of socio-spatial differentiation in Europe associated with shifting patterns of boundaries and identities, while Tickell (Chapter 3) demonstrated that present trends within one branch of the economy (the financial services sector) are only likely to intensify existing patterns of inter-regional inequality. More generally, the emphasis in this volume has been on the interplay *between* social and spatial questions, so that space is seen as neither determinant of, nor being determined by, social change, but instead the two are mutually constitutive.

To suggest that social and spatial division is integral to current models of political–economic organization, and to accept that the dominant tendency is for differentiation of the latter rather than convergence, is not to indicate that social science should abdicate from the search for ways of conceptualizing the linkages *between* political, economic and social change, however. Acceptance of complexity in a polycentric world need not lead to a celebration of difference, but can rather inspire attempts to conceptualize the ways in which the different parts of that European mosaic fit together (Hadjimichalis and Sadler, 1995). In thinking about that task, several of the contributors to this volume explicitly recognized that an alternative European future would require an alternative macro-economic framework. For Perrons, this would involve jettisoning the principles enshrined in the Treaty of Maastricht, while for Hudson a new model would require greater attention to be paid to social and environmental sustainability. Such questions of political choice form the context for this chapter's final remarks.

Before that however, the chapter explores several further questions which have only been touched upon in this volume. The first of these concerns the nature of economic change, for there are issues here which require further consideration, including the notion of economic transition. Grabher and Stark (1997) criticized the extent to which change in central and eastern Europe during the 1990s was grounded in the neo-liberal notion that a necessary condition for growth was rapid privatization and marketization, based upon a wholesale adoption of a particular western European model and the destruction of previous institutional forms. They argued that while this might have fostered *adaptation* in the short run, the consequent loss of indigenous institutions would impede *adaptability* – in the sense of the capacity to adapt to continuously changing macro-economic conditions – in the long run. Thus there are conflicting views on the necessity, let alone desirability, of the implantation of one model of 'development' in central and eastern Europe during the 1990s, and on the extent to which the changes that have taken place there can be conceptualized as some kind of seamless transition or flow from one condition to another. This raises important questions which are relevant to the whole of Europe, and indeed more generally.

Secondly, the question of time scale was hinted at by Leontidou and Afouxenidis in this volume. Great stress is placed in many contemporary accounts on the economic impacts of political reform in central and eastern Europe after 1989. As well as introducing considerable uncertainty into the economic environment there and more generally, it is also likely to prove to be the case that many of the economic effects across the whole of Europe of this concentrated burst of political change will take several decades to emerge (Nilsson and Schamp, 1996). This introduces another element into the equation of the search for a less-divided European future. An additional factor relates to that heightened state of uncertainty, for it is also argued (for instance Thrift, 1996) that one consequence of it is an acceptance on the part of capital that the economy is no longer 'knowable'. As such, a new kind of managerialism has emerged, one geared to the development of looser organizational forms. In this way, it is suggested, new ways of exercising corporate power have risen to prominence. Challenging such notions of radical indeterminacy requires new visions of alternative forms of organization.

Thirdly, consideration of the likely future shape of a divided Europe demands attention to a slightly different question: what might a less-divided European future look like? Central to an answer to this are questions to do with the creation of identity and the idea of Europe, as identified by White (Chapter 9). Laffan (1996) argued that, historically, the 'European project' was embraced by many of the member states as a means of strengthening their existing state identities, and as an arena within which to project themselves. More recently, it is argued, identity politics have become increasingly important in Europe partly because the European Union is moving from issues of instrumental problem-solving to fundamental questions about its nature as a part-formed polity. Thus shared values and identities are significant if the EU is to become a focus for genuine legitimacy in the future (see also Therborn, 1995). The creation of identity within Europe today cannot be set aside from the broader issue of Europe's past and present role in a wider world. A less-divided Europe would be a more *inclusive* project, which would in turn involve recognizing the legitimacy of the voices and experiences previously marginalized as 'others'. The power of margins to be a source of critique and inspiration is fundamental to such a process of change. A less-divided Europe would also be more sensitive to local, regional and national identity: in other words, a project that works *with* rather than *against* difference.

Finally, however, the question that remains to be addressed is one identified by Perrons: why have there been so few alternatives emerging from so-called 'radical' channels of urban and regional debate? In part this issue has been addressed elsewhere. For instance, Graham (1992) demonstrated how the rhetoric of post-Fordism was associated with a foreclosing of debate on political alternatives, while Hallin and Malmberg (1996) debunk the myth that regional competitiveness (now the touchstone of many development efforts and analyses) is anything other than a zero-sum game, and go on to ask why this search for competitiveness should have become so prominent. This volume has provided ample evidence of the unequal nature of contemporary growth paths; of the non-inevitability of those directions and of the sheer human costs involved in the creation of a divided Europe. As Europe (in the sense of the European Union and its antecedents) approaches the half-century mark, and as a new millennium beckons, perhaps the time has come for a fuller exposition of the alternatives?

References

Aasland, A. (1996), 'Russians outside Russia: the new Russian diaspora', in G. Smith (ed.), *The Nationalities Question in the Post-Soviet States*, Harlow: Longman, pp. 477–97.

Abercrombie, N. and Warde, A. (1988), *Contemporary British Society*, Cambridge: Polity Press.

Acherman, A. (1995), *Schengen Agreement and its Consequences: the Removal of Border Controls in Europe*, Bern: Stampfli.

Afouxenidis, A. (1990), 'Industrial relations and workers' participation issues: a case study of the Greek telecommunications sector', unpublished PhD Thesis, University of Durham.

Afouxenidis, A. (1998) (in Greek), 'The Greek welfare state: in transition, restructuring or deconstruction', in M. Matsaganis (ed.), *Prospects of the Welfare State in Southern Europe*, Athens: Themelio.

Aglietta, M. (1979), *A Theory of Capitalist Regulation*, London: New Left Books.

Agnew, J. (ed.) (1997), *Political Geography: a Reader*, London: Arnold.

Agnew, J.A. and Corbridge, S. (1995), *Mastering Space*, London: Routledge.

Allen, J. and Henry, N. (1997), 'Ulrich Beck's *Risk Society* at work: labour and employment in the contract services industries', *Transactions of the Institute of British Geographers, New Series*, 22 (2).

Altunbas, Y., Molyneux, P. and Thornton, J. (1997), 'Big bank mergers in Europe: an analysis of the cost implications', *Economica*, 64: 317–29.

Amersfoort, H. van (1988), 'Ethnic residential patterns in Dutch cities: class, race or culture?', in T. Gerholm and Y.G. Lithman (eds), *The New Islamic Presence in Western Europe*, London: Mansell, pp. 219–38.

Amersfoort, H. van (1991), 'Nationalities, citizens and ethnic conflicts: towards a theory of ethnicity in the modern state', in H. van Amersfoort and H. Knippenberg (eds), *States and Nations: the Rebirth of the 'Nationalities Question' in Europe*, Nederlandse Geografische Studies 137, pp. 12–29.

Amin, A. (1985), 'Restructuring in Fiat and the decentralisation of production into southern Italy', in R. Hudson and J. Lewis (eds), *Uneven Development in Southern Europe*, London: Methuen, pp. 155–91.

Amin, A. (ed.) (1994), *Post-Fordism: a Reader*, Oxford: Blackwell.

Amin, A. (1996), 'Beyond associative democracy', *New Political Economy*, 1 (3): 309–33.

Amin, A. (1997a), 'Placing globalisation', University of Durham, Department of Geography (mimeo).

Amin, A. (1997b), 'Globalisation and regional development: a relational perspective', University of Durham, Department of Geography (mimeo).

Amin, A. and Thrift, N.J. (1992), 'Neo-Marshallian nodes in global networks', *International Journal of Urban and Regional Research*, 16: 571–87.

Amin, A. and Thrift, N.J. (1994), *Globalization, Institutions and Regional Development in Europe*, Oxford: Oxford University Press.

Anderson, B. (1991), *Imagined Communities: Reflections on the Origin and Spread of Nationalism*, London: Verso.

Anderson, B. (1997), *Imagined Communities: Reflections on the Origins and Spread of Nationalism*, Greek trans., Athens: Nefeli.

Anderson, J. (1995), 'The exaggerated death of the nation state', in J. Anderson, C. Brook and A. Cochrane (eds), *A Global World?*, Oxford: Oxford University Press, pp. 65–112.

Anderson, J. (1996), 'The shifting stage of politics: new medieval and postmodern territorialities', *Environment and Planning D: Society and Space*, 14: 133–53.

Arfé, G. (1981), 'On a community charter of regional languages and cultures and on a charter of rights of ethnic minorities'. Resolution adopted by the European Parliament, Strasbourg.

Aronowitz, S., Martinsons, B. and Menser, M. (eds) (1996), *Technoscience and Cyberculture*, London: Routledge.

Ash, T., Hare, P. and Canning, A. (1994), 'Privatisation in the former centrally planned economies', in P.M. Jackson and C.M. Price (eds), *Privatisation and Regulation: a Review of the Issues*, Harlow: Longman, pp. 213–36.

Atkinson, A.B. (1995), *Incomes and the Welfare State*, Cambridge: Cambridge University Press.

Atkinson, A.B. and Micklewright, J. (1992), *Economic Transformation in Eastern Europe and the Distribution of Income*, Cambridge: Cambridge University Press.

Atkinson, J. (1985), Flexibility, Uncertainty and Management, IMS report no. 89, Brighton: Institute of Manpower Studies, University of Sussex.

Aviss, L. (1995), 'Micro-chips or *pommes frites* – can you tell them apart?', paper presented to the Conference on Education and Training for the Future Labour Markets of Europe, Durham, University of Durham, 21–24 September.

Ayubi, N. (1991), *Political Islam: Religion and Politics in the Arab World*, London: Routledge.

Azzam, M. (1994), 'Islam: political implications for Europe and the Middle East', in P. Ludlow (ed.), *Europe and the Mediterranean*, London: Brassey's, pp. 89–104.

Bachtler, J. and Michie, R. (1995), 'A new era in EU regional policy evaluation? The appraisal of structural funds', *Regional Studies*, 29 (8): 745–51.

Bagnasco, A. (1977), *Tre Italie: la problematica territoriale dello sviluppo Italiano*, Bologna: Il Mulino.

Bailly, A., Jensen-Butler, C. and Leontidou, L. (1996), 'Changing cities: restructuring, marginality and policies in urban Europe', *European Urban and Regional Studies*, 3 (2): 161–76.

Baimbridge, M., Burkitt, B. and Macey, M. (1994), 'The Maastricht Treaty: exacerbating racism in Europe?', *Ethnic and Racial Studies*, 18: 420–41.

Baker, C. (1996), *Foundations of Bilingual Education and Bilingualism*, 2nd edn, Clevedon, Avon: Multilingual Matters.

Bakker, T. (1992), 'All aboard for allfinanz', *The Banker*, September: 65–7.

Balakrishnan, G. (ed.) (1996), *Mapping the Nation*, London: Verso.

Balaz, V. (1996), 'The Wild East? Capital markets in the V4 countries', *European Urban and Regional Studies*, 3 (3): 251–7.

Balcells, A. (1996), *Catalan Nationalism*, London: Macmillan.

Banfield, E.C. (1958), *The Moral Basis of a Backward Society*, Glencoe, Ill.: The Free Press.

Bank of England (1993), 'Cross-border alliances in banking and financial services', *Bank of England Quarterly Bulletin*, 33: 372–78.

The Banker (1990), '1992: the key directives', November, supplement.

The Banker (1991), 'Foreign banks', November: 58–85.

The Banker (1996a), 'Foreign banks in London', November: 42–57.

The Banker (1996b), 'Italy: dash for the Euro', August: 25–29.

Banks, A., Day, A. and Muller, T. (eds) (1996), *Political Handbook of the World: 1995–1996*, Binghampton, NY: CSA Publications.

Barcelona Declaration (1995), *Bulletin of the European Union*, 11–1995: 136–45.

Barnes, I. and Barnes, P. (1995), *The Enlarged European Union*, Harlow: Longman.

Barrientos, S. and Perrons, D. (1996), 'Fruit of the vine: linkages between flexible women workers in the production and retailing of winter fruit', paper presented to the Conference on the Globalization of Production and the Regulation of Labour, University of Warwick, September.

Batt, J. (1994), 'Political dimensions of privatization in Eastern Europe', in S. Estrin (ed.), *Privatization in Central and Eastern Europe*, Longman: Harlow, pp. 83–91.

Bazo, M-T. (1997), 'New social policies on living arrangements in Spain to promote elder independence', paper presented to the World Congress of Gerontology, Adelaide, August (Faculty of Economic Science, Universidad del País Vasco, Bilbao).

Beck, U. (1992), *Risk Society: Towards a New Modernity*, London: Sage.

Bedem, R. van den (1994), 'Towards a system of plural nationality in the Netherlands: changes in regulations and perceptions', in R. Baubock (ed.), *From Aliens to Citizens: Redefining the Status of Immigrants in Europe*, Aldershot: Avebury, pp. 95–110.

Begg, I. and Green, D. (1995), 'Banking supervision in Europe and economic and monetary union', *South Bank European Papers* 3/95 (available from the European Institute, South Bank University, 103 Borough Road, London SE1 0AA).

Belka, M. (1994), 'Financial restructuring of banks and enterprises in Poland', *Moct-Most*, 4 (3): 71–84.

Benhabib, S. (ed.) (1996), *Democracy and Difference: Contesting the Boundaries of the Political*, Princeton, NJ: Princeton University Press.

Benjamin, W. (1973), *Illuminations*, London: Fontana.

Best, M.H. (1990), *The New Competition: Institutions of Industrial Restructuring*, Cambridge: Polity Press.

Beynon, H. (1984), *Working for Ford*, Harmondsworth: Penguin.

Beynon, H. (1995), 'The changing experience of work: Britain in the 1990s', paper presented to the Conference on Education and Training for the Future Labour Markets of Europe, University of Durham, 21–24 September.

Beynon, H., Hudson, R. and Sadler, D. (1994), *A Place Called Teesside: a Locality in a Global Economy*, Edinburgh: Edinburgh University Press.

Bissoondath, N. (1994), *The Cult of Multiculturalism in Canada*, Toronto: Penguin.

Blacksell, M. and Williams, A. (eds) (1994), *The European Challenge: Geography and Develoment in the European Community*, Oxford: Oxford University Press.

Bloomfield, J. (1993), 'The new Europe: a new agenda for research?', in M. Fulbrook (ed.), *National Histories and European History*, London: UCL Press.

Böhning, W.R. (1972), *The Migration of Workers in the United Kingdom and the European Community*, London: Oxford University Press.

Bosch, G. (1995), *Flexibility and Work Organisation*, Brussels: Social Europe, Commission of the European Communities.

Bourcier, V. (1996), 'EMU: the impact of transition on the French financial market and the MATIF', *ECU-Europe* 37 (http://www.ecu-activities.be).

Boyer, R. (1987), 'Labour flexibilities: many forms, uncertain effects', *Labour and Society*, 12 (1): 107–29.

Boyer, R. (1996), 'State and market: a new engagement for the twenty-first century?', in R. Boyer and D. Drache (eds), *States against Markets: the Limits of Globalization*, London: Routledge, pp. 84–116.

Boyer, R. and Drache, D. (eds) (1996), *States against Markets: the Limits of Globalization*, London: Routledge.

Braudel, F. (1972), *The Mediterranean and the Mediterranean World in the Age of Philip II*, London: Collins.

Braverman, H. (1974), *Labors and Monopoly Capital*, New York: Monthly Review Press.

Brown, B. (1996), 'EMU: implications for international debt markets', paper presented at Euromoney Seminar, London, 17 October.

Brown, J.F. (1994), *Hopes and Shadows: Eastern Europe after Communism*, Harlow: Longman.

Brown, J.M. (1997), 'Dublin: brokers see key role within Europe', *Financial Times*, 16 November: 8.

Brown, M. (1997), *Replacing Citizenship: AIDS Activism and Radical Democracy*, New York, NY: Guilford Press.

Brown, P. and Crompton, R. (eds) (1994), *A New Europe? Economic Restructuring and Social Exclusion*, London: UCL Press.

Brubaker, W.R. (1992a), 'Citizenship struggles in Soviet successor states', *International Migration Review*, 26: 269–91.

Brubaker, W.R. (1992b), *Citizenship and Nationhood in France and Germany*, Cambridge, Mass.: Harvard University Press.

Bruegel, I. and Perrons, D. (1995), 'Where do the costs of unequal treatment fall? An analysis of the incidence of the costs of unequal pay and sex discrimination in the UK', *Gender, Work and Organization*, 2 (3): 113–24.

Bruegel, I. and Perrons, D. (1998), 'Deregulation and women's employment: the diverse experiences of women in Britain', *Feminist Economics*, 4 (1): 103–25.

Brunn, S. and Leinbach, T.R. (eds) (1991), *Collapsing Space and Time: Geographic Aspects of Communication and Information*, London: HarperCollins.

Buckler, A. (1991), *Illegal Border Crossers: Experiences from the Years 1947–1961 on the German–German Frontier*, Leipzig: Thomas-Verlag.

Bugajski, J. (1995), *Nations in Turmoil: Conflict and Cooperation in Eastern Europe*, Boulder, Col.: Westview Press.

Bugie, S. (1996), 'EMU and the banks: costs vs implications', *Standard and Poor's Creditweek*, 13 November.

Burgat, F. (1988), *L'Islamisme au Maghreb*, Paris: Karthala.

Business in the Community (1993), *Corporate Culture and Caring: the Business Case for Family-friendly Provision*, London: Business in the Community/Institute of Personnel Management.

Butterwegge, C. and Jäger, S. (eds) (1992), *Rassismus in Europa*, Cologne: Bund-Verlag.

Buzan, B. (1991), *People, States and Fear*, New York: Harvester Wheatsheaf.

Callahan, D. (1987), *Setting Limits: Medical Goals in an Aging Society*, Washington, DC: Georgetown University Press.

Callahan, D. (1994), *What Kind of Life? The Limits of Medical Progress*, Washington, DC: Georgetown University Press.

Caritas di Roma (1996), *Immigrazione dossier statistico '96*, Rome: Anterem.

Carnevali, F. (1996), 'Between markets and networks: regional banks in Italy', *Business History*, 38: 84–100.

Carter, E., Donald, J. and Squires, J. (eds) (1993), *Space and Place: Theories of Identity and Location*, London: Lawrence and Wishart.

Carter, F.W., Hall, D.R., Turnock, D. and Williams, A.M. (1996), *Interpreting the Balkans*, London: Royal Geographical Society, Geographical Intelligence paper no. 2.

Caselli, G. (1994), *Long-term Trends in European Mortality*, Studies on Medical and Population Subjects no. 56, London: OPCS.

Caselli, G. and Lopez, A.D. (eds) (1996), *Health and Mortality among European Populations*, Oxford: Clarendon Press.

Casey, B. (1996), 'The German pension reform of 1989', in P. Lloyd-Sherlock and

P. Johnson (eds), *Ageing and Social Policy: Global Comparisons*, STICERD occasional paper 19, London: London School of Economics, pp. 36–43.

Castells, M. (1983), *The City and the Grassroots*, London: Arnold.

Castells, M. (1996), *The Rise of the Network Society*, Oxford: Blackwell.

Castles, S. (1993), 'Migrations and minorities in Europe. Perspectives for the 1990s: eleven hypotheses', in J. Wrench and J. Solomos (eds), *Racism and Migration in Western Europe*, Oxford: Berg, pp. 17–34.

Castles, S. (1995), 'How nation-states respond to immigration and ethnic diversity', *New Community*, 21: 293–308.

Castles, S. and Kosack, G. (1973), *Immigrant Workers and Class Structure in Western Europe*, London: Oxford University Press.

Castles, S. and Miller, M.J. (1993), *The Age of Migration*, Basingstoke: Macmillan.

Castles, S., Booth, H. and Wallace, T. (1984), *Here for Good: Western Europe's New Ethnic Minorities*, London: Pluto Press.

Castro, B. (ed.) (1996), *Business and Society*, Oxford: Oxford University Press.

Cecchini, P. (1988), *The European Challenge 1992: the Benefits of a Single Market*, Aldershot: Wildwood House.

Central Statistical Office (1997), *Ireland: Statistical Bulletin*, vol. 72, no. 1, Dublin: Central Statistical Office.

Centre for Contemporary Cultural Studies (1982), *The Empire Strikes Back: Race and Racism in 70s Britain*, London: Hutchinson.

Centre for the Study of Public Policy (1996), 'Support for democracy and market system rising in Central and Eastern Europe', Glasgow: University of Strathclyde (http://www.strath.ac.uk/Departments/CSPP/ndb4pr2.html).

Cerny, P. (1990), *The Changing Architecture of Politics: Structure, Agency and the Future of the State*, London: Sage.

Cerny, P. (1993), 'The deregulation and re-regulation of financial markets in a more open world', in P. Cerny (ed.), *Finance and World Politics: Markets, Regimes and States in the Post-hegemonic Era*, Aldershot: Edward Elgar, pp. 51–85.

Chesnais, J-C. (1986), *La transition démographique: etapes, formes, implication économiques*, Paris: Presses Universitaires de France.

Churchill, W. (1947), Speech to the House of Commons, November.

Clark, R.L. and Spengler, J.J. (1980), *The Economics of Individual and Population Aging*, Cambridge: Cambridge University Press.

Clark, R.P. (1979), *The Basques: the Franco Years and Beyond*, Reno, NV: University of Nevada Press.

Clark, R.P. (1984), *The Basque Insurgents, ETA, 1952–1980*, Madison, Wisc.: The University of Wisconsin Press.

Clerc, D. (1995), 'Production of marginals', *Le Monde diplomatique*, 6: 14–16.

Clifford, J. and Marcus, G.E. (eds) (1986), *Writing Culture. The Poetics and Politics of Ethnography*, Chicago: University of Chicago Press.

Coale, A. and Watkins, S.C. (eds) (1986), *The Decline in Fertility in Europe*, Princeton, NJ: Princeton University Press.

Cockfield, F.A.B. (1985), *Completing the Internal Market*, Brussels: Commission of the European Communities.

Coe, N. (1997), 'US transnationals and the Irish software industry: assessing the nature, quality and stability of a new wave of foreign direct investment', *European Urban and Regional Studies*, 4 (3): 211–30.

Cole, J. and Cole, F. (1993), *The Geography of the European Community*, London: Routledge.

Commissariat Général du Plan (1988), *Immigration: Le devoir d'insertion*, Paris: La Documentation Française.

Commission of the European Communities (1988), 'The Economics of 1992', *European Economy*, 35.

Commission of the European Communities (1993a), *Social Protection in Europe*, Brussels: Commission of the European Communities, Directorate General V.

Commission of the European Communities (1993b), *Growth, Competitiveness and Employment: the Challenges and Ways Forward into the 21st Century*, White Paper and White Paper Part C, Luxembourg: Commission of the European Communities.

Commission of the European Communities (1993c), *European Social Policy: Options for the Union*, Green Paper, Luxembourg: Directorate-General for Employment, Industrial Relations and Social Affairs, Office for Official Publications of the EC.

Commission of the European Communities (1994a), *Employment in Europe, 1994*, Brussels: Commission of the European Communities.

Commission of the European Communities (1994b), *Competitiveness and Cohesion: Trends in the Regions*, Fifth Periodic Report on the Social and Economic Situation and Development of the Regions in the Community, Luxembourg: Commission of the European Communities.

Commission of the European Communities (1994c), *Democracy at Work in the European Union*, Brussels: Commission of the European Communities.

Commission of the European Communities (1994d), *Growth, Competitivity and Employment*, White Paper, Luxembourg: Directorate-General for Employment, Industrial Relations and Social Affairs, Office for Official Publications of the EC.

Commission of the European Communities (1995a), *Competitiveness and Cohesion: Trends in the Regions*, Brussels: Commission of the European Communities.

Commission of the European Communities (1995b), *Social Protection in Europe*, Brussels: Directorate General Employment, Industrial Relations and Social Affairs.

Commission of the European Communities (1995c), *Eastern and Central Europe 2000+, Final Report, Studies 2*, Brussels: Commission of the European Communities.

Commission of the European Communities (1995d), *Intergovernmental Conference 1996 – Commission Report for the Reflection Group*, Luxembourg: Office for Official Publications of the Commission of the European Communities.

Commission of the European Communities (1995e), *Employment in Europe*, Luxembourg: Commission of the European Communities.

Commission of the European Communities (1996a), *Employment in Europe*, Luxembourg: Commission of the European Communities.

Commission of the European Communities (1996b), *Reform Processes and Spatial Development in Central and Eastern Europe*, Brussels: Commission of the European Communities.

Commission of the European Communities (1996c), *Environmental Taxes, Implementation and Environmental Effectiveness*, Copenhagen: European Environmental Agency.

Commission of the European Communities (1996d), *The Demographic Situation in the European Union 1995*, COM(96), 60 final, Brussels: Commission of the European Communities .

Commission of the European Communities (1996e), *Central and Eastern Euro-barometer*, no. 6, Brussels: Commission of the European Communities (http://europe.eu.int/en/comm/dg10/infcom/epo/ceeb6/en/chap1-4.htm).

Commission of the European Communities (1996f), *First Report on Economic and Social Cohesion*, preliminary edition, Luxembourg: Commission of the European Communities.

Commission of the European Communities (1996g), *Incorporating Equal Opportunities for Women and Men into all Community Policies and Activities*, Luxembourg: European Commission Com (96), 67 final CB-CO-96-083-En-C.

Commission of the European Communities (1997a), *Commission Opinion on Poland's Application for Membership of the European Union*, Brussels: European Commission Com (97), 2002.

Commission of the European Communities (1997b), *Equal Opportunities for Women and Men in the European Union, 1996*, Brussels: Directorate General for Employment, Industrial Relations and Social Affairs.

Commission of the European Communities (1997c), *Agenda 2000 – Summary and Conclusions of the Opinions of Commission Concerning the Applications for Membership to the European Union Presented by the Candidates Countries*, Brussels: Commission of the European Communities.

Commission of the European Communities (1997d), *Agenda 2000: for a Stronger and Wider Union*, Brussels: European Commission Com (97), 2000.

Commission of the European Communities (1997e), *A New Treaty for Europe*, Brussels: Commission of the European Communities.

Commission of the European Communities (1997f), *Commission Opinion on Hungary's Application for Membership of the European Union*, Brussels: European Commission Com (97), 2001.

Commission of the European Communities (1997g), *Commission Opinion on Romania's Application for Membership of the European Union*, Brussels: European Commission Com (97), 2003.

Commission of the European Communities (1997h), *Commission Opinion on Slovakia's Application for Membership of the European Union*, Brussels: European Commission Com (97), 2004.

Commission of the European Communities (1997i), *Commission Opinion on Latvia's Application for Membership of the European Union*, Brussels: European Commission Com (97), 2005.

Commission of the European Communities (1997j), *Commission Opinion on Estonia's Application for Membership of the European Union*, Brussels: European Commission Com (97), 2006.

Commission of the European Communities (1997k), *Commission Opinion on Lithuania's Application for Membership of the European Union*, Brussels: European Commission Com (97), 2007.

Commission of the European Communities (1997l), *Commission Opinion on Bulgaria's Application for Membership of the European Union*, Brussels: European Commission Com (97), 2008.

Commission of the European Communities (1997m), *Commission Opinion on the Czech Republic's Application for Membership of the European Union*, Brussels: European Commission Com (97), 2009.

Commission of the European Communities (1997n), *Commission Opinion on Slovenia's Application for Membership of the European Union*, Brussels: European Commission Com (97), 2010.

Commission of the European Communities (1997o), *First Cohesion Report*, Brussels: Commission of the European Communities.

Commission of the European Communities (1998), *Flexible Working and the Reconciliation of Work and Family Life or a New Form of Precariousness*, Brussels: Commission of the European Communities.

Conti, S. and Enrietti, A. (1995), 'The Italian automobile industry and the case of Fiat: one country, one company, one market', in R. Hudson and E.W. Schamp (eds), *Towards a New Map of Automobile Manufacturing in Europe? New Production Concepts and Spatial Restructuring*, Berlin: Springer, pp. 117–46.

Cooke, P. (1989), 'Ethnicity, economy and civil society: three theories of political regionalism', in C.H. Williams and E. Kofman (eds), *Community Conflict, Partition and Nationalism*, London: Routledge.

Cooke, P. (1995), 'Keeping to the high road: learning, reflexivity and associative governance in regional economic development', in P. Cooke (ed.), *The Rise of the Rustbelt*, London: University of London Press, pp. 231–46.

Cortie, C. and Kesteloot, C. (1998), 'Housing Turks and Moroccans in Brussels and Amsterdam: the difference between private and public markets', *Urban Studies*, 35.

Costa-Lascoux, J. (1993), 'Continuité ou rupture dans la politique française de l'immigration: les lois de 1993', *Revue Européenne des Migrations Internationales*, 9: 233–61.

Costa-Lascoux, J. (1994), 'French legislation against racism and discrimination', *New Community*, 20: 371–9.

Costa-Lascoux, J. (1995), 'La lutte contre le racisme en Europe. I: Les instruments internationaux', *Revue Européenne des Migrations Internationales*, 11: 205–20.

Council of Europe (1985), *Changing Age Structure of the Population and Future Policy*, Population Studies no. 18, Strasbourg: Council of Europe.

Council of Europe (1992), *European Charter for Regional or Minority Languages*, Strasbourg: Council of Europe.

Council of Europe (1996), *Recent Demographic Developments in Europe, 1996*, Strasbourg: Council of Europe.

Cram, L. (1993), 'Calling the tune without paying the piper? Social policy regulation: the role of the Commission in European Community social policy', *Policy and Politics*, 21: 135–46.

Crewe, L. and Davenport, E. (1992), 'The puppet show: changing buyer–supplier relations within clothing retailing', *Transactions of the Institute of British Geographers*, n.s. 17: 183–97.

Cross, M. and Keith, M. (eds) (1993), *Racism, the City and the State*, London: Routledge.

Czech Statistical Office (1997), *Statistical Yearbook*, Prague: Czech Statistical Office.

Czechoslovakia Federal Statistical Office (1990), *The 1990 Yearbook of the Czech and Slovak Federative Republic*, Prague: Federal Statistical Office.

Daatland, S.O. (1992), 'Ideals lost? Current trends in Scandinavian welfare policies on ageing', *Journal of European Social Policy*, 2: 33–47.

Dahrendorf, R. (1990), *Reflections on the Revolution in Europe*, London: Chatto and Windus.

Dalby, S. (1992), 'Security, modernity, ecology: the dilemmas of post-Cold War security discourse', *Alternatives*, 17 (1): 95–134.

Dallago, B. (1995), 'Privatization in Europe: a comparison', in R. Daviddi (ed.), *Property Rights and Privatization in the Transition to a Market Economy*, Maastricht: European Institute of Public Administration, pp. 231–65.

Dangerfield, M.V. (1995), 'The economic opening of Central and Eastern Europe: continuity and change in foreign economic relations', *Journal of European Integration*, 19 (1): 5–42.

Danthine, J-P. (1993), 'Comment', in J. Dermine (ed.), *European Banking in the 1990s*, 2nd edn, Oxford: Blackwell, pp. 370–2.

Davidson, C. (ed.) (1994) *Anyway*, New York: Anyone Corporation and Rizzoli International.

Davis, J. (1962), *Land and Family in Pisticci*, London: Athlone Press.

Dean, M. (1994), *Critical and Effective Histories: Foucault's Methods and Historical Sociology*, London: Routledge.

Declaració de Barcelona (1996), *Declaració universal de drets lingüístics*, Barcelona: International PEN and CIEMEN.

De la Serre, F. (1981), 'The Community's Mediterranean policy after the Second Enlargement', *Journal of Common Market Studies*, 19 (4), 377–87.

Delors, J. (1989), *Report on Economic and Monetary Union in the European*

Community, Luxembourg: Office for the Official Publications of the European Communities.

Dermine, J. (1996), 'European banking with a single currency', Wharton Financial Institutions Centre working paper 96–54.

Devetak, S., Flere, S. and Seewann, G. (eds) (1993), *Small Nations and Ethnic Minorities in an Emerging Europe*, Munich: Slavica Verlag Dr Anton Kovac.

Dex, S. and McCulloch, A. (1995), *Flexible Employment in Britain: a Statistical Analysis*, Research Discussion Series 15, Manchester: Equal Opportunities Commission.

Diaz, V.M. (1993), *The Return of Civil Society*, Cambridge, Mass.: Harvard University Press.

Dicken, P. (1994), *Global Shift*, London: Paul Chapman.

Dicken, P. and Oberg, S. (1996), 'The global context: Europe in a world of dynamic economic and population change', *European Urban and Regional Studies*, 3 (2): 101–20.

Dicken, P., Forsgren, M. and Malmberg, A. (1994), 'The local embeddedness of transnational corporations', in A. Amin and N.J. Thrift (eds), *Globalization, Institutions and Regional Development in Europe*, Oxford: Oxford University Press.

Dijk, T.A. van (1993), 'Denying racism: elite discourse and racism', in J. Wrench and J. Solomos (eds), *Racism and Migration in Western Europe*, Oxford: Berg, pp. 179–94.

Dixon, S. (1996), 'The Russians and the Russian question', in G. Smith (ed.), *The Nationalities Question in the Post-Soviet States*, Harlow: Longman, pp. 47–74.

Dooghe, G. (1991), *The Ageing of the Population in Europe: Socio-Economic Characteristics of the Elderly Population*, Brussels: Centrum voor Bevolkigns en Gezinsstudien, Ministerie Vlaamse Gemeenschap.

Dow, S.C. (1994), 'European monetary integration and the distribution of credit availability', in S. Corbridge, N.J. Thrift and R. Martin (eds), *Money, Power and Space*, Oxford: Blackwell, pp. 149–64.

Drèze, J. and Malinvaud, E. (1994), 'Growth and employment: the scope for a European initiative', *European Economy*, 43: 77–106.

Drobizheva, L., Gottemoeller, R., Kelleher, C.M. and Walker, L. (1996), *Ethnic Conflict in the Post-Soviet World*, New York, M.E. Sharpe.

Duke, V. and Grime, K. (1997), 'Inequality in post-communism', *Regional Studies*, 31 (9): 883–90.

Duncan, S. (1996), 'The diverse worlds of European patriarchy', in M.D. García-Ramon and J. Monk (eds), *Women of the European Union: the Politics of Work and Daily Life*, London: Routledge.

Dunford, M. (1994), 'Winners and losers: the new map of economic inequality in the European Union', *European Urban and Regional Studies*, 1 (2): 95–114.

Dunford, M. (1997), 'Mediterranean economies: the dynamics of uneven development', in R. King, L. Proudfoot and B. Smith (eds), *The Mediterranean: Environment and Society*, London: Arnold, pp. 126–54.

Dunford, M. (1998), 'Differential development, institutions, modes of regulation and comparative transitions to capitalism: Russia, CIS and the GDR', in J. Pickles and A. Smith (eds), *Theorising Transition*, London: Routledge. pp. 76–111.

Dunford, M. and Hudson, R. (1996), *Successful European Regions: Northern Ireland Learning from Others?*, Belfast: Northern Ireland Economic Council.

Dunford, M. and Perrons, D. (1994), 'Regional inequality, regimes of accumulation and unequal development in contemporary Europe', *Transactions of the Institute of British Geographers*, n.s., 19: 163–82.

Dworkin, D.L. and Roman, L.G. (1993), *Views Beyond the Border Country: Raymond Williams and Cultural Politic*, New York: Routledge.

Dwyer, C. and Meyer, A. (1995), 'The institutionalisation of Islam in the Netherlands and in the UK: the case of Islamic schools', *New Community*, 21: 37–54.

Dyson, K., Featherstone, K. and Michalopoulos, G. (1995), 'Strapped to the mast: EC central bankers between global financial markets and regional integration', *Journal of European Public Policy*, 2: 465–87.

East, R. and Pontin, J. (eds) (1997), *Revolution and Change in Central and Eastern Europe*, London: Pinter.

Edwards, J. (1994), *Multilingualism*, London: Longman.

Ehrlich, C. (1997), 'Federalism, regionalism, nationalism: a century of Catalan political thought and its implications for Scotland in Europe', *Space and Polity*, 1: 205–24.

Ehrlich, E. and Révész, G. (1997), 'The economy and some key fields of the social and human dimension in Hungary', in R. Weichhardt (ed.), *Economic Developments and Reforms in Cooperation Partner Countries: the Social and Human Dimension*, Brussels: NATO, pp. 289–309.

EKKE (National Centre of Social Research) (1996), *Dimensions of Social Exclusion in Greece*, 2 vols (in Greek), Athens: Report for the European Social Fund.

Emerson, M. (1990), 'One market, one money', *European Economy*, 44.

Engels, F. (1984 edn), *The Condition of the Working Class in England*, London: Lawrence and Wishart.

Enjeux: Les Echos (1997), *Le reveil des regions*, special issue no. 127, July–August.

Enoch, Y. (1994), 'The intolerance of a tolerant people: ethnic relations in Denmark', *Ethnic and Racial Studies*, 18: 282–300.

Entzinger, H. (1994), 'Y a-t-il un avenir pour le modèle néerlandais des "minorités ethniques"?', *Revue Européenne des Migrations Internationales*, 10: 73–94.

Escott, K. and Whitfield, D. (1995), *The Gender Impact of CCT in Local Government*, Manchester: Equal Opportunities Commission.

Esping-Andersen, G. (1990), *The Three Worlds of Welfare Capitalism*, Cambridge: Polity Press.

Esping-Andersen, G. (1994), *After the Golden Age: the Future of the Welfare State in the New Global Order*, Occasional Paper 7, Geneva: World Summit for Social Development.

Esping-Andersen, G. (ed.) (1996a), *Welfare States in Transition: National Adaptations in Global Economies*, London: Sage.

Esping-Andersen, G. (1996b), 'After the golden age? Welfare state dilemmas in a global economy', in G. Esping-Andersen (ed.), *Welfare States in Transition: National Adaptations in Global Economies*, London: Sage, pp. 1–31.

Esping-Andersen, G. (1996c), 'Welfare states without work: the impasse of labour shedding and familialism in continental European social policy', in G. Esping-Andersen (ed.), *Welfare States in Transition: National Adaptations in Global Economies*, London: Sage, pp. 66–87.

Esposito, J.L. (1992), *The Islamic Threat: Myth or Reality?*, New York: Oxford University Press.

Estrin, S. (1994), 'Economic transition and privatization: the issues', in S. Estrin (ed.), *Privatization in Central and Eastern Europe*, Longman: Harlow, pp. 3–30.

Estrin, S., Hughes, K. and Todd, S. (1997), *The Nature and Scope of Foreign Direct Investment in Central and Eastern Europe*, European Programme Working Papers no. 3, London: Royal Institute for International Affairs.

Eurolink Age (1997), *Public Policy Options to Assist Older Workers*, Brussels and London: Eurolink Age.

Eurostat (1981), *Review 1970–1979*, Brussels: Commission of the European Communities.

Eurostat (1996), *Yearbook '96*, Brussels: Commission of the European Communities.

Eurostat (1997a), *Statistics in Focus: Population and Social Conditions*, Brussels: Commission of the European Communities.

Eurostat (1997b), 'Part-time work in the European Union', Luxembourg: Eurostat.

Evans, G. (1996), *For the Sake of Wales: the Memoirs of Gwynfor Evans*, Bridgend: Welsh Academic Press.

Evans, M. (1996), 'Languages of racism in contemporary Europe', in B. Jenkins and S.A. Sofos (eds), *Nation and Identity in Contemporary Europe*, London: Routledge, pp. 33–53.

Faist, T. (1993), 'From school to work: public policy and underclass formation among young Turks in Germany during the 1980s', *International Migration Review*, 27: 306–31.

Faist, T. (1995), 'Ethnicization and racialization of welfare-state politics in Germany and the USA', *Ethnic and Racial Studies*, 18: 219–50.

Fakiolas, R. and King, R. (1996), 'Emigration, return, immigration: a review and evaluation of Greece's postwar experience of international migration', *International Journal of Population Geography*, 2 (2): 171–90.

Federal Trust (1995), 'Towards the single currency', London: Federal Trust Working Papers no. 2.

Ferrão, J. and Vale, M. (1995), 'Multi-purpose vehicles: a new opportunity for the periphery?', in R. Hudson and E.W. Schamp (eds), *Towards a New Map of Automobile Manufacturing in Europe? New Production Concepts and Spatial Restructuring*, Berlin: Springer.

Ferri, E. and Smith, K. (1996), *Parenting in the 1990s*, London: Family Studies Centre/Joseph Rowntree Foundation.

Fijalkowski, J. (1994), 'Solidarités intra-communautaires et formations d'associations au sein de la population étrangère d'Allemagne', *Revue Européenne des Migrations Internationales*, 10: 33–57.

Financial Times (1997), 'LIFFE: link up threat to market's supremacy', *Financial Times*, 18 September: 3.

Fleras, A. and Elliott, J.L. (1992), *Multiculturalism in Canada*, Scarborough: Nelson Canada.

Fogel, D.S. and Etcheverry, S. (1994), 'Reforming the economies of Central and Eastern Europe', in S.D. Fogel (ed.), *Managing in Emerging Market Economies: Cases from the Czech and Slovak Republics*, Boulder, Col.: Westview Press, pp. 3–33.

Folch-Serra, M. and Nogué-Font, J. (1996), 'Mapping contemporary Catalanism: the production of meaning in municipal and district publications', London, Ontario: UWO (mimeo).

Fothergill, S. (1998), 'The premature death of EU regional policy', *European Urban and Regional Studies*, 5: 183–8.

Fouéré, Y. (1984), *L'Europe des Régions*, St Brieuc: Les Cahiers de l'Avenir, 14.

Fraser, D. (1982), *The Evolution of the British Welfare State*, London: Macmillan.

Freeman, C. (1987), *Technology Policy and Economic Performance: Lessons from Japan*, London: Frances Pinter.

Freeman, C. and Soete, L. (1994), *Work for All or Mass Unemployment: Computerised Technical Change into the 21st Century*, London: Pinter.

Fröbel, F., Heinrichs, J. and Kreye, O. (1980), *The New International Division of Labour*, Cambridge: Cambridge University Press.

Fukuyama, F. (1992), *The End of History and the Last Man*, London: Hamish Hamilton.

Fuller, G.E. and Lesser, I.O. (1995), *A Sense of Siege: the Geopolitics of Islam and the West*, Boulder, Col.: Westview/RAND.

Galtung, J. (1995), 'Rethinking the future of cultural identity: nationalism reconsidered', symposium honouring Sven Tägil, Lund: Lund University (mimeo).

Gardener, E.P.M. and Molyneux, P. (1990), *Changes in Western European Banking*, London: Unwin Hyman.

Garofoli, G. (1986), 'Le development peripherique en Italie', *Economie et Humanisme*, 289: 30–6.

Garrahan, P. and Stewart, P. (1992), *The Nissan Enigma: Flexibility at Work in a Local Economy*, London: Mansell.

Gates, H. (1995), 'Goodbye, Columbus? Notes on the culture of criticism', in T.D. Goldberg (ed.), *Multiculturalism: a Critical Reader*, Oxford: Blackwell.

Gellner, E. (1983), *Nations and Nationalism*, Oxford: Basil Blackwell.

Gellner, E. (1992), *Postmodernism, Reason and Religion*, London: Routledge.

Gellner, E. (1994), *Encounters with Nationalism*, Oxford: Blackwell.

Gerholm, T. and Lithman, Y.G. (eds) (1988), *The New Islamic Presence in Western Europe*, London: Mansell.

Gibson-Graham, J.K. (1995), 'Identity and economic plurality: rethinking capitalism and capitalist hegemony', *Society and Space*, 13: 275–82.

Giddens, A. (1985), *The Nation-state and Violence*, Cambridge: Polity Press.

Giddens, A. (1990), *The Consequences of Modernity*, Cambridge: Polity Press.

Giddens, A. (1994), *Beyond Left and Right: the Future of Radical Politics*, Cambridge: Polity Press.

Giddens, A. (1996), 'Affluence, poverty and the idea of a post-scarcity society', *Development and Change*, 27: 365–77.

Gillespie, R. (ed.) (1994), *Mediterranean Politics*, London: Pinter.

Girard, A. (1977), 'Opinion publique, immigration et immigrés', *Ethnologie Française*, 7: 219–28.

Glazer, N. (1977), 'Individual rights against group rights', in E. Kamenka (ed.), *Human Rights*, London: E. Arnold.

Glenny, M. (1993), *The Fall of Yugoslavia*, London: Penguin.

Glover, J. and Arber, S. (1995), 'Polarisation in mother's employment', *Gender, Work and Organisation*, 2 (4): 165–79.

Goddard, V.A. (1994), 'From the Mediterranean to Europe: honour, kinship, and gender', in V.A. Goddard, J.R. Llobera and C. Shore (eds), *The Anthropology of Europe: Identities and Boundaries in Conflict*, Oxford: Berg, pp. 57–92.

Godfrey, M. (1986), *Global Unemployment*, Sussex: Wheatsheaf Books.

Goldberg, T.D. (ed.) (1995), *Multiculturalism: a Critical Reader*, Oxford: Blackwell.

Golini, A., Gesano, G. and Heins, F. (1991) 'South–north migration with special reference to Europe', *International Migration*, 29 (2): 253–77.

Gonas, L. (1998), 'Has equality gone too far? Changing labour market regimes and new employment patterns in Sweden', *European Urban and Regional Studies*, 5 (1): 41–54.

Gordon, M.S. (1988), *Social Security Policies in Industrial Countries: a Comparative Analysis*, Cambridge: Cambridge University Press.

Gorzelak, G. (1996), *The Regional Dimension of Transformation in Central Europe*, London: Jessica Kingsley/Regional Studies Association.

Göting, U. (1996), In Defence of Welfare: Social Protection and Social Reform in Eastern Europe, working paper 96/42, Florence: European University Institute.

Gow, J.W. (1991), 'Deconstructing Yugoslavia', *Survival*, 33 (4): 291–311.

Gowan, P. (1995), 'Neo-liberal theory and practice for Eastern Europe', *New Left Review*, 213: 3–60.

Gowan, P. (1996), 'Eastern Europe, Western power and neo-liberalism', *New Left Review*, 216: 129–40.

Grabher, G. and Stark, D. (eds) (1997), *Restructuring Networks in post-Socialism: Legacies, Linkages and Localities*, Oxford: Oxford University Press.

Graham, B. (1997), 'The Mediterranean in the Medieval and Renaissance world', in R. King, L. Proudfoot and B. Smith (eds), *The Mediterranean: Environment and Society*, London: Arnold, pp. 75–93.

Graham, B. (ed.) (1998), *Modern Europe: Place, Culture, Identity*, London: Arnold.

Graham, G. and Martinson, J. (1997), 'The $600 bn question', *Financial Times*, 8 December: 19.

Graham, J. (1992), 'Post-Fordism as politics: the political consequences of narratives on the left', *Environment and Planning D: Society and Space*, 10: 393–410.

Gregory, A. and O'Reilly, J. (1996), 'Checking out and cashing up: the prospects and paradoxes of regulating part-time work in Europe', in R. Crompton, D. Gallie and K. Purcell (eds), *Changing Forms of Employment: Organisation, Skills and Gender*, London: Routledge, pp. 207–34.

Grenon, M. and Batisse, M. (1989), *Futures for the Mediterranean Basin*, Oxford: Oxford University Press.

Gruen, E. (ed.) (1970), *Imperialism in the Roman Republic*, New York: Holt, Rinehart and Winston.

Gruffydd, P. (1994), 'Back to the land: historiography, rurality and the nation in interwar Wales', *Transactions of the Institute of British Geographers*, 19: 61–77.

Guibernau, M. (1995), 'Spain: a federation in the making?', in G. Smith (ed.), *Federalism: the Multiethnic Challenge*, London: Longman, pp. 239–54.

Guoth, J. (1994), *Issuing Activities of Investment 'Privatization' Funds* (in Czech), Prague: Národní hospodárství, 3–99, pp. 16–17.

Habermas, J. (1975), *Legitimation Crisis*, London: Heinemann.

Habermas, J. (1996), 'The European nation-state – Its achievements and its limits', in G. Balakrishnan and B. Anderson (eds), *Mapping the Nation*, London: Verso.

Hadjimichalis, C. (1987), *Uneven Development and Regionalism: State, Territory and Class in Southern Europe*, London: Croom Helm.

Hadjimichalis, C. (1994), 'The fringes of Europe and EU integration: a view from the South', *European Urban and Regional Studies*, 1 (1): 19–29.

Hadjimichalis, C. and Sadler, D. (eds) (1995), *Europe at the Margins: New Mosaics of Inequality*, Chichester: Wiley.

Hagendorn, L. and Hraba, J. (1989), 'Foreign, different, deviant, seclusive and working class: anchors to an ethnic hierarchy in The Netherlands', *Ethnic and Racial Studies*, 2: 441–68.

Halfmann, J. (1997), 'Immigration and citizenship in Germany: contemporary dilemmas', *Political Studies*, 45: 260–74.

Hall, D. and Danta, D. (eds) (1996), *Reconstructing the Balkans: a Geography of the New Southeast Europe*, Chichester: John Wiley.

Hall, J.B. (1991), 'Sectoral transformation and economic decline: views from Berkeley and Cambridge', *Cambridge Journal of Economics*, 15 (2): 229–37.

Hall, R. (1998), 'Agenda 2000 and European cohesion policies', *European Urban and Regional Studies*, 5: 176–83.

Hallin, G. and Malmberg, A. (1996), 'Attraction, competition and regional development in Europe', *European Urban and Regional Studies*, 3: 323–37.

Hamill, J. (1993), 'Cross-border mergers, acquisitions and strategic alliances', in P. Bailey, A. Parisotto and G. Renshaw (eds), *Multinationals and Employment: the Global Economy of the 1990s*, Geneva: International Labour Office, pp. 95–123.

Hammar, T. (1990), *Democracy and the Nation State: Aliens, Denizens and Citizens in a World of International Migration*, Aldershot: Gower.

Hamnett, C. (1994), 'Social polarisation in global cities: theory and evidence', *Urban Studies*, 31 (3): 401–24.

Hansen, H. (1997), *Elements of Social Security in Six European Countries*, Copenhagen: Danish National Institute of Social Research.

Hantrais, L. (1995), *Social Policy in the European Union*, Basingstoke: Macmillan.

Hargreaves, A.G. (1995), *Immigration, 'Race' and Ethnicity in Contemporary France*, London: Routledge.

Harvey, D. (1989a), *The Condition of Postmodernity: an Enquiry into the Origins of Cultural Change*, Oxford: Oxford University Press.

Harvey, D. (1989b), 'From managerialism to entrepreneurialism: the transformation in urban governance in late capitalism', *Geografiska Annaler*, 71B (1): 3–17.

Haussermann, H. (1993), 'Regional perspectives of East Germany after the unification of the two Germanies', in P. Getimis and G. Kafkalas (eds), *Urban and Regional Development in the New Europe*, Athens: Topos, pp. 77–92.

Haussermann, H. and Mingione, E. (eds) (1996), *Urban Poverty and the Underclass: a Reader*, Oxford; Blackwell.

Havel, V. (1991), 'Otváraci prejav na konferencii "Minorities in Politics"', Bratislava: The Czechoslovak Committee of the European Cultural Foundation, 13–16 November.

Hebbert, M. (1990), 'The new map of Europe', in M. Hebbert and J.C. Hansen (eds), *Unfamiliar Territory: the Reshaping of European Geography*, Aldershot: Avebury, pp. 1–7.

Held, D. (1995), *Democracy and the Global Order: from the Modern State to Cosmopolitan Governance*, Cambridge: Polity Press.

Held, D. (1996), *Models of Democracy*, 2nd edn, Cambridge: Polity Press.

Héraud, G. (1963), *L'Europe des ethnies*, Paris: Presse d'Europe.

Héraud, G. (1971), *Federalisme et communautés ethniques*, Charleoi: Edition J. Destrée.

Herdegen, G. (1989), 'Aussiedler in der Bundesrepublik Deutschland: Einstellen und aktuelle Ansichten der Bundesbürger', *Informationen zur Raumentwicklung*, 5: 331–41.

Héthy, L. (1997), 'Central and Eastern Europe: economic transformation, social cohesion and social dialogue', in R. Weichhardt (ed.), *Economic Developments and Reforms in Cooperation Partner Countries: the Social and Human Dimension*, Brussels: NATO, pp. 9–16.

Hills, J. (1993), *The Future of Welfare: a Guide to the Debate*, York: Joseph Rowntree Foundation.

Hinrichs, K. (1991), 'Public pensions and demographic change', *Society*, 28 (6): 32–7.

Hirst, P. (1994), *Associative Democracy: New Forms of Economic and Social Governance*, Cambridge: Polity Press.

Hirst, P. and Thompson, G. (1992), 'The problem of "globalization": international economic relations, national economic management and the formation of trading blocs', *Economy and Society*, 21 (4): 357–95.

Hirst, P. and Thompson, G. (1996a), 'Globalisation: ten frequently asked questions and some surprising answers', *Soundings*, 4: 47–66.

Hirst, P. and Thompson, G. (1996b), *Globalization in Question*, Cambridge: Polity Press.

Hobsbawm, E. (1984), *Industry and Empire*, London: Penguin.

Holland, S. (1980), *Uncommon Market*, London: Macmillan.

Holtermann, S. (1995), 'The economics of equal opportunities', in J. Shaw and D. Perrons (eds), *Making Gender Work: Managing Equal Opportunities*, Buckingham: Open University Press.

Hope, K. (1997), 'Greek bankers succumb to the lure of big bucks', *Financial Times*, 13 September: 4.

Hoschka, T. (1993), *Cross Border Entry in European Retail Financial Services*, New York: St Martin's Press.

Hudson, R. (1983), 'Capital accumulation and chemicals production in Western Europe in the post-war period', *Environment and Planning A*, 15: 105–22.

Hudson, R. (1988), 'Uneven development in capitalist societies: changing spatial divisions of labour, forms of spatial organisation of production and service provision, and their impact upon localities', *Transactions of the Institute of British Geographers*, n.s., 13: 484–96.

Hudson, R. (1989), 'Labour market changes and new forms of work in industrial regions: maybe flexibility for some but not flexible accumulation', *Society and Space*, 7: 5–30.

Hudson, R. (1994a), 'New production concepts, new production geographies? Reflections on changes in the automobile industry', *Transactions of the Institute of British Geographers*, n.s., 19: 331–45.

Hudson, R. (1994b), 'Institutional change, cultural transformation and economic regeneration: myths and realities from Europe's old industrial regions', in A. Amin and N.J. Thrift (eds), *Globalization, Institutions and Regional Development in Europe*, Oxford: Oxford University Press, pp. 331–45.

Hudson, R. (1995a), 'The role of foreign investment', in A. Darnell, L. Evans, P. Johnson and B. Thomas (eds), *The Northern Region Economy: Progress and Prospects*, London: Mansell, pp. 79–95.

Hudson, R. (1995b), 'Towards sustainable industrial production: but in what sense sustainable?', in M. Taylor (ed.), *Environmental Change: Industry, Power and Place*, Avebury: Winchester, pp. 37–56.

Hudson, R. (1996), 'The learning economy, the learning firm and the learning region: a sympathetic critique of the limits to learning', paper prepared for an International Seminar on Learning and Territoriality, Université Montesquieu-Bordeaux XIV, 28–30 November.

Hudson, R. (1997a), 'Regional futures: industrial restructuring, new production concepts and spatial development strategies in Europe', *Regional Studies*, 31 (5): 467–78.

Hudson, R. (1997b), 'The end of mass production and of the mass collective worker? Experimenting with production, employment and their geographies', in R. Lee and J. Wills (eds), *Geographies of Economies*, London: Edward Arnold, pp. 302–10.

Hudson, R. and Lewis, J. (eds) (1985), *Uneven Development in Southern Europe: Studies of Accumulation, Class, Migration and the State*, London: Methuen.

Hudson, R. and Sadler, D. (1983), 'Region, class and the politics of steel closures in the European Community', *Society and Space*, 1: 405–28.

Hudson, R. and Sadler, D. (1989), *The International Steel Industry*, London: Routledge.

Hudson, R. and Schamp, E.W. (eds) (1995), *Towards a New Map of Automobile Manufacturing in Europe? New Production Concepts and Spatial Restructuring*, Berlin: Springer.

Hudson, R. and Weaver, P. (1997), 'In search of employment creation via environmental valorisation: exploring a possible eco-Keynesian future for Europe', *Environment and Planning A*, 29, 1647–61.

Hudson, R. and Williams, A.M. (1995), *Divided Britain*, 2nd edn, Chichester: Wiley.

Humbeck, E. (1996), 'The politics of cultural identity: Thai women in Germany', in M.D. García-Ramon and J. Monk (eds), *Women of the European Union: the Politics of Work and Daily Life*, London: Routledge, pp. 186–201.

Huntington, S.P. (1993), 'The clash of civilizations?', *Foreign Affairs*, 72 (3): 22–49.

Hutchinson, J. and Smith, A.D. (eds) (1996), *Ethnicity*, Oxford: Oxford University Press.

Hutton, W. (1995), *The State We're In*, London: Cape.

Icduygu, A. (1996), 'Becoming a new citizen in an immigration country: Turks in

Australia and Sweden and some comparative implications', *International Migration Review*, 34: 257–72.

Ilbery, B.W. (1984), 'Core–periphery contrasts in European social well-being', *Geography*, 69 (4): 289–302.

Ilbery, B.W. (1986), *Western Europe: a Systematic Human Geography*, Oxford: Oxford University Press.

Illeris, S. (1989), *Services and Regions in Europe*, Aldershot: Avebury.

Isaksen, A. (1994), 'New industrial spaces and industrial districts in Norway: productive concepts in explaining regional development?', *European Urban and Regional Studies*, 1: 31–48.

Jackson, M. (1997), 'Labor markets and income maintenance: a survey of transition', in R. Weichhardt (ed.), *Economic Developments and Reforms in Cooperation Partner Countries: the Social and Human Dimension*, Brussels: NATO, pp. 89–109.

Jackson, P. (ed.) (1987), *Race and Racism: Essays in Social Geography*, London: Allen and Unwin.

Jäger, S. (1992), 'Wie die Deutschen die "Fremden" sehen: Rassismus im Alltagsdiskurs', in C. Butterwegge and S. Jäger (eds), *Rassismus in Europa*, Cologne: Bund-Verlag, pp. 248–61.

Jeffrey, C. (1996), 'Regional identity and economic development', in *Regional Identity and Economic Development*, Leuven: Institute for European Policy, pp. 2–10.

Jessop, B. (1990), 'Regulation theories in retrospect and prospect', *Economy and Society*, 19 (2): 153–216.

Jessop, B. (1994), 'Post-Fordism and the state', in A. Amin (ed.), *Post-Fordism: a Reader*, Oxford: Blackwell, pp. 57–84.

Joffe, G. (1996), 'Integration or peripheral dependence: the dilemma facing the South Mediterranean states', in A. Bin (ed.), *Cooperation and Security in the Mediterranean: Prospects after Barcelona*, Malta: Malta University Press, pp. 175–97.

Johnson, C. (1996), *In with the Euro, out with the Pound: the Single Currency for Britain*, Harmondsworth: Penguin.

Johnston, N. (1995), 'Cast in stone: monuments, geography and nationalism', *Environment and Planning D: Society and Space*, 13: 51–65.

Johnston, R. (1979), *Political, Electoral and Spatial Systems*, Oxford: Clarendon Press.

Johnston, R., Shelley, F. and Taylor, P. (eds) (1990), *Developments in Electoral Geography*, London: Routledge.

Jones, A. (1997), 'The European Union's Mediterranean policy: from pragmatism to partnership', in R. King, L. Proudfoot and B. Smith (eds), *The Mediterranean: Environment and Society*, London: Arnold, pp. 155–63.

Jones, C. (1997), 'The centre of attention', *The Banker*, September: 65–7.

Jonung, C. and Persson, I. (1993), 'Women and market work: the misleading tale of participation rates in international comparisons', *Work Employment and Society*, 7 (2): 259–74.

Jouve, B. (1997), 'France: from the regionalized state to the emergence of regional governance?', in M. Keating and J. Loughlin (eds), *The Political Economy of Regionalism*, London: Cass, pp. 347–69.

Kaldor, M. (1996), 'Nation-states, European institutions and citizenship', in B. Einhorn, M. Kaldor and Z. Kavan (eds), *Citizenship and Democratic Control in Contemporary Europe*, Cheltenham: Edward Elgar, pp. 9–23.

Kaldor, N. (1970), 'The case for regional policies', *Scottish Journal of Political Economy*, 17: 337–48.

Kallen, E. (1995), *Ethnicity and Human Rights in Canada*, Don Mills: Oxford University Press.

Kamen, H. (1971), *The Iron Century: Social Change in Europe 1550–1660*, London: Cardinal.

Kaufman, F. (1996), 'The welfare states', in X-F. Merrien (ed.), *Facing Poverty* (Greek trans.), Athens: Katarti.

Kazepov, Y. (1995), 'Urban poverty patterns in Italy: the case of Milan', *Espace, Populations, Société*, 4: 329–40.

Keane, J. (ed.) (1988), *Civil Society and the State*, London: Verso.

Keating, M. (1988), *State and Regional Nationalism*, Brighton: Harvester Wheatsheaf.

Keating, M. (1992), 'Regional autonomy in the changing state order: a framework of analysis', *Regional Politics and Policy*, 2 (3): 45–61.

Keating, M. (1997), 'Culture, collective identities and development', in *Regional Identity and Economic Development*, Brussels: The Foundation of Europe of the Cultures, pp. 16–30.

Keating, M. and J. Loughlin (eds) (1996), *The Political Economy of Regionalism*, London: Frank Cass.

Keeble, D. (1989), 'Core–periphery disparities, recession and new regional dynamisms in the European Community', *Geography*, 74 (1): 1–11.

Kelleher, C.M. (1996), 'Indicators, implications, and policy choices', in L. Drobizheva, R. Gottemoeller, C. McArdle Kelleher and L. Walker (eds), *Ethnic Conflict in the Post-Soviet World*, Armonk, NY: M.E. Sharpe, pp. 337–52.

Killilea, M. (1994), 'On linguistic and cultural minorities in the European Community'. Resolution adopted by the European Parliament, Strasbourg.

Kim, T.J. (1997), 'Emerging market derivatives: the central European option', *Euromoney*, August (http://www.euromoney.com).

Kindleberger, C.P. (1965), 'Emigration and economic growth', *Banca Nazionale del Lavoro Quarterly Review*, 75: 235–54.

Kindleberger, C.P. (1974), *The Formation of Financial Systems: a Study in Comparative Economic History*, Princeton, NJ: University of Princeton, Princeton Studies in International Finance, 36.

King, R. (1976), 'The evolution of international labour migration movements concerning the EEC', *Tijdschrift voor Economische en Sociale Geografie*, 67: 66–82.

King, R. (1982), 'Southern Europe: dependency or development?' *Geography*, 67 (3): 221–34.

King, R. (1984), 'Population mobility: emigration, return migration and internal migration', in A.M. Williams (ed.), *Southern Europe Transformed*, London: Harper and Row, pp. 145–78.

King, R. (1993a), 'Italy reaches zero population growth', *Geography*, 78 (1): 63–9.

King, R. (ed.) (1993b), *The New Geography of European Migrations*, London: Belhaven.

King, R. (ed.) (1993c), *Mass Migrations in Europe: the Legacy and the Future*, London: Belhaven Press.

King, R. (1996), 'Migration and development in the Mediterranean region', *Geography*, 81 (1): 3–14.

King, R. (1997), 'Population growth: an avoidable crisis?', in R. King, L. Proudfoot and B. Smith (eds), *The Mediterranean: Environment and Society*, London: Arnold, pp. 164–80.

King, R. (1998), 'The Mediterranean – Europe's Rio Grande?', in M. Anderson and E. Bort (eds), *Frontiers of Europe*, London: Pinter, pp. 109–34.

King, R., Proudfoot, L. and Smith, B. (eds) (1997), *The Mediterranean: Environment and Society*, London: Arnold.

Kinsella, K. and Gist, Y.J. (1995), *Older Workers, Retirement and Pensions: a Comparative International Chartbook*, Washington, DC: Bureau of the Census, Department of Commerce.

Klein, R. and O'Higgins, M. (eds) (1985), *The Future of Welfare*, Oxford: Blackwell.

Kleyman, P. (1997), 'Social security debate to move ahead', *Aging Today*, 18 (6): 1–2.

Kliot, N. (1997), 'Politics and society in the Mediterranean Basin', in R. King, L. Proudfoot and B. Smith (eds), *The Mediterranean: Environment and Society*, London: Arnold, pp. 108–25.

Klug, F., Starmer, K. and Weir, S. (1996), *The Three Pillars of Liberty: Political Rights and Freedoms in the United Kingdom*, London: Routledge.

Knell, M. (1996), 'Structural adjustments and growth: is Eastern Europe catching up?', in M. Knell (ed.), *Economics of Transition: Structural Adjustments and Growth Prospects in Eastern Europe*, Cheltenham: Edward Elgar, pp. 1–24.

Knight, K. (1997), 'Students of nationalism', *West European Politics*, 20 (2): 173–9.

Knights, M. (1996), 'Bangladeshi immigrants in Italy: from geopolitics to micropolitics', *Transactions of the Institute of British Geographers*, 21: 105–203.

Knox, P. and Taylor, P. (eds) (1995), *World Cities in a World-System*, Cambridge: Cambridge University Press.

Kromschröder, E. (1983), *Als ich ein Türke war: Reportagen*, Frankfurt-am-Main: Eichborn.

Kuijpers, W. (1987), 'On the languages and cultures of regional and ethnic minorities in European Community'. Resolution adopted by the European Parliament, Strasbourg.

Kurthen, H. (1995), 'Germany at the crossroads: national identity and the challenges of imperialism', *International Migration Review*, 29: 914–38.

Kymlicka, W. (1995a), *Multicultural Citizenship*, Oxford: Oxford University Press.

Kymlicka, W. (ed.) (1995b), *The Rights of Minority Cultures*, Oxford: Oxford University Press.

Laclau, E. and Mouffe, C. (1985), *Hegemony and Socialist Strategy: Towards a Radical Democratic Politics*, London: Verso.

Lacoste, Y. and Lacoste, C. (1991), *L'Etat du Maghreb*, Paris: La Découverte.

Laffan, B. (1996), 'The politics of identity and political order in Europe', *Journal of Common Market Studies*, 34: 81–102.

Landesmann, M. and Székely, I.P. (1995), 'Industrial structural change in Central and Eastern European economies', in M. Landesmann and I.P. Székely (eds), *Industrial Restructuring and Trade Reorientation in Eastern Europe*, Cambridge: Cambridge University Press, pp. 23–68.

Lane, D. (1994), 'Italy: from bad to worse', *The Banker*, 19–23 August.

Langan, M. and Ostner, I. (1991), 'Gender and welfare: towards a comparative framework', in G. Room (ed.), *Towards a European Welfare State*, Bristol: School of Advanced Urban Studies, pp. 127–44.

Langer, W. (1996), 'The marginalized', in B. Castro (ed.), *Business and Society*, Oxford: Oxford University Press.

Lanno, K. (1996), 'The implications of EMU for European banking', *ECU-Europe* 37, (http://www.ecu-activities.be/1996_4/lanoo.htm).

Lash, S. and Urry, J. (1987), *Disorganised Capitalism*, Cambridge: Polity Press.

Lash, S. and Urry, J. (1994), *Economies of Signs and Space*, London: Sage.

Latawski, P. (ed.) (1995), *Contemporary Nationalism in East Central Europe*, London, Macmillan.

Leborgne, D. and Lipietz, A. (1991), 'Two social strategies in the production of new industrial spaces', in G. Benko and M. Dunford (eds), *Industrial Change and Regional Development*, London: Belhaven, 27–50.

Lee, P. (1996), 'European monetary union: a technical ascent' *Euromoney* September (http://www.euromoney.com).

Leontidou, L. (1989) (in Greek), *Cities of Silence*, Athens: ETBA & Themelio.

Leontidou, L. (1990), *The Mediterranean City in Transition: Social Change and Urban Development*, Cambridge: Cambridge University Press.

Leontidou, L. (1993), 'Informal strategies of unemployment relief in Greek cities:

the relevance of family, locality and housing', *European Planning Studies*, 1 (1): 43–68.

Leontidou, L. (1995), 'Repolarization in the Mediterranean: Spanish and Greek cities in neoliberal Europe', *European Planning Studies*, 3 (2): 155–72.

Leontidou, L. (1996), 'Alternatives to modernism in (Southern) urban theory: exploring in-between spaces', *International Journal of Urban and Regional Research*, 20 (2): 180–97.

Leontidou, L. (1997), 'Five narratives for the Mediterranean city', in R. King, L. Proudfoot and B. Smith (eds), *The Mediterranean: Environment and Society*, London: Arnold, pp. 181–92.

Les, E. (1992), 'Poland', in J. Dixon and D. Macarov (eds), *Social Welfare in Socialist Countries*, London: Routledge.

Leško, M. (1997), 'Predseda HZDS dal, aby mohol dostat (The HZDS chairman gave in order too receive)', *The SME Daily*, 17 October.

Lesser, I.O. (1995), 'Growth and change in Southern Europe', in J.W. Holmes (ed.), *Maelstrom: the United States, Southern Europe, and the Challenges of the Mediterranean*, Cambridge, Mass.: World Peace Foundation, pp. 11–28.

Levesque, J. (1997), *The Enigma of 1989: the USSR and the Liberation of Eastern Europe*, Berkeley, Calif.: University of California Press.

Lewis, B. (1993), *Islam and the West*, New York: Oxford University Press.

Lewis, P. (1997a), 'Democratization in Eastern Europe', in D. Potter et al. (eds), *Democratization*, Cambridge: Polity Press, pp. 399–420.

Lewis, P. (1997b), 'Political participation in post-community democracies', in D. Potter et al. (eds), *Democratization*, Cambridge: Polity Press, pp. 443–65.

Lewis, S. (1926), *Egwyddorion Cenedlaetholdeb*, Caernarfon: Plaid Cymru.

Leyshon, A. and Thrift, N.J. (1997), *Money/Space*, London: Routledge.

Leyshon, A. and Tickell, A. (1994), 'Money order? The discursive constitution of Bretton Woods and the making and breaking of regulatory space', *Environment and Planning A*, 26: 1861–90.

Liebfried, S. (1993), 'Conceptualising European social policy: the EC as a social actor', in L. Hantrais and S. Mangen (eds), *The Policy Making Process and the Social Actors*, Loughborough: Loughborough University European Research Centre.

LIFFE (1997), *LIFFE's Strategy for Economic and Monetary Union*, London: London International Financial Futures Exchange.

Lipietz, A. (1985), *The Enchanted World: Inflation, Credit and the World Crisis*, London: Verso.

List, F. (1909), *The National System of Political Economy*, London: Longman.

Llewellyn, D.T. (1992), 'Banking and financial services', in D. Swann (ed.), *The Single European Market and Beyond*, London: Routledge, pp. 106–45.

Lloyd, J. (1996), 'Eastern reformers and neo-Marxist reviewers', *New Left Review*, 216: 119–28.

Lloyd-Sherlock, P. and Johnson, P. (1996), 'Introduction', in P. Lloyd-Sherlock and P. Johnson (eds), *Ageing and Social Policy: Global Comparisons*, STICERD occasional paper 19, London: London School of Economics, pp. 1–4.

Loughlin, J. (1996a), 'Nationalism, regionalisation and regionalism in Ireland', in G. Färber and M. Forsyth (eds), *The Regions-Factors: Integration or Disintegration in Europe?* Baden-Baden: Nomos Verlagsgesellschaft, pp. 79–90.

Loughlin, J. (1996b), 'Representing regions in Europe: the Committee of the Regions, *Regional and Federal Studies: an International Journal*, 6 (3): 147–65.

Loughlin, J. (1996c), 'Wales in Europe: Welsh regional actors and European integration', Cardiff: University of Wales (mimeo).

Loughlin, J. (1996d), '"Europe of the Regions" and the federalization of Europe', *Publius: the Journal of Federalism*, 26: 407–16.

Lowe, S. (1986), *Urban Social Movements*, London: Macmillan.

McAll, C. (1990), *Class, Ethnicity and Social Inequality*, Montreal and Kingston: McGill-Queen's University Press.

McCauley, R.N. and White, W.R. (1997), 'The Euro and European financial markets', Bank for International Settlements Working Paper 41 (http://www.bis.org).

McIntyre, J., Porter, L. and Wendelova, P. (1994), 'Ex-émigré entrepreneur Viktor Kozený and the Harvard Group', in S.D. Fogel (ed.), *Managing in Emerging Market Economies: Cases from the Czech and Slovak Republics*, Boulder, Col.: Westview Press, pp. 149–65.

McIvor, G. (1997), 'Swedes loosen purse-strings', *Financial Times*, September/October.

MacKay, R.R. and Molyneux, P. (1996), 'Bank credit and the regions: a comparison within Europe', *Regional Studies*, 30: 757–63.

MacLaughlin, J. (1986), 'The political geography of nation-building and nationalism in the social sciences: structural versus dialectical accounts', *Political Geography Quarterly*, 3 (4): 299–329.

MacLaughlin, J. (1993), 'Defending the frontiers: The political geography of race and racism in the European Community', in C.H. Williams (ed.), *The Political Geography of the New World Order*, London: Wiley, pp. 20–46.

Magas, B. (1993), *The Destruction of Yugoslavia: Tracing the Break-up, 1980–92*, London: Verso.

Magosci, P.R. (1993), *Historical Atlas of East Central Europe*, Seattle: University of Washington Press.

Mair, A., Florida, R. and Kenney, M. (1988), 'The new geography of automobile production: Japanese transplants in North America', *Economic Geography*, 64 (4): 352–73.

Maleki, E. (1997), *Technology and Economic Development: the Dynamics of Local, Regional and National Competitiveness*, Harlow: Longman.

Mandel, E. (1975), *Late Capitalism*, London: New Left Books.

Mandel, E. (1978), *The Second Slump*, London: New Left Books.

Mann, M. (1993), 'Nation-states in Europe and other continents: diversifying, developing, not dying', *Proceedings of the American Academy of Arts and Sciences*, 122 (3): 115–40.

Mann, M. (1996), 'Nation-states in Europe and other continents: diversifying, developing, not dying', in G. Balakrishnan (ed.), *Mapping the Nation*, London: Verso Books, pp. 295–316.

Maret, X. and Schwartz, G. (1994), 'Poland: social protection and the pension system during the transition', *International Social Security Review*, 47 (2): 51–69.

Mar-Molinero, C. (1994), 'Linguistic nationalism and minority language groups in the "New" Europe', *Journal of Multilingual and Multicultural Development*, 15 (3): XX.

Marquand, D. (1997), 'Blair's split personality', *Guardian* 16 July: 17.

Marshall, T. (1950), *Citizenship and Social Class*, Cambridge: Cambridge University Press.

Marston, S. and Staeheli, L. (eds) (1994), 'Citizenship theme issue', *Environment and Planning*, 26 (6), 840–955.

Martin, R. (1995), 'Undermining the financial basis of regions: the spatial structure and implications of the UK pension fund system', *Regional Studies*, 29: 125–44.

Massey, D. (1995), 'Thinking radical democracy spatially', *Society and Space*, 13: 283–8.

Matsaganis, M. (ed.) (1998) (in Greek), *Prospects of the Welfare State in Southern Europe*, Athens: Themelio.

Mayes, D. (1995), 'Conflict and cohesion in the single European market: a reflection', in A. Amin and J. Tomaney (eds), *Behind the Myth of the European Union: Prospects for Cohesion*, London: Routledge, pp. 1–9.

Meadows, P. (1997), 'On the unexpected legacy of Lawson's "feminist" measure', *Guardian*, 24 June: 19.

Meidner, R. (1994), 'The rise and fall of the Swedish model', in W. Clement and R. Mahon (eds), *Swedish Social Democracy: a Model in Transition*, Toronto: Canada Scholars Press.

Memmi, A. (1987), *Rassismus*, Frankfurt-am-Main: Athenaeum Verlag.

Mernissi, F. (1993), *Islam and Democracy: Fear of the Modern World*, London: Virago.

Merrien, X-F. (ed.) (1996), *Facing Poverty* (Greek trans.), Athens: Katarti.

Mertlik, P. (1995), 'Transformation of the Czech and Slovak economies 1990–92: design, problems, costs', in J. Hausner, B. Jessop and K. Nielsen (eds), *Strategic Choice and Path Dependency in Post Socialism: Institutional Dynamics in the Transformation Process*, Aldershot: Edward Elgar, pp. 218–29.

Meulders, D., Plasman, O. and Plasman, R. (1994), *Atypical Employment in the EC*, Aldershot: Dartmouth.

Meulders, D., Plasman, O. and Plasman, R. (1997), 'Atypical labour market relation in the European Union', in A.G. Dijkstra and J. Plantenga (eds), *Gender and Economics: a European Perspective*, London: Routledge, pp. 75–85.

Meulders, D., Plasman, R. and Vander Stricht, V. (1993), *Position of Women in the Labour Market in the EU*, Hampshire: Aldershot.

Michel, A-M. (1995), 'Poverty in Europe', *Le Monde Diplomatique*, 6: 10–13.

Miguel, J.M. (1996), 'Desarrollo o desigualdad? Análisis de una polémica sociológica de medio siglo en España', *Revista Española de Investigaciones Sociológicas*, 75: 55–108.

Miles, D. (1994), 'Fixed and floating rate finance in the UK and abroad', *Bank of England Quarterly Bulletin*, 34: 34–45.

Miles, R. (1989), *Racism*, London: Routledge.

Miles, R. (1993), *Racism after 'Race Relations'*, London: Routledge.

Miles, R. (1994), 'A rise of racism and fascism in contemporary Europe?: Some sceptical reflections on its nature and extent', *New Community*, 20: 547–62.

Miller, D. (1992), 'Deliberative democracy and public choice', in D. Held (ed.), *Prospects for Democracy*, Political Studies special issue, 40.

Mingione, E. (1991), *Fragmented Societies: a Sociology of Economic Life Beyond the Market Paradigm*, Oxford: Basil Blackwell.

Mingione, E. (ed.) (1993), 'The new urban poverty', *International Journal of Urban and Regional Research*, special issue, 17 (3).

Mingione, E. (1995), 'Labour market segmentation and informal work in Southern Europe', *European Urban and Regional Studies*, 2 (2): 121–43.

Minority Rights Group (1991), *Minorities and Autonomy in Western Europe*, London: Minority Rights Group.

Misiti, M., Muscarà, C., Pumares, P., Rodriguez, V. and White, P. (1995), 'Future migration into southern Europe', in R. Hall and P. White (eds), *Europe's Population: Towards the Next Century*, London: UCL Press, pp. 161–87.

Mitchell, M. and Russell, D. (1996), 'Immigration, citizenship and the nation-state in the new Europe', in B. Jenkins and S.A. Sofos (eds), *Nation and Identity in Contemporary Europe*, London: Routledge, pp. 54–80.

Mlinar, Z. (ed.) (1992), *Globalization and Territorial Identities*, Aldershot: Avebury.

Mlinar, Z. (1994), 'Transnational flows and language identity of a small nation (the case of Slovenia)', paper presented to the Conference on Nation and Languages and the Construction of Europe, Katholieke Universiteit, Leuven, November.

Molle, W. and Cappellin, R. (eds) (1988), *Regional Impact of Community Policies in Europe*, Aldershot: Avebury.

Moore, P. (1996), 'Italy: the quest for a risorgimento', *Euromoney*, September (http://www.euromoney.com).

Moors, H. and Palomba, R. (eds) (1995), *Population, Family and Welfare: a Comparative Survey of European Attitudes*, Oxford: Clarendon Press.

Moragas Moragas, R. (1991), *Gerontología social: envejecimiento y calidad de vida*, Barcelona: Herda.

Moran, M. (1991), *The Politics of the Financial Services Revolution: London, New York, Tokyo*, Basingstoke: Macmillan.

Moreno, L. (1995), 'Multiple ethnoterritorial concurrence in Spain', *Nationalism and Ethnic Politics*, 1 (1): 11–32.

Morokvasic, M. (1991), 'Fortress Europe and migrant women', *Feminist Review*, 39: 69.

Morris, L. (1994), *Dangerous Classes: the Underclass and Social Citizenship*, London: Routledge.

Mouffe, C. (1992), 'Democratic citizenship and the political community', in C. Mouffe (ed.), *Dimensions of Radical Democracy*, London: Verso, pp. 225–39.

Mouffe, C. (1995a), 'Post-Marxism: democracy and identity', *Society and Space*, 13: 259–65.

Mouffe, C. (1995b), 'Politics, democratic action and solidarity', *Inquiry*, 38: 99–108.

Mouffe, C. (1996), 'Radical democracy or liberal democracy', in D. Trend (ed.), *Radical Democracy: Identity, Citizenship and the State*, New York: Routledge, pp. 19–26.

Moulaert, F. and Leontidou, L. (1995), 'Localiteé desintegrées et strategies de lutte contre la pauvrété: une reflexion methodologique post-moderne', *Espaces et Sociétés*, 78: 35–53.

Moulaert, F. and Tödtling, F. (1995), 'The geography of advanced producer services in Europe – conclusions and prospects', *Progress in Planning*, 43 (2/3): 1261–74.

Munchau, W. (1997a), 'Pensions: big project imperfectly remodelled', *Financial Times*, 18 November: Germany Survey.

Munchau, W. (1997b), 'EMU unlikely to dislodge City', *The Banker*, supplement: International Banking in London, 23–4 November.

Musterd (ed.) (1994), 'A rising European underclass?', *Built Environment*, special issue, 20 (3).

Nath, R. and Jirásek, J. (1994), 'Transformation management in Czechoslovakia', in S. D. Fogel (ed.), *Managing in Emerging Market Economies: Cases from the Czech and Slovak Republics*, Boulder, Col.: Westview Press, pp. 69–84.

NATO (1996), 'Partnership for peace', *Basic Factsheet 9* (http://www.nato.int/docu/facts/fs9.htm).

Natter, W. (1995), 'Radical democracy: hegemony, reason, time and space', *Society and Space*, 13: 283–8.

Neathey, F. and Hurstfield, J. (1995), *Flexibility in Practice: Women's Employment and Pay in Retail and Finance*, Discussions Series, no. 16, Equal Opportunities Commission/Institute of Racial Studies.

Nelde, P., Labrie, N. and Williams, C.H. (1992), 'The principles of territoriality and personality in the solution of linguistic conflicts', *Journal of Multilingual and Multicultural Development*, 13 (5): 387–406.

Nielsen, K., Jessop, B. and Hausner, J. (1995), 'Institutional change in post-socialism', in J. Hausner, B. Jessop and K. Nielsen (eds), *Strategic Choice and Path Dependency in Post Socialism: Institutional Dynamics in the Transformation Process*, Aldershot: Edward Elgar, pp. 3–44.

Nilsson, J-E. and Schamp, E. (1996), 'Restructuring of the European production system: processes and consequences', *European Urban and Regional Studies*, 3: 121–32.

Nugent, N. (1989), *The Government and Politics of the European Community*, London: Macmillan.

Nuti, D.M. (1995), 'Mass privatization: costs and benefits of instant capitalism', in R. Daviddi (ed.), *Property Rights and Privatization in the Transition to a Market Economy*, Maastricht: European Institute of Public Administration, pp. 103–32.

OECD (1987), *Financing and Delivering Health Care*, Paris: OECD.

OECD (1988a), *Reforming Public Pensions*, Paris: OECD.

OECD (1988b), *Ageing Populations: the Social Policy Implications*, Paris: OECD.

OECD (1990), *Health Care Systems in Transition*, Paris: OECD.

OECD (1994), *New Orientations for Social Policy*, Paris: OECD.

OECD (1996), *Beyond 2000: the New Social Policy Agenda*, Paris: OECD.

OECD (1998), *Maintaining Prosperity in an Ageing Society*, Paris: OECD.

Offe, C. (1975), 'The theory of the capitalist state and the problem of policy formation', in L.N. Lindberg, R. Alford, C. Crouch and C. Offe (eds), *Stress and Contradiction in Modern Capitalism*, Lexington: D.C. Heath, pp. 125–44.

Office of Population Censuses and Surveys (OPCS) (1993), *National Population Projections: 1991-Based*, Series PP2, no. 18, London: HMSO.

Ohmae, K. (1990), *The Borderless World*, London: Collins.

Ohmae, K. (1995), *The End of the Nation State*, Glencoe, Ill.: Free Press.

O'Riagain, D. (1989), 'The EBLUL: its role in creating a Europe united in diversity', in T. Veiter (ed.), *Federalisme, regionalisme et droit des groupes ethnique en Europe*, Vienna: Braümuller.

Orloff, A. (1996), 'Gender in the welfare state', *Annual Review of Sociology*, 22: 51–78.

Orridge, A.W. and Williams, C.H. (1982), 'Autonomist nationalism: a theoretical framework for spatial variations in its genesis and development', *Political Geography Quarterly*, 1 (1): 18–42.

Pace, R. (1996), 'Peace, stability and prosperity in the Mediterranean region', in A. Bin (ed.), *Cooperation and Security in the Mediterranean: Prospects after Barcelona*, Malta: Malta University Press, pp. 105–26.

Painter, J. (1995), *Politics, Geography and 'Political Geography': a Critical Perspective*, London: Arnold.

Painter, J. (1998), 'Entrepreneurs are made, not born: knowledge, learning and regime in the production of entrepreneurial cities', in T. Hall and P. Hubbard (eds), *The Entrepreneurial City*, Chichester: Wiley, pp. 259–73.

Painter, J. and Philo, C. (eds) (1995), 'Spaces of citizenship', *Political Geography*, special issue, 14 (2).

Pampel, F. and Williamson, J.B. (1989), *Age, Class, Politics and the Welfare State*, Cambridge: Cambridge University Press.

Paparela, I. (1997), 'Some random reflections concerning financing of social measures in countries in transition', in R. Weichhardt (ed.), *Economic Developments and Reforms in Cooperation Partner Countries: the Social and Human Dimension*, Brussels: NATO, pp. 61–72.

Pearson, P. (1985), *Twilight Robbery – Trade Unions and Low Paid Workers*, London: Pluto Press.

Peck, J.A. and Tickell, A. (1994), 'Jungle law breaks out: neoliberalism and global–local disorder', *Area*, 26: 317–26.

Periwal, S. (1995), *Notions of Nationalism*, Budapest: Central European University.

Perrons, D. (1995), 'Economic strategies, welfare regimes and gender equality in European employment', *European Urban and Regional Studies*, 2 (2): 99–120.

Perrons, D. (1998), 'Maps of meaning: gender inequality in the regions of Western Europe', *European Urban and Regional Studies*, 5 (1): 13–26.

Perrons, D. and Gonas, L. (1998), 'Perspectives on gender inequality in European employment', *European Urban and Regional Studies*, 5 (1): 5–12.

Pfau-Effinger, B. (1995), 'Social change in the gendered division of labour in cross-national perspective', Second European Conference of Sociology,

Working Group, 'Gender Relations and the Labour Market in Europe', Budapest, September.

Pickvance, C. (1997), 'Decentralization and democracy in Eastern Europe: a sceptical approach', *Environment and Planning: Government and Policy*, 15 (2): 129–42.

Piper, N. (1997), 'The relationship between racism, nationalism and citizenship: a comparative analysis of Britain and Germany', unpublished PhD thesis, University of Sheffield.

Pirani, M., Yolles, M. and Bassa, E. (1992), 'Ethnic pay differentials', *New Community*, 19: 31–42.

Pitt-Rivers, J. (1954), *People of the Sierra*, London: Weidenfeld and Nicolson.

Plantenga, J. (1995), 'Part-time work and equal opportunities: the case of the Netherlands', in J. Humphries and J. Rubery (eds), *The Economics of Equal Opportunities*, Manchester: Equal Opportunities Commission, pp. 277–89.

Plantenga, J. (1997), 'European constants and national particularities: the position of women in the European labour market', in A.G. Dijkstra and J. Plantenga (eds), *Gender and Economics: a European Perspective*, London: Routledge.

Plasser, F. and Ulram, P. (1996), 'Measuring political culture in East Central Europe: political trust and system support', in F. Plasser and A. Pribersky (eds), *Political Culture in East Central Europe*, Aldershot: Avebury.

Plichtova, J. (ed.) (1992), *Minorities in Politics: Cultural and Language Rights*, Bratislava: Czechoslovak Committee of the European Cultural Foundation.

Pohl, G., Anderson. R., Clasessens, S. and Djankov, S. (1997), *Privatisation and Restructuring in Central and Eastern Europe*, technical paper 386, Washington: World Bank.

Poznanski, K. (1995), 'Institutional perspectives on post-communist recession in Eastern Europe', in K. Poznanski (ed.), *The Evolutionary Transition to Capitalism*, Boulder, Col.: Westview Press, pp. 3–30.

Pratt, J. (1998), 'Re-placing money: the evolution of the British banking industry', *Environment and Planning A*, forthcoming.

Pratt, J., Leyshon, A. and Thrift, N.J. (1996), 'Financial exclusion in the 1990s: the changing geography of UK retail financial services', Working Papers on Producer Services 34, Universities of Bristol and Birmingham.

Preston, S.H. (1984), 'Children and the elderly: divergent paths for America's dependants', *Demography*, 21: 435–57.

Pridham, G. and Vanhanen, T. (eds) (1994), *Democratization in Eastern Europe*, London: Routledge.

Pugliese, E. (1993), 'Restructuring of the labour market and the role of Third World migrations in Europe', *Society and Space*, 11 (4): 513–22.

Pugliese, E. (1995), 'New international migrations and the "European Fortress"', in C. Hadjimichalis and D. Sadler (eds), *Europe at the Margins: New Mosaics of Inequality*, Chichester: Wiley, pp. 51–68.

Quinn, J.A. (1996), *Entitlements and the Federal Budget: Securing our Future*, Washington: National Academy on Aging.

Raento, P. (1997), 'Political mobilisation and place specificity: radical nationalist street campaigning in the Spanish Basque country', *Space and Polity*, 1: 191–204.

Ramoneda, J. (1994), 'Anyway: geopolitics and architecture', in C. Davidson (ed.), *Anyway*, New York: Anyone Corporation and Rizzoli International.

Ramsay, H. (1990), *1992: the Year of the Multinational?*, Warwick Papers in Industrial Relations, Warwick: University of Warwick.

Rees, P., Stillwell, J., Convey, A. and Kupiszewski, M. (eds) (1996), *Population Migration in the EU*, Chichester: John Wiley.

Reeves, P. (1984), *Round about a Pound a Week*, London: Virago Press.

Reichs, R. (1996), Paper presented to the Conference on the Globalization of Production and the Regulation of Labour, University of Warwick, September.

Rex, J. (1994), 'Ethnic mobilisation in Britain', *Revue Européenne des Migrations Internationales*, 10: 7–31.

Rex, J. (1997), 'The problematic of multinational and multicultural societies', *Ethnic and Racial Studies*, 20 (3): 455–73.

Rhodes, M. (1995), 'A regulatory conundrum: industrial relations and the social dimension', in S. Liebfried and P. Pierson (eds), *European Social Policy*, Washington: Brookings.

Rhodes, M. (1996), *Globalisation and West European Welfare States*, working paper 96/43, Florence: European University Institute.

Richardson, R. and Marshall, J.N. (1996), 'The growth of telephone call centres in peripheral areas of Britain: evidence from Tyne and Wear', *Area*, 28 (3): 308–17.

Ridder, C. and Zajicek, E. (1995), 'Mass privatisation in Poland', *International Journal of Politics, Culture and Society*, 9 (1): 133–48.

Rix, S.E. and Fisher, P. (1982), *Retirement-age Policy: an International Perspective*, Oxford: Pergamon.

Robertson, A. (1997), 'Beyond apocalyptic demography: towards a moral economy of interdependence', *Ageing and Society*, 17 (4): 425–46.

Robinson, W. (1996), 'Globalisation: nine theses on our epoch', *Race and Class*, 38 (2): 13–31.

Roche, M. (1992), *Rethinking Citizenship*, Cambridge: Polity Press.

Rodrigues, P.R. (1994), 'Racial discrimination and the law in the Netherlands', *New Community*, 20: 381–91.

Roelandt, T. and Veenman, J. (1992), 'An emerging ethnic underclass in The Netherlands? Some empirical evidence', *New Community*, 19: 129–41.

Rondinelli, D.A. and Yurkiewicz, J. (1996), 'Privatisation and economic restructuring in Poland: an assessment of transition policies', *American Journal of Economics and Sociology*, 55 (2): 145–60.

Rosenau, J.N. (1993), 'Notes on the servicing of triumphant sub-groupism', *International Sociology*, 8 (1): 77–92.

Ross, C.J. (1997), *Contemporary Spain: a Handbook*, London: Arnold.

Roux, C. (1991), 'Même école, même ambition: étude comparée des aspirations scolaires et professionnelles de jeunes filles d'origine maghrébine et de jeunes filles françaises de souche en classe de 3ème', *Revue Européenne des Migrations Internationales*, 7: 151–61.

Rowlands, M. (1996), 'Memory, sacrifice and the nation', *New Formations*, 30: 8–18.

Ruggie, J.G. (1993), 'Territoriality and beyond: problematizing modernity in international relations', *International Organization*, 47 (1): 139–74.

Rutherford, F. (1997), 'The contribution of the European Commission to combating social exclusion', in K. Duffy (ed.), *Partnership and Participation: the Experience of Poverty 3 in the UK*, London: Department of Social Security, pp. 53–62.

Sachs, J. (1990), 'What is to be done?', *Economist*, 13 January.

Sachs, J. (1995), 'Consolidating capitalism', *Foreign Policy*, 98.

Safran, W. (1995), 'Nationalism, ethnic groups and politics: a preface and an agenda', *Nationalism and Ethnic Politics*, 1 (1): 1–10.

Sainsbury, D. (ed.) (1994), *Gendering Welfare States*, London: Sage.

Sainsbury, D. (1996), *Gender Equality and Welfare States*, Cambridge: Cambridge University Press.

Salomon Brothers (1996), 'What EMU might mean for European banks', *European Equity Research*, 29 October.

Satava, L. (1994), *Národnostni mensiny u Europe*, Prague: Ivo Zelezny.

Sayer, A. and Walker, R. (1992), *The New Social Economy*, Oxford: Blackwell.

Schamp, E.W. (1995), 'The German automobile industry going European', in R. Hudson and E.W. Schamp (eds), *Towards a New Map of Automobile Manufacturing in Europe? New Production Concepts and Spatial Restructuring*, Berlin: Springer, pp. 93–116.

Scharf, F. (1996), 'Negative and positive integration in the political economy of European welfare states', in G. Marks, F. Scharpf, P. Schmitter and W. Streeck (eds), *Government in the European Union*, London: Sage, pp.15–39.

Schmähl, W. (1993), 'The "1992 reform" of public pensions in Germany: main elements and some effects', *Journal of European Social Policy*, 3 (1): 39–51.

Schor, R. (1996), 'L'extrême-droite française et les immigrés en temps de crise: années trente – années quartre vingt', *Revue Européenne des Migrations Internationales*, 12: 241–60.

Schumpeter, J. (1934), *The Theory of Economic Development*, Oxford: Oxford University Press.

Scott, A. (1988), *New Industrial Spaces*, London: Pion.

Seers, D. (1979), 'The periphery of Europe', in D. Seers, B. Schaffer and M.L. Kiljunen (eds), *Underdeveloped Europe: Studies in Core–Periphery Relations*, Hassocks: Harvester Press, pp. 3–34.

SEIES (1995), 'Women lead social cohesion and innovation', Autumn University and Conference Setúbal, SEIES working paper, Lisboa.

Seifert, W. (1996), 'Occupational and social integration of immigrant groups in Germany', *New Community*, 22: 417–36.

Sellers, M. (ed.) (1996), *The New World Order*, Oxford: Berg.

Shields, R. (1991), *Places on the Margin: Alternative Geographies of Modernity*, London: Routledge.

Shirreff, D. (1997), 'Regulation: naughty Germans', *Euromoney*, April (http://www.euromoney.com).

Sibley, D. (1995), *Geographies of Exclusion*, London: Routledge.

Siegfried, A. (1948), *The Mediterranean*, London: Jonathan Cape.

Silver, H. (1994), 'Social exclusion and social solidarity: three paradigms', *International Labour Review*, 133 (5–6): 531–78.

Simon, G. (1987), 'Migration in Southern Europe: an overview', in *The Future of Migration*, Paris: OECD, pp. 258–91.

Slovak Statistical Office (1997), *The 1996 Statistical Yearbook of the Slovak Republic*, Bratislava: Slovak Statistical Office.

Smith, A. (1994), 'Uneven development and the restructuring of the armaments industry in Slovakia', *Transactions of the Institute of British Geographers*, 19 (4): 404–24.

Smith, A. (1996a), 'From convergence to fragmentation: uneven regional development, industrial restructuring, and the "transition to capitalism" in Slovakia', *Environment and Planning A*, 28: 135–56.

Smith, A. (1996b), 'Industrial restructuring and uneven development in Slovakia: a regulationist approach to "the transition" in Central and Eastern Europe', unpublished Dphil thesis, University of Sussex.

Smith, A. (1997), 'Breaking the old and constructing the new? Geographies of uneven development in Central and Eastern Europe', in R. Lee and J. Wills (eds), *Geographies of Economies*, London: Arnold, pp. 331–44.

Smith, A.D. (ed.) (1976), *Nationalist Movements*, London: Macmillan.

Smith, A.D. (1982), 'Nationalism, ethnic separatism and the intelligentsia', in C.H. Williams (ed.), *National Separatism*, Cardiff: The University of Wales Press, pp. 17–41.

Smith, A.D. (1991), *National Identity*, London: Penguin.

Smith, A.D. (1993), 'Ties that bind', *LSE Magazine*, Spring, pp. 8–13.

Smith, D. and Blanc, M. (1997), 'Grass-roots democracy and participation: a new analytical and practical approach', *Environment and Planning D: Society and Space*, 15: 281–303.

Smith, G. (1993), 'Transitions to liberal democracy', in S. Whitefield (ed.), *The New Institutional Architecture of Eastern Europe*, London: Macmillan, pp. 1–13.

Smith, G. (ed.) (1996), *The Nationalities Question in the Post-Soviet States*, London: Longman.

Snyder, J. (1993), 'Nationalism and the crisis of the post-Soviet state', in M.E. Brown (ed.), *Ethnic Conflict and International Security*, Princeton, NJ: Princeton University Press.

Soininen, M. and Bäck, H. (1993), 'Electoral participation among immigrants in Sweden: integration, culture and participation', *New Community*, 20: 111–30.

Solé, C. (1995), 'Racial discrimination against foreigners in Spain', *New Community*, 21: 95–102.

SOPEMI (1997), *Trends in International Migration: Annual Report 1996*, Paris: OECD.

Sowell, T. (1994), *Race and Culture*, New York: Basic Books.

Soysal, N.Y. (1996), 'Changing citizenship in Europe: remarks on postnational membership and the national state', in D. Cesarini and M. Fulbrook (eds), *Citizenship, Nationality and Migration in Europe*, London: Routledge, pp. 17–29.

Standing, G. (1996), 'Social protection in central and eastern Europe', in G. Esping-Andersen (ed.), *Welfare States in Transition: National Adaptations in Global Economies*, London: Sage, pp. 225–55.

Staniszkis, J. (1991), 'Political capitalism in Poland', *East European Politics and Societies*, 5: 127–41.

Stephens, J.D. (1996), 'The Scandinavian welfare states: achievements, crisis and prospects', in G. Esping-Andersen (ed.), *Welfare States in Transition: National Adaptations in Global Economies*, London: Sage, pp. 32–66.

Stichter, S. and Parpart, J. (1990), *Women, Employment and the Family in the International Division of Labour*, London: Macmillan.

Storper, M. (1995), 'The resurgence of regional economies, ten years later: the region as a nexus of untraded dependencies', *European Urban and Regional Studies*, 2 (3): 191–223.

Storper, M. and Scott, A. (1989), 'The geographical foundations and social regulation of flexible production complexes', in M. Dear and J. Wolch (eds), *The Power of Geography*, London: Unwin Hyman.

Strange, S. (1987), *Casino Capitalism*, Oxford: Blackwell.

Sugar, P.F. (ed.) (1995), *Eastern European Nationalism*, Lanham, Maryland: The American University Press.

Sunley, P. (1992), 'Marshallian industrial districts: the case of the Lancashire cotton industry in the inter-war years', *Transactions of the Institute of British Geographers*, n.s., 17 (3): 306–20.

Swain, A. (1996), 'A Geography of transformation: the automotive industry in Eastern Germany and Hungary, 1989–94', unpublished PhD thesis, University of Durham.

Swain, N. (1992), *Hungary: the Rise and Fall of Feasible Socialism*, London: Verso.

Swoboda, A.K. (1993), 'Swiss banking after 1992', in J. Dermine (ed.), *European Banking in the 1990s*, 2nd edn, Oxford, Blackwell, pp. 351–69.

Symes, V. (1995), *Unemployment in Europe: Problems and Policies*, London: Routledge.

Szucs, J. (1988), 'Three historical regions of Europe', in J. Keane (ed.), *Civil Society and the State*, London: Verso.

Tamadonfar, M. (1989), *The Islamic Policy and Political Leadership: Fundamentalism, Sectarianism and Pragmatism*, Boulder, Col.: Westview Press.

Tapinos, G., Cogneau, D., Lacroix, P. and de Rugy, A. (1994), *Libre-echange et migration internationale au Maghreb*, Brussels: Rapport pour la Commission de la Communauté Européenne.

Taylor, C. (1991), *The Malaise of Modernity*, Concord: Anansi.

Taylor, C. (1992), *Multiculturalism and 'the Politics of Recognition'*, Princeton, NJ: Princeton University Press.

Taylor, P. and Johnston, R. (1979), *Geography of Elections*, London: Penguin.

Tett, G. (1997), 'Japan: borrowing costs fall for banks', *Financial Times*, 9 December: 4.

Therborn, G. (1995), *European Modernity and Beyond: the Trajectory of European Societies, 1945–2000*, London: Sage.

Thomas, R. (1997), 'Pensions panel fails to agree on a solution', *Guardian*, 1 January: 7.

Thompson, G. (1993), *The Economic Emergence of a New Europe? The Political Economy of Cooperation and Competition in the 1990s*, Aldershot: Edward Elgar.

Thomson, D. (1996), *Selfish Generations? How Welfare States Grow Old*, 2nd edn, Cambridge: White Horse.

Thomson, G. and Taylor, M. (1994), 'The developing single market in financial services', *Bank of England Quarterly Bulletin*, 34: 341–6.

Thorslund, M. (1993), 'The increasing number of very old people will change the Swedish model of the welfare state', *Social Science and Medicine*, 32: 455–64.

Thrift, N.J. (1994), 'On the social and cultural determinants of international financial centres', in S. Corbridge, N.J. Thrift and Martin, R. (eds), *Money, Power and Space*, Oxford: Blackwell, pp. 327–55.

Thrift, N.J. (1996), 'Shut up and dance, or, is the world economy knowable?', in P. Daniels and W. Lever (eds), *The Global Economy in Transition*, London: Longman, pp. 11–23.

Thrift, N.J. and Leyshon, A. (1988), 'The gambling propensity: banks, developing country debt exposures and the new international financial system', *Geoforum*, 19: 55–69.

Thurow, L.C. (1996a), *The Future of Capitalism*, London: Breckley.

Thurow, L.C. (1996b), 'Today's elderly are bringing down the social welfare state and threatening the nation's economic future', *New York Times Magazine*, 19 May.

Tickell, A. (1997), 'Restructuring the financial system into the twenty-first century', *Capital and Class*, 62: 13–19.

Tickell, A. (1998a), 'Dangerous derivatives? Controlling and creating risk in international money', paper presented at the annual conference of the RGS-IBG, Guildford, January (available from the Department of Geography, University of Southampton, Southampton SO17 1BJ, UK).

Tickell, A. (1998b), 'Questions about globalisation', *Geoforum*, 29: 1–5.

Tickell, A. (forthcoming), 'Unstable futures', in C. Leys and L. Panitch (eds), *Socialist Register 1999*, London: Merlin Press.

Tickell, A. and Peck, J.A. (1992), 'Accumulation, regulation and the geographies of post-Fordism: missing links in regulation theory', *Progress in Human Geography*, 16: 190–218.

Titmuss, R.M. (1955), 'Pensions systems and population change', *Political Quarterly*, 26 (2), reprinted in R.M. Titmuss, *Essays on the Welfare State*, 3rd edn, London: Allen and Unwin, 1976.

Titmuss, R.M. (1979), *Commitment to Welfare*, London: Unwin.

Tödtling, F. (1997), 'Strategic reactions of firms in Austrian regions to change in Central and Eastern Europe', *European Urban and Regional Studies*, 4 (2): 171–82.

Townsend, A.R. (1997), *Making a Living in Europe*, London: Routledge.

Treibel, A. (1994), 'Le "sentiment de nous" en Allemagne', *Revue Européenne des Migrations Internationales*, 10: 57–71.

Tribalat, M., Simon, P. and Riandey, B. (1996), *De l'immigration à l'assimilation: enquête sur les populations d'origine etrangère en France*, Paris: La Découverte/INED.

Tsardanidis, C. (1996), 'The southern EU member states' policy towards the Mediterranean: regional or global cooperation', *Journal of Area Studies*, 9: 53–69.

Tsoukalis, L. (1997), *The New European Economy Revisited*, Oxford: Oxford University Press.

Tucker, E. and Jack, A. (1997), 'Brussels irked by French evasions over bank rescue', *Financial Times*, 27 November.

van Tulder, R. and Ruigrok, W. (1993), 'Regionalisation or glocalisation: the case of the world car industry', in M. Humbert (ed.), *The Impact of Globalisation on Europe's Firms and Industries*, London: Pinter, pp. 22–33.

Turok, I. (1993), 'Inward investment and local linkages: how deeply embedded is Silicon Glen?', *Regional Studies*, 27 (5): 401–18.

Tyden (1996), 'Top 50 richest Czechs and Eastern Europeans (50 nejbohatsich obcanu CR i nejbohatsich Vychodoevropanu)', *Tyden Weekly*, 37.

UNDP (1996), *Human Development Report 1996*, New York and Oxford: Oxford University Press.

UNECE (United Nations Economic Commission for Europe) (1996), *International Migration Bulletin*, no. 9, November.

UNICEF (1997), *Children at Risk in Central and Eastern Europe: Perils and Promises*, London: UNICEF.

UNO (1956), *The Ageing of Populations and its Economic and Social Implications*, New York: United Nations.

UNO (1985), *The World Aging Situation*, New York: United Nations.

UNO (1993), *Demographic Yearbook 1993: Special Issue, Population Aging and the Situation of Elderly Persons*, New York: United Nations.

UNO (1996), *Economic Survey of Europe I: 1994–5*, Geneva: United Nations.

Úspešná privatizácia (1997), 'A successful privatisation', a survey by Slovak Statistical Office published in *Hospodárske noviny*, 24 October.

Vaiou, D. (1995), ' Women of the south after, like before, Maastricht', in C. Hadjimichalis and D. Sadler (eds), *Europe on the Margins: New Mosaics of Inequality*, London: Wiley, pp. 35–50.

Van Boeschoten, R. (1997), 'Euromarches: a powerful initiative, but where are the women?', *European Forum of Left Feminists Newsletter*, April 1997.

Vandermotten, C. and Vanlaer, J. (1993), 'Immigrants and the extreme-right vote in Europe and in Belgium', in R. King (ed.), *Mass Migrations in Europe: the Legacy and the Future*, London: Belhaven, pp. 136–55.

Vaughan, R.E. (1996), 'Procurement and capital projects', paper presented to the Conference on Supply Chain Management: the Challenges for the 21st Century, Durham Business School, University of Durham, 9–10 May.

Velkoff, V.A. and Kinsella, K. (1993), *Aging in Eastern Europe and the Former Soviet Union*, Washington, DC: Bureau of the Census.

Vercernik, J. (1995), 'Changing earnings distribution in the Czech Republic', *The Economics of Transition*, 3 (3): 355–71.

Verkuyten, M., de Jong, W. and Masson, K. (1994), 'Similarities in anti-racist and racist discourse: Dutch local residents talking about ethnic minorities', *New Community*, 20: 253–67.

Vermeylen, G. (1993), 'Elementen van het juridisch statuut van de vreemdeling in West-Europa', in R. Deslé, R. Lesthaeghe and E. Witte (eds), *Denken over Migranten in Europa*, Brussels: VUB Press, pp. 211–28.

Vertovec, S. (1996a), 'Berlin Multikulti: Germany, "foreigners" and "world-openness"', *New Community*, 22: 381–99.

Vertovec, S. (1996b), 'Multiculturalism, culturalism and public incorporation', *Ethnic and Racial Studies*, 19: 49–69.

Vieillard-Baron, H. (1994), *Les Banlieues Françaises ou le Ghetto Impossible*, La Tour d'Aigues: Editions de l'Aube.

Vincent, D. (1991), *Poor Citizens: the State and the Poor in Twentieth Century Britain*, London: Longman.

Vintrová, R. (1993), 'The general recession and the structural adaptation crisis', *Eastern European Economics*, 31 (3): 78–94.

Vives, X. (1991), 'Banking competition and European integration', in A. Giovannini and C. Mayer (eds), *European Financial Integration*, Cambridge: Cambridge University Press, pp. 9–34.

Voszka, É. (1995), 'Centralization, re-nationalization, and redistribution: government's role in changing Hungary's ownership structure', in J. Hausner, B. Jessop and K. Nielsen (eds), *Strategic Choice and Post-dependency in Post-socialism: Institutional Dynamics in the Transition Process*, Aldershot: Edward Elgar, pp. 287–308.

Wacquant, L.J.D. (1993), 'Urban outcasts: stigma and division in the black American ghetto and the French urban periphery', *International Journal of Urban and Regional Research*, 17: 366–83.

Walby, S. (1994), 'Methodological and theoretical issues in the comparative analysis of gender relations in Western Europe', *Environment and Planning A*, 26 (9): 1339–54.

Walby, S. (1997), *Gender Transformations*, London: Routledge.

Walker, A. (1990), 'The economic "burden" of ageing and the prospect of inter-generational conflict', *Ageing and Society*, 10: 377–96.

Walker, A. and Maltby, T. (1997), *Ageing Europe*, Buckingham: Open University Press.

Walker, L. (1996), 'Nationalism and ethnic conflict in the post-Soviet transition', in L. Drobizheva, R. Gottemoeller, C.M. Kelleher and L. Walker (eds), *Ethnic Conflict in the Post-Soviet World*, New York: M.E. Sharpe, pp. 3–13.

Walraff, G. (1988), *Lowest of the Low*, London: Methuen.

Walsh, J.H. (1992), 'Migration and European nationalism', *Migration World*, 20 (4): 19–22.

Warner, A. (1997), 'Borders blur as mergers mount', *The Banker*, September: 35–9.

Warner, E.W. (1992), '"Mutual recognition" and cross-border financial services in the European Community', *Law and Contemporary Problems*, 55: 7–28.

Warnes, A.M. (1993a), 'Demographic ageing: trends and policy responses', in D. Noin and R.I. Woods (eds), *The Changing Population of Europe*, Oxford: Blackwell, pp. 82–99.

Warnes, A.M. (1993b), 'Being old, old people and the burdens of burden', *Ageing and Society*, 13 (3): 297–338.

Webster, P. (1997), 'Britain sank Jospin's EU jobs scheme', *Guardian*, 21 June: 16.

Weiner, M. (1995), *The Global Migration Crisis: Challenge to States and to Human Rights*, New York: HarperCollins.

Weiss, I. (1997), 'Globalization and the myth of the powerless state', *New Left Review*, 225: 3–27.

Werth, M. and Körner, H. (1991), 'Immigration of citizens from third countries into the southern member states of the EEC', *Social Europe*, supplement 1/91: 1–134.

White, P. (1998), 'Urban life and social stress', in D. Pinder (ed.), *The New Europe: Economy, Society and Environment*, Chichester: Wiley, pp. 305–21.

White, S. (1997), 'Russia's troubled transition', in D. Potter et al. (eds), *Democratization*, Cambridge: Polity, pp. 421–42.

Whitehorn, K. (1997), 'Health of the nation: a prescription to tackle all of society's ills', *Observer*, 11 May.

Wickens, P. (1986), *The Road to Nissan*, London: Macmillan.

Wihtol de Wenden, C. (1995), 'Generational change and political participation in French suburbs', *New Community*, 21: 69–78.

Willems, H. (1995), 'Right-wing extremism, racism or youth violence? Explaining violence against foreigners in Germany', *New Community*, 21: 501–24.

Williams, A.M. (1984), 'Introduction', in A.M. Williams (ed.), *Southern Europe Transformed*, New York: Harper and Row, pp. 1–29.

Williams, A.M. (1987), *The Western European Economy: a Geography of Postwar Development*, London: Hutchinson.

Williams, A.M. (1994), *The European Community: the Contradictions of Integration*, Oxford: Blackwell.

Williams, A.M. (1997), 'Tourism and uneven development in the Mediterranean', in R. King, L. Proudfoot and B. Smith (eds), *The Mediterranean: Environment and Society*, London: Arnold, pp. 208–26.

Williams, A.M. and Balaz, V. (1998), 'Privatisation, markets and globalisation in Central Europe', University of Exeter, Department of Geography (mimeo).

Williams, A.M., Balaz, V. and Zajac, S. (1998), 'The EU and Central Europe: the remaking of economic relationships', *Tijdschrift voor Economische en Sociale Geografie*, forthcoming.

Williams, A.M., King, R. and Warnes, A.M.. (1997), 'A place in the sun: international retirement migration from Northern to Southern Europe', *European Urban and Regional Studies*, 4 (2): 115–34.

Williams, C.H. (1980), 'Ethnic separatism in Western Europe', *Tijdschrift voor Economische and Sociale Geografie*, 71 (3): 42–58.

Williams, C.H. (ed.) (1982), *National Separatism*, Cardiff: University of Wales Press; Vancouver: University of British Columbia Press.

Williams, C.H. (1986), 'The question of national congruence', in R.J. Johnston and P.J. Taylor (eds), *A World in Crisis?*, Oxford: Blackwell, pp. 196–231.

Williams, C.H. (ed.) (1988), *Language in Geographic Context*, Clevedon, Avon: Multilingual Matters.

Williams, C.H. (ed.) (1991), *Linguistic Minorities, Society and Territory*, Clevedon, Avon: Multilingual Matters.

Williams, C.H. (1992), 'Identity, autonomy and the ambiguity of technological development', in Z. Mlinar (ed.), *Globalization and Territorial Identities*, Aldershot: Avebury.

Williams, C.H. (1993a), 'Towards a new world order: European and American perspectives', in C.H. Williams (ed.), *The Political Geography of the New World Order*, London: Wiley, pp. 1–19.

Williams, C.H. (1993b), 'The rights of autochthonous minorities in contemporary Europe', in C.H. Williams (ed.), *The Political Geography of the New World Order*, London: Wiley, pp. 74–99.

Williams, C.H. (1993c), 'The European Community's lesser used languages', *Rivista Geografica Italiana*, 100: 531–64.

Williams, C.H. (1994), *Called unto Liberty: On Language and Nationalism*, Clevedon, Avon: Multilingual Matters.

Williams, C.H. (1996), 'Ethnic identity and language issues in development', in D. Dwyer and D. Drakakis-Smith (eds), *Ethnicity and Development: Geographical Perspectives*, Chichester: Wiley, pp. 45–85.

Williams, C.H. and Evas, J.E. (1997), *The Community Language Project*, Cardiff: The Welsh Language Board.

Williams, C.H. and Smith, A.D. (1983), 'The national construction of social space', *Progress in Human Geography*, 7: 502–18.

Williams, R. (1983), *Keywords*, London: Fontana.

Wills, J. (1997), *Harbingers of a Social Economy in Europe? Preliminary Investigations into the Operations of European Works Councils*, Re-scaling Workplace Solidarity working paper no. 2, Southampton: University of Southampton, Department of Geography.

Wilson, P-L. (1996), 'Boundary violations', in S. Aronowitz, B. Martinsons and M. Menser (eds), *Technoscience and Cyberculture*, London: Routledge.

Wilson, T. and Donnan, H. (eds) (1998), *Border Identities: Nation and State at International Frontiers*, Cambridge: Cambridge University Press.

Witte, R. (1995), 'Racist violence in western Europe', *New Community*, 21: 489–500.

Wolf, M. (1995), 'Cooperation or conflict? The European Union in a liberal global economy', *International Affairs*, 71 (3): 325–37.

World Bank (1994), *Averting the Old Age Crisis: Policies to Protect the Old and Promote Growth*, Oxford: Oxford University Press.

World Bank (1996), *World Development Report 1996: from Plan to Market*, Oxford: Oxford University Press.

Zelinsky, W. (1984), 'O say, can you see? Nationalistic emblems in the landscape', *Winterthur Portfolio*, 19 (4): 277–86.

Zelinsky, W. (1988), *Nation into State: the Shifting Symbolic Foundations of American Nationalism*, Chapel Hill, NC: University of North Carolina.

Zimmer, S.A. and McCauley, R. N. (1991), 'Bank cost of capital and international competition', *Federal Reserve Bank of New York Quarterly Review*, 16: 33–59.

Zincone, G. (1993), 'The political rights of immigrants in Italy', *New Community*, 20: 191–206.

Zon, H. van (1996), *The Future of Industry in Central and Eastern Europe*, Aldershot: Avebury Press.

Index